THERMO-HYDRO-MECHANICAL PROCESSING OF WOOD

Engineering Sciences — *Materials*

THERMO-HYDRO-MECHANICAL PROCESSING OF WOOD

Parviz Navi and Dick Sandberg

EPFL Press
A Swiss academic publisher distributed by CRC Press

CRC Press
Taylor & Francis Group

Taylor and Francis Group, LLC
6000 Broken Sound Parkway, NW, Suite 300,
Boca Raton, FL 33487

Distribution and Customer Service
orders@crcpress.com

www.crcpress.com

Library of Congress Cataloging-in-Publication Data
A catalog record for this book is available from the Library of Congress.

This book is published under the editorial direction of
Professor Michel Rappaz (EPFL).

ÉCOLE POLYTECHNIQUE FÉDÉRALE DE LAUSANNE The author and publisher express their thanks to the Swiss Federal Institute of Technology in Lausanne (EPFL) for the generous support towards the publication of this book.

About the cover: A photo taken by D. Sandberg of a chair-based sculpture created by the benders/craftsmen at the company Gemla, Inc. of Sweden.

EPFL Press
is an imprint owned by Presses polytechniques et universitaires romandes, a Swiss academic publishing company whose main purpose is to publish the teaching and research works of the Ecole polytechnique fédérale de Lausanne (EPFL) and other universities and institutions of higher learning.

Presses polytechniques et universitaires romandes
EPFL – Rolex Learning Center
Post office box 119
CH-1015 Lausanne, Switzerland
E-mail: ppur@epfl.ch
Phone: 021 / 693 21 30
Fax: 021 / 693 40 27

www.epflpress.org

ISBN 978-2-940222-41-1 (EPFL Press)
ISBN 978-1-4398-6042-7 (CRC Press)

Printed in USA

PREFACE

The fabrication of wood-derived products is widespread and growing, thanks to its unique advantages: widespread availability, natural renewability, favorable ecological assessment, and flexibility of implementation. Moreover, the polymeric components of wood, together with its porous structure, confer upon it a faculty for transformation exceeding that of other materials.

Since the dawn of civilization wood has been used in its natural state. Only recently has wood been developed to form a range of products that are increasingly functional, based on a combination of performance and sustainability requirements. Indeed, since the beginning of the last century, knowledge on this topic has advanced constantly, mainly through the efforts of systematic scientific research and new types of applications.

The early 1900s, preliminary work on the science and technology of wood was first published. Our current understanding of the material properties, chemistry, and physics of wood – coupled with advances in materials science and modeling techniques – has provided the tools to allow engineers and researchers to fully exploit wood as a material and to produce new components and products under controlled processing conditions.

In order to overcome the inherent difficulties in wood-processing technologies, an interdisciplinary approach is required. Close co-operation between scientific disciplines – such as the anatomy of wood, physics, chemistry and mechanics – allows for mutual and constructive progress to be made. Thus the wood-treatment technology evolves with our understanding of the basic science in such areas as conservation, drying, machining, shaping and joining, etc.

One of the emerging eco-friendly treatment methods is the combined use of temperature, moisture and mechanical action – the so-called Thermo-Hydro-Mechanical (THM) treatment. THM processing is implemented to improve the intrinsic properties of wood, to produce new kinds of materials, and to acquire a form and functionality desired by engineers without changing the eco-friendly nature of the material. These processes can be divided into two major categories; Thermo-Hydro (TH) treatments and Thermo-Hydro-Mechanical (THM) treatments.

As mentioned, wood consists of natural polymeric chains, connecting to each other by hydrogen bonds and, in other zones, by covalent bonds. The hydrogen bond is at the origin of its transformation properties. For example, when wood is put under thermo-hydrous conditions, allowing for the softening of its amorphous components, it can easily deform, making it possible to apply a range of industrial processes such as molding, densification, surface densification, bending, shaping, and drying at high temperature. However, the application of high temperature – with or without moisture – can mechanically damage or chemically modify the polymeric components of wood.

In this book, we bring together the key elements of the chemical degradation of wood constituents under TH and THM processing; the behavior of wood under these conditions; as well as a selection of the principal technologies that can be considered

TH/THM treatments. This work is intended for researchers, professionals of timber construction, as well as students in the fields of material science, wood technology and processing, civil engineering, and architecture. Our goal is not to provide exhaustive coverage of the subject, but rather to highlight the scientific disciplines necessary to comprehend THM technologies, as well as the behavior of wood during treatment and later in product application.

This book consists of 11 chapters. The first is devoted to the justification for TH/THM processing. Ancient treatments of wood by THM processing are discussed in the second chapter. It is shown that different heat-treatment processes were already in use thousands of years ago in order to improve the performance of wood. The description of the structure and the chemical composition of the components of wood are given in the third chapter. In the fourth chapter, the explanation and modeling of certain THM behavior of wood is presented. In the first part, the small and large deformations of wood are described, and the constitutive equations of elastic linear and elastic nonlinear behavior of wood, are derived. In the second part the viscoelastic behavior of wood under ambient temperature, constant and variable humidity is described. In the chapter five, the behavior of THM of wood under variable moisture and temperatures (as high as 200 °C), is examined by considering that during the processing, at high temperatures, the components of wood undergo certain chemical modifications. In this chapter the effects of the processing parameters – temperature, moisture content and time – on the THM wood characteristics are discussed. The sixth chapter is devoted to the process of wood densification by THM treatment. In the first part of this chapter, various THM wood-densification processing methods are discussed, with illustrations of the machines developed in different countries corresponding to open, closed, and mixed processing. Later in the chapter, the origin and mechanisms of shape memory and elimination of the compression-set recovery by THM treatment are discussed. Chapters seven and eight are devoted to techniques of wood welding by friction and wood-surface densification. In each of these chapters, different techniques are discussed and the problems related to these different 'open systems' are explained.

In the recent decades, heat-treatment techniques have advanced considerably. At present, many countries have developed their own national version of wood TH treatments. In chapter nine, most of these processing methods are presented and discussed. In chapter ten, a number of wood-bending processes – including bending of solid-wood, laminated-wood bending, green-wood bending and kerfing – are presented and different techniques discussed. In addition, the reader will find a presentation of the theory of solid-wood bending and a demonstration of solid-wood bending in the laboratory and at industrial levels. Finally, the eleventh chapter provides a selection of technologies on the fabrication of reconstituted wood, namely fiberboards, particle boards and panels made of veneers.

For the benefit of wood engineers and others with an interest in this fascinating industry, we hope that the convenient availability of this material will lead to a deeper understanding of some of the fundamentals involved in TH and THM processing of wood.

Parviz Navi & Dick Sandberg

Acknowledgements

The authors wish particularly to thank Dr. Brenhard Stamm (wood circular friction welding), Lauri Rautkari (surface densification), Professor Preben Hoffmayer (creep and mechano-sorption), Professor Alain Curnier (on large nonlinear deformations), Fred Girardet (conception of THM closed reactor), Professor Frederick Kamke (conception of THM mixed reactor), David Anguish and Professor Ove Söderström for his constructive remarks.

The financial support of the Swiss National Foundation on the field of wood THM processing is gratefully acknowledged.

A special thank is directed to the Swedish Knowledge Foundation and Ikea of Sweden for their long-term support in the areas of solid wood bending and laminated bending.

The authors would also express their thanks to the Editor, Dr. Frederick Fenter, as well as to the entire helpful team of the Presses polytechniques et universitaires romandes (PPUR) for dealing with all the numerous details through the book's final production.

CONTENTS

CHAPTER 1

WOOD MODIFICATION

1.1 INTRODUCTION

Wood is the ultimate renewable material. It possesses qualities that have made it a material of choice for millennia and these qualities are further enhanced by its recognized carbon sequestration. Moreover, the polymeric components of wood and its porous structure confer on it a noble, versatile and general-purpose character and a faculty for transformation exceeding that of all other materials. The unique advantages of this material, its widespread availability, sustainable renewal, favorable ecological assessment and its flexibility of implementation, give this worthy material its letters patent of nobility in the eyes of scientists and engineers. Nevertheless, the fact that wood is a natural product originating from different individual trees imposes limits on its use. This natural material may need to be transformed in order to acquire the desired functionality.

There are ways in which the properties of wood can be enhanced by modification through eco-friendly methods. One emerging eco-friendly method is the combined use of temperature, moisture and mechanical action, the so-called thermo-hydro-mechanical treatment. There are numerous thermo-hydro-mechanical processing techniques and the number of these processes is growing continuously. Thermo-hydro-mechanical processing is implemented to improve the intrinsic properties of wood, to produce new materials and to acquire a form and functionality desired by engineers without changing the eco-friendly characteristics of the material. These processes can be divided into two major categories; Thermo-Hydro treatments (TH) and Thermo-Hydro-Mechanical (THM) treatments.

Craftsmen have modified wood for centuries by TH/THM treatments, but it was in the 19th century, with the construction of the Vienna Chair by Michael Thonét (1796-1871), that the industrialized process of wood molding technology was born, allowing its products to be commercialized all over the world. Besides greatly improving the understanding of basic material properties, chemistry and physics as well as giving rise to advances in materials and wood science, modeling techniques provided the means for engineers and researchers to engineer wood as a material and produce new materials under controlled processing conditions.

1.2 REASONS FOR THERMO-HYDRO-MECHANICAL PROCESSING

Never before have the forestry and forestry-related industries been so sharply in the focus of discussions concerning the major challenges for the future. The challenges being discussed are of great significance: instead of consuming the Earth's non-renewable resources, the use of renewable materials is being encouraged in all quarters; fossil fuels must be phased out and human consumption must increasingly reflect a concern for the climate and the environment. Whilst in service, the carbon having been fixed in wood by photosynthesis can act as a long-term carbon store. Wood-based products such as solid wood can also act as short-term carbon stores. Carbon accounts for about half of the mass of a wood product. For example, with a wood product stock of about 60 million tonnes in Europe alone, the carbon storage effect of wood products plays a significant role in mitigating ecological and societal effects. The forest industry uses renewable raw materials and manufactures eco-friendly products. Carbon is stored in the forests and in their products. This means that both forests and forest products are part of the answer to the climate challenge.

Concern for the environment and the climate has put pressure on wood researchers and the wood industry. The demand for investigations and the development of new treatments for eco-friendly products, for improving wood properties and finding alternatives to tropical hardwoods as well as to energy-consuming material processes and fossil-fuel-based materials is greater than ever before, and this is the reason why some new modification technologies are being commercialized.

The TH process is based on wood, water and heat, and the THM treatment incorporates an additional mechanical force. This is an environment-friendly process that uses a recyclable and renewable natural resource that decreases the CO_2 in the atmosphere. At the same time, TH/THM processes consume relatively little energy. The result of studies in different fields of TH/THM processing also indicates that the properties of wood such as its dimensional stability, strength, surface hardness and durability are improved.

Enhancing the competitiveness of timber by using an innovative upgrading method with a low environmental impact can increase the use of local materials instead of imported rare tropical hardwoods, creating added-value for local timber resources, increasing profitability, providing regional income and providing employment in regional/convergence areas. People are also starting to recognize the advantages of combining renewable sources such as wood with an environmentally acceptable processing technique in order to achieve new high-quality materials. Thermo-hydro-mechanical processing as a base for product improvement and the development of new product-market combinations is now of significant interest in the industry. Potential areas of application for the thermo-hydro-mechanical processing include the building industry, furniture manufacture, improving service lives of wood products through increased durability and stability, wood-finish compatibility and in new market areas now being identified.

1.3 MODIFICATION OF WOOD

The purpose of any technological development involving wood is to standardize, homogenize and produce new wood-based materials with dimensions that are independent of those of the tree, to decrease anisotropy, overcome the problems of dimensional instability and improve the material's durability and resistance to fire. In order to obtain semi-finished and finished products with or without added value, wood can be transformed

- by transformation, i.e., sawing, chipping, pulping;
- by reconstitution, i.e., in the manufacture of engineered wood products from wood, wood chips and particles; and
- by modification.

This book presents the principal tools and instruments used in the TH/THM wood modification, the rich experience gained during several millennia of wood TH/THM treatment and especially the role of wood science, chemistry, mechanics and physics on the recent techniques of wood thermo-hydro-mechanical processing.

To modify wood, two treatment types are implemented: chemical treatments and TH/THM treatments. The former are much more numerous and the range of chemical agents is very broad, whereas for the TH/THM treatments, only heat, water and mechanical forces are employed. Figure 1.1 displays a simplified synoptic diagram of the current chemical and THM treatments.

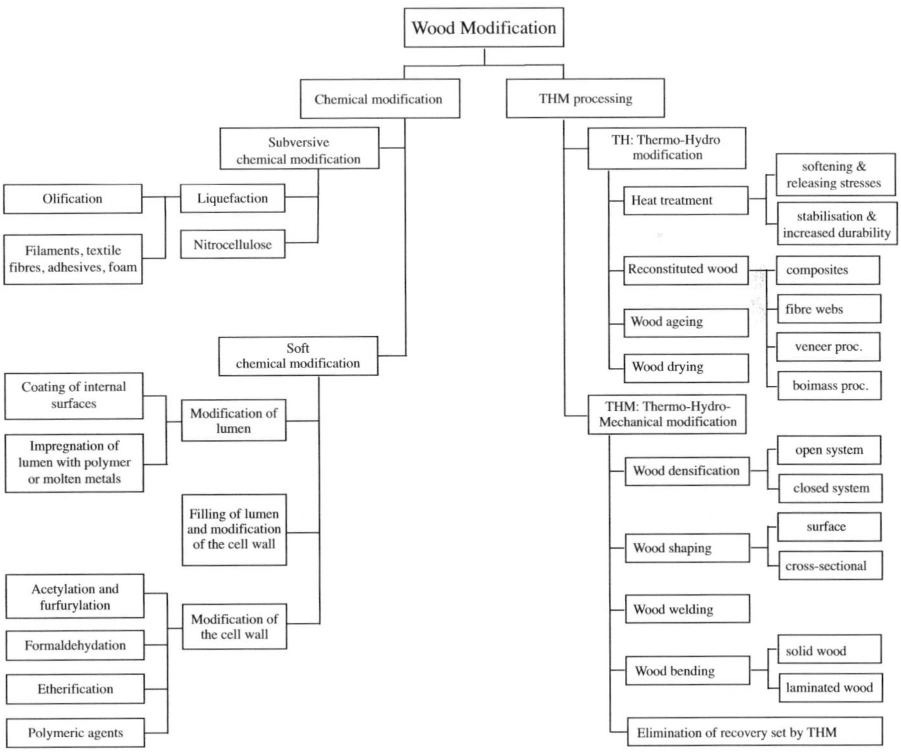

Fig. 1.1 A diagram of the various processes of chemical modification and thermo-hydro-mechanical processing of wood.

1.3.1 Chemical modification

Chemical modification of wood is defined as any chemical reaction between a reactive wood cell component and a chemical reagent, with or without a catalyst, giving rise to a covalent bond between the two components (Hon & Shiraishi, 1991; Hill, 2006). The most promising application of this technology is developed in the area of reconstituted wood products, e.g., particle boards, plywood and veneer-based products and in the enhancement of the dimensional instability, resistance to biological degradation (durability) and resistance to fire of the wood.

In Figure 1.1, two classes of chemical treatments can be distinguished: subversive and soft. The subversive treatments are those that enter into the core of the cellulose fibrils. They break down the crystalline structure of cellulose and eliminate the multilevel and hierarchical structure of wood. These treatments radically modify the chemical components of the wood, and consequently the material produced lacks practically all the intrinsic characteristics of untreated natural wood. Examples of subversive treatments are liquefaction and olification of lignocelluloses (Yao *et al.*, 1994), where liquefaction is adopted mainly to produce oil from biomass under very severe conditions. Appel *et al.* (1969, 1975) converted celluloses to oil using reactive chemicals, under high temperature and high pressure. This type of liquefaction is called olification of lignocelluloses. Other recent types of liquefaction involve the dissolution of chemically modified wood or even untreated wood in a solvent. Applications of these types of liquefaction have been developed in the preparation of adhesives, moldings, foams etc. Nitrocellulose (cellulose nitrate) is prepared by treating cellulose with a mixture of nitric acid and sulfuric acid. It is employed as a propellant or low-order explosive (guncotton) and as a film-forming resin in the inks and coatings market.

The soft chemical treatments, on the other hand, leave the crystalline structure of wood intact and touch only the amorphous part and the side groups (functional groups) of the wood components at the molecular level. These treatments allow the microstructure of the modified wood to remain more or less unchanged. In this category, some typical treatments include: formaldehydation, acetylation, etherification, impregnation by carbinol and maleic acid, impregnation by vinyl resins, impregnation by polyethylene glycol (PEG), impregnation by phenol formaldehyde resin, filling by metal alloys of low melting point, etc. These treatments can lead to improvements in certain physical and mechanical properties of wood, such as hygroscopicity, resistance to micro-organisms, and fire resistance, as well as in its mechanical behavior with respect to the moisture variations. Norimoto and Grill (1993) have investigated the influence of various soft chemical treatments on the mechanical and physical charachteristics of wood, and they have shown that some treatments improve certain properties of wood but are likely to lead to deterioration in others. For example, the impregnation of wood with a polyethylene glycol (PEG) resin reducees the hygroscopicity of wood but at the same time increases the mechano-sorptive creep deformation. In contrast, acetylation can simultaneously reduce both the hygroscopicity and the mechano-sorptive creep behavior of wood.

The chemical treatments that involve a radical modification of the wood structure, i.e., subversive chemical wood treatment, constitute a delicate field, which requires a detailed study of the organic chemistry. The soft chemical treatments of wood are mainly implemented to reduce the hygroscopicity and mechano-sorptive creep and increase the resistance to micro-organisms.

1.3.2 Thermo-hydro and thermo-hydro-mechanical modification of wood

The TH/THM treatments have an important advantage over the chemical treatments since the products of the TH/THM wood are particularly environment-friendly. During thermo-hydro-mechanical processing, no chemical agents are introduced into the wood and no subversive chemical degradation affects any of its components. Under TH/THM treatments, many types of processes are developed to produce different products. Some of the results achieved at the laboratory level or in industrialized-commercialized processes are shown in Figure 1.1. The TH treatments include many types of processes, such as those to

- improve dimensional stability and reduce the wood hygroscopicity;
- enhance the resistance to micro-organisms;
- achieve an accelerated aging of the wood; and
- release internal stresses (reduce the recovery set) and to soften the wood through wet heating.

One can also perform a TH treatment of reconstituted wood products, e.g., wooden composites, paper, fiber-boards, plywood and wood plastic composites. The THM treatments include

- densification, i.e., THM densification in a closed or open system;
- wood shaping by surface densification and embossing;
- wood shaping by wood cross-sectional transformation;
- wood welding;
- wood bending, i.e., solid wood bending and laminated bending; and
- THM treatments to eliminate the shape memory of densified wood.

The fact that wet wood can be shaped under the action of mechanical and thermal loads has been known for a very long time. Figure 1.2 shows the principles used in the construction of a canoe by an elementary and ancient THM process. The various stages of the construction include the preparation of the hull starting from a trunk, heating of the hull of humidified wood, giving the desired shape to the canoe and mechanical fixing to maintain the given shape.

The shaping of steamed wood has been employed since antiquity and has been carried out in numerous ways. Plasticization of the solid wood in order to make it possible to bend the piece without fracture is the most common way of shaping, e.g., for furniture-making and ship-construction. The bending can be achieved in one or in two planes as well as in different directions in the same plane. Wood is normally bent after pre-steaming in various types of strapping devices that minimize the effect of tension on the convex surface and maximize the compressive yield on the concave surface, thus limiting the risk of breakage.

Another process involving the shaping of wood is the technique of laminated bending that is primarily directed towards the bending of veneers for other than

(a)

(b)

Fig. 1.2 The construction of a canoe (Freed & Baby, 1983). (a) Mallets and wedges are used to hollow out the interior of the trunk. The trunk is then scraped to give it the desired thickness; one finger thickness for the upper part and two fingers for the bottom. (b) In order to widen the canoe and give it the final shape, it is filled with water to humidify the wood. The water is then heated to boiling by heated stones. The fire lit at the side also heats the outside of the canoe. Once the wood has been softened by the hot water and fire, the wooden beams are set up to widen the hull, fixing and maintaining the desired shape.

structural purposes, such as in sports goods, furnishing details etc. (Stevens & Turner, 1970). The need to increase the efficiency of the process of manufacturing wood veneer products and the increasing importance of laminated bends in structural applications have led to an increasing interest in a basic understanding of the wood material and the process (Ormarsson & Sandberg, 2007). The fundamental mechanisms that

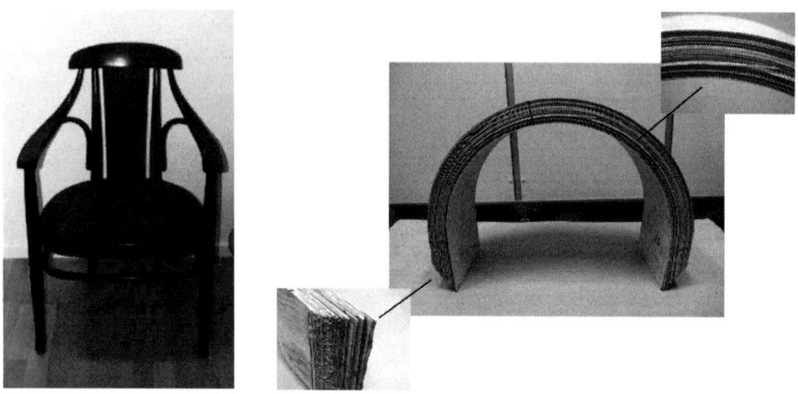

(a) (b)

Fig. 1.3 Examples of products made by (a) solid wood bending (chair) and (b) laminated bending (window frame).

must be understood and developed for bending laminated wood for structural purposes and for non-structural members are however similar, especially when the radius of curvature is small. Figure 1.3 shows examples of products made by solid wood bending and laminated bending.

Wood can be densified and its properties modified not only by filling its void volume with polymers, molten natural resins, waxes, sulphur and even low fusion metals, but also by compressing it under conditions such that the structure does not become fractured. Densification is the process whereby the wood density is increased by reducing the void volume of the lumens in the wood material.

In wood shaping, the wood can be processed either at the surface or in the cross-section of the wood. Embossment is a process where the surface of the wood is partially densified, normally by a steel tool, to create a decorative pattern in the surface, as exemplified Figure 1.4. This process is comparable with the methods of densification of wood, but its main purpose is not to increase the density of the wood, but only to shape the wood surface. Embossment has a long tradition. For example, the art of pressing, engraving and fretting patterns into wood has been a part of Iran's ancient history and goes as far back as 5000 years when it comes to the interior design of palaces, boxes laid over tombstones, pulpits and book-racks (Khamouski, 1999).

Fig. 1.4 Example of embossment work.

The advances achieved during the 20th century in wood material science and engineering provided the means for the engineers and researchers to develop new eco-friendly materials from wood. One of the most interesting areas is wood processed in a closed system under controlled processing conditions. The most remarkable examples of science-based technologies in thermo-hydro-mechanical processing are briefly described in the next section.

Thermo-hydro wood processing

The thermo-hydro treatment of wood constitues a modification technique since the wood constituents undergo chemical changes during the treatment. The reason for heat-treating the wood is to improve some of its intrinsic properties such as the dimensional stability, resistance to micro-organisms, stress relaxation during the forming of wood-based composites, veneer production, wood cutting and wood fracture resistance. Currently, this technique is implemented to make artificially aged wood, so-called accelerated wood ageing.

Stamm *et al.* (1946) investigated the effect of heat and processing time on the dimensional stability of wood. Wood was immersed in baths of molten metal at temperatures between 140 and 320 °C, and it was reported that heating for seven hours at 260 °C reduced the swelling of Sitka spruce by 60%. However, Stamm (1964) reported that, although the dimensional stability and set recovery of compressed wood were improved by heating, certain parameters related to strength were reduced to an unacceptable level, particularly the surface strength. Hence, it was predicted that the dimensional stabilization of wood by heat had no future.

In the 1970s, Burmester (1973) studied the effects of temperature, pressure and moisture content to find the optimum conditions for the heat-stabilization of different species of wood, and a heat-treatment method called FWD (Feuchte-Wärme-Druck) was launched. The optimum conditions for pine wood for instance were stated to be a moisture content of 20-30%, a temperature of 160 °C and a pressure of 0.7 MPa. A considerable resistance towards brown rot fungus was reported, and the strength reduction was stated to be insignificant.

During the 1980s, French and Japanese industries began to study the heat treatment of wood in a closed system in order to increase the microbial durability. Since then, the interest for heat treatment has increased all over the world. The process essentially involves a controlled degradation of wood, primarily resulting in the destruction of hemicelluloses. Several different methods are included. The basic diversity of the different processes is indicated by their oxygen-excluding and heat-transporting media. A substantial similarity is that these processes run in a temperature range between 180 and 240 °C to change the chemical composition of the cell wall. Due to a severe loss of strength, TH treatments of wood at elevated temperatures above 300 °C are limited (Hill, 2006). The temperature is achieved with superheated steam in vacuum, or with an inert gas such as nitrogen. Pre-heated oil can also be used in the process. A simplified picture of the results obtained from the heat-treatment methods indicates that heat treatment increases the stability and durability but also the brittleness of wood and leads to a loss of certain strength properties including impact toughness, modulus of rupture and the work to failure.

The commercializing of industrial heat-treated wood products is a recent advance. This type of industry aims at improving the biological durability of less durable wood species and enhancing the dimensional stability of wood or wood-based products, e.g., particle boards. The properties of industrially produced heat-treated wood in general have been intensively investigated in recent years.

On the European market, several industrial heat-treatment processes have during the last few years been introduced using closed systems. Figure 1.5 shows one heat-treatment chamber for controlling the temperature and humidity of a two-stage treatment plant.

The most wide-spread industrial heat-treatment process of wood is the seasoning sawn wood in kilns. This process is performed at temperatures below 100 °C, i.e., low-temperature convective drying, or above 100 °C, i.e., high-temperature drying. Thermal modification of wood may be viewed as an extended seasoning process,

Fig. 1.5 Heat treatment according to the Plato process in a thermo-hydro vessel, 20 m long and 2.4 m in diameter, with a controlled temperature and humidity (by courtesy of the Plato International BV, the Netherlands).

which ends with an increased temperature period when all the water in the wood has vaporized.

In a TH process, wood is heated in water to soften it and to release its internal stresses. A common industrial TH process is steaming and boiling logs before slicing or rotary peeling them to veener. Actually, the boiling temperature of water is seldom reached. Almost always, the temperature is lower than 100 °C. TH processes have also been extensively studied as a technology for the processing of biomass for the production of sugars (Garrote *et al.*, 1999).

For several decades, friction welding has found broad applications in many metal and plastic industries. Suthoff *et al.* (1996) made the first attempts to join wood by means of pressure and frictional movement. They stated that two pieces of wood can be welded by means of an oscillating or linear frictional movement.

Early work in joining pieces of wood by friction welding was carried out by Gfeller *et al.* (2004), Stamm *et al.* (2005) and Stamm (2006). It was demonstrated that during the welding process the contact zones between two wooden pieces were joined together by "molten wood" through the heat produced during the frictional movement. The movement causes a thermal alteration of the wood cells, which leads to the formation of a viscous layer or "molten wood". After stopping the friction movement and cooling, the material in the contact zone forms the connection between the pieces, as displayed in Figure 1.6.

Investigations of the welded wood show that the strength achieved by this method is close to that obtained using conventional glues, but that the strength of friction-welded joints decreases at high humidity. The development of a relatively high initial resistance also permits for instance the continuous welding of multilayered laminates. In this technique, the temperature of the wood between the welded surfaces may rise to more than 400 °C during welding. Moreover, the total welding processing time is short, about 10-20 seconds. These techniques have attracted industrial interest.

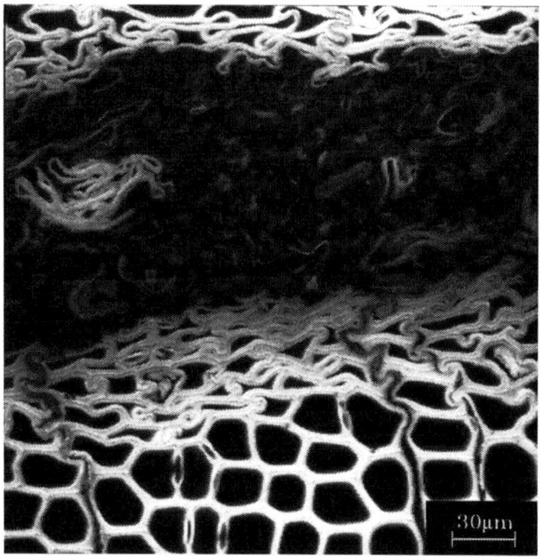

Fig. 1.6 Cross-sectional view of a confocal laser-scanning micrograph picture of the heat-affected zone and the adjacent regions of friction-welded Norway spruce (Stamm *et al.*, 2005).

Mechanically induced vibration wood-welding techniques can also be utilized to obtain wood surface finishes with a greater surface hardness and performance in the presence of polymerizing unsaturated oils, such as sunflower oil or other polymerizing finishes (Pizzi *et al.*, 2005).

Thermo-hydro-mechanical wood processing

Compressing wood in the transverse direction reduces the void volume of the lumens in the wood material and increases its density. This process is commonly called densification. In 1886, the idea of wood densification by compressing the wood in its radial direction existed (Vorreiter, 1949). In Austria, the Plumfes brothers developed a method in 1922 for densifying wood by impregnation with rubber. This type of wood was used in the aviation industry until 1945 when it was replaced by aluminum. Compressed solid wood has also been made in Germany since the early 1930s and commercialized under the trade name of Lignostone, a corresponding laminated compressed wood under the trade name of Lignofol, and a resin-treated laminated compressed wood under the name of Kunstharzschichtholz (Kollmann, 1936; Stamm, 1964). The first aim of the densification was to improve the mechanical behavior and moisture sorption of wood. At present, the objective of wood densification is manifold and includes numerous processes, such as wood shaping, enhancing the intrinsic mechanical, physical properties of the wood, producing high-quality wood, wood surface densification. One of the reasons for densifying wood in the transverse direction is to produce high quality timber from timbers of low quality. A process developed in Japan called "Compressed Lumber Processing System" transfers a local timber with a low density of about 300 kg/m^3 to a timber with a density greater than 900 kg/m^3. Figure 1.7 shows two elements fabricated by this THM process at a temperature of 180 °C.

Fig. 1.7 Photograph of two elements of densified wood fabricated by the "Compressed Lumber Processing System" from low-density wood.

However, densified wood shows an undesired behavior, i.e., a tendency to recover all or part of its compression-set and return to its initial dimensions when subjected to heat and humidity. This phenomenon is called "shape memory" or "compression-set recovery". Various wood products (bent wood, densified wood, wood surface densified wood, molded wood and compressed wood fibers) suffer from shape memory. To overcome this issue, Stamm and Seaborg (1941) impregnated the wood after compression with a phenol formaldehyde resin. The wood was then heated for 10 to 20 minutes to polymerize the resin. This type of compressed and impregnated wood was called "Compreg". Stamm *et al.* (1946) had thus succeeded in partially eliminating the shape memory of densified wood by a TH treatment.

Advanced investigations of the densified wood treatment in THM processing have been reported by Tanahashi (1990), Inoue *et al.* (1993), and Ito *et al.* (1998a). They have shown that the THM post-treatment of densified thin wood specimens at 200 °C for only 4 minutes was sufficient to eliminate almost totally the shape memory of densified wood. On the other hand, the thermal treatment of wood was limited to 300 °C, due to severe degradation of the strength of the material (Hon & Shiraishi, 1991; Hon, 1996; Navi & Heger, 2005; Hill, 2006).

The purpose of recent research into molding and wood densification as well as the development of new techniques for THM post-treatment (to eliminate set recovery) is to implement wooden elements with large dimensions. Currently, in Japan, various investigations into the THM densification of wood are being carried out. These processes are based on a two-directional transverse densification of wood elements and involve four stages: wood plasticization by high temperature steam, compressive molding, THM post-treatment and cooling. Two techniques are being developed: one by Ito *et al.* (1998b), in which small trunks of wood with a circular sections are transformed into trunks with square sections; and one known as the "Compressed Lumber Processing System". The differences between the two technique lies in the stage of post-treatment. In the first technique, the molding and post-treatment form a continuous process, whereas in the second technique these two stages are separated. The problems related to the dimension of the wood elements (size effects), such as the development of cracks and exfoliation during shaping, cooling and drying have apparently yet to be completely solved. Shigematsu *et al.* (1998), Kyomori *et al.* (2000), and Tanahashi *et al.* (2001) have reported that to overcome the problems of "compressive molding of wood under a high-pressure steam technique" requires much fundamental knowledge on wood THM behavior is required.

From the 1990s, as in Japan, various investigations and research studies on the molding, densification of wood and THM post-treatments as well as on the thermal heat treatment of wood were started in Europe, in the United States and recently in Canada. For example, in Denmark, besides the work of Morsing (2000), a machine was developed for pre-compression of wood in the longitudinal direction. The procedure involves:

1. compressing the steamed softened wood element longitudinally about 20%;
2. releasing the applied compression, after which the wood partially recovers the compressed set, so that the residual deformation is about 4-5%; and
3. then allowing the wood to cool at room temperature and dry to a moisture content of 12%. This type of densified wood element can be bent without steaming and jigs.

In Switzerland, Schrepfer and Schweingruber (1998) have studied the deformation of wood cells during densification. Navi and Girardet (2000), Heger (2004), Navi and Heger (2005) and Navi *et al.* (2007) have investigated the origin of set recovery and have shown that the hydrolysis of wood hemicelluloses is sufficient to almost totally eliminate the shape memory through relaxation of the stored strain energy. In Germany, Rapp and Sailer (2001), Rapp *et al.* (2006), have developed a system for wood heat treatment based on oil at a high temperature (OHT). In this open system, rape-seed oil is used as a medium as it can be heated above 200 °C. The same system was investigated by Haller and Wehsener (2004) and Welzbacher *et al.* (2008) for eliminating the set-recovery of large industrial densified wood elements. The results have shown that oil-heating at a temperature above 200 °C can completely eliminate the compression-set recovery of large densified spruce panels. Such a product has shown improved resistance against micro-organisms. Because of its improved durability and the impregnation of densified wood by oil during the post-treatment, this type of product can be used for outdoor purposes.

In USA, due to the harvesting of rapidly grown wood with low density, wood researchers have shown an interest in the possible opportunities of using densified wood in the production of composite layered materials (Kamke, 2004; Kutnar *et al.*, 2008a,b). To densify small low-density hybrid poplar specimens, Kamke and Sizemore (2005) have developed a semi-closed THM reactor. This system might have had some advantages for the closed system developed in Japan or in Switzerland, but unfortunately the dimensions of specimens used for densification are small and it would be difficult to evaluate the interest of this system when employed for large-sized wood elements for building and construction purposes.

At the present time, the technology of TH wood treatments has been industrialized and their products are being commercialized. The research in the field of wood THM molding and wood densification as well as the post-treatment of compressed wood by THM actions aims to widen this field to large-sized wood elements for applications in the building industry. Moreover, many other THM treatments are advancing at the laboratory level. However, the main characteristics of all recent TH/THM treatments are based on scientific knowledge, wood physics, chemistry, mechanics, material and wood sciences. Understanding these treatments clearly requires the appropriate scientific knowledge related to wood. It is essential to know the structure and composition of wood, the elasto-viscoplastic behavior of the wood according to the temperature and the wood moisture content, the effects of TH/THM processing parameters (heat, humidity and processing time) on the mechanical properties, the chemical kinetics of reaction and degradation of wood, heat and mass transfer in wood during TH/THM processing and the mechanisms of eliminating recovery-set in THM woods.

1.4 REFERENCES

Appel, H.R., Wender, I. & Miller, R.D. (1969). *Conversion of urban refuse to oil.* U.S. Bureau of Mines, Technical Progress Report 25.

Appel, H.R., Fu, Y.C., Illing, E.G., Steffgen, F.W. & Miller, R.D. (1975). *Conversion of cellulosic wastes to oil.* U.S. Bureau of Mines, Technical Progress Report RI 8013,27.

Burmester, V.A. (1973). Effect of heat-pressure treatments of semi-dry wood on its dimensional stability. *Holz als Roh- und Werkstoff, 31*(6):237-243.

Freed, S.A. & Baby, R.S. (eds.) (1983). *La grande aventure des Indiens d'Amérique du Nord.* (The great adventure of Indians in North America.) Sélection du Reader's Digest, Cop. Paris, ISBN 2-7098-0074-8.

Garrote, G., Domínguez, H. & Parajó, J.C. (1999). Hydrothermal processing of lignocellulosic materials. *Holz als Roh- und Werkstoff, 57*(3):191-202.

Gefeller, B., Pizzi, A., Zanetti, M., Properzi, M., Pichelin, F., Lehmann, M. & Delmotte, L. (2004). Solid wood joints by in situ welding of structural wood constituents. *Holzforschung, 58*(1):45-52.

Haller, P. & Wehsener, J. (2004). Festigkeitsuntersuchungen an Fichtenpressholz (FPH). (Investigation of the strength of compressed wood of pine.) *Holz als Roh- und Werkstoff, 62*(6):452-454.

Heger, F. (2004). *Etude du phénomène de l'élimination de la mémoire de forme du bois densifié par post-traitement thermo-hydro-mécanique.* (Study of the phenomenon of the elimination of the shape memory by thermo-hydro-mechanical post-processing of densified wood.) PhD. Thesis No. 3004, EPFL, Lausanne, Switzerland.

Hill, C. (2006). *Wood modification – Chemical, thermal and other processes.* John Wiley & Sons, Ltd. West Sussex, England, ISBN 0-470-02172-1.

Hon, D.N.-S. (ed.) (1996). *Chemical modification of lignocellulosic materials.* Marcel Dekker, Inc., New York and Basel, ISBN 0-8247-9472-9.

Hon, D.N.-S. & Shiraishi, N. (eds.) (1991). *Wood and cellulosic chemistry.* Marcel Dekker, Inc., New York and Basel, ISBN 0-8247-8304-2.

Inoue, M., Norimoto, M., Tanahashi M. & Rowell, R.M. (1993). Steam or heat fixation of compressed wood. *Wood and Fiber Science, 25*(3):224-235.

Ito, Y., Tanahashi, M., Shigematsu, M. & Shinoda, Y. (1998a). Compressive-molding of wood by high-pressure steam-treatment: Part 2. Mechanism of permanent fixation. *Holzforschung, 52*(2):217-221.

Ito, Y., Tanahashi, M., Shigematsu, M., Shinoda, Y. & Otha, C. (1998b). Compressive-molding of wood by high-pressure steam-treatment: Part 1. Development of compressively molded squares from thinnings. *Holzforschung, 52*(2):211-216.

Kamke, F.A. (2004). A novel structural composite from low density wood. In: *Proceeding of the 7th Pacific Rim Bio-Based Composites Symposium.* Nanjing, China, pp. 176-185.

Kamke, F.A. & Sizemore, H. (2005). *Viscoelastic thermal compression of wood.* US Patent Application No. US2005/0006004A1.

Khamouski, A.N. (1999). Handicrafts, a means of expanding jobs, social justice. *Iran Commerce Magazine, 6*(1).

Kollmann, F. (1936). *Technologies des Holzes*. (Wood technology.) Verlag von Julius Springer, Berlin.

Kutnar, A., Kamke, F.A. & Sernek, M. (2008a). The mechanical properties of densified VTC wood relevant for structural composites. *Holz als Roh- und Werkstoff, 66*(6):439-446.

Kutnar, A., Kamke, F.A., Nairn J.A. & Sernek, M. (2008b). Mode II fracture behavior of bonded viscoelastic thermal compressed wood. *Wood and Fibre Science, 40*(3):362-373.

Kyomori, K., Shigematsu, M., Onwona-Agyeman, S. & Tanahashi, M., (2000). Virtual models of the molding process of wood compressed by the high-pressure steam technique: towards the development of an aid-system for designing and manufacturing. In: *Proceedings of the 6th International Symposium on Virtual System and Multimedia (VSMM 2000)*, Thawahites, H., (ed.). Gifu, Japan, pp. 403-408.

Morsing, N. (2000). *Densification of wood: The influence of hygrothermal treatment on compression of beech perpendicular to the grain*. PhD. Thesis No. R79, Department of Structural Engineering and Materials, Technical University of Denmark.

Navi, P. & Girardet, F. (2000). Effects of thermo-hydro-mechanical treatment on the structure and properties of wood. *Holzforschung, 54*(3):287-293.

Navi, P. & Heger, F. (2005). *Comportement thermo-hydromécanique du bois*. (Behaviour of thermo-hydro-mechanical processed wood.) Presses polytechniques et universitaires romandes, Lausanne, ISBN 2-88074-620-5.

Navi, P., Girardet, F., Vulliemin, P., Spycher, M. & Heger, F. (2007). Effects of post-treatment parameters on densified wood set-recovery. In: *Proceedings of Third International Symposium on Wood Machining*, Navi, P. & Guidoum, A., (eds.). PPUR, Lausanne, Switzerland, pp. 69-72.

Norimoto, M. & Gril, J. (1993). Structure and properties of chemically treated woods. In: *Recent Research on Wood and Wood-based Materials*, Shiraishi, N., Kajita, H. & Norimoto, M., (eds.). Elsevier, Barking, UK, pp. 135-154.

Ormarsson, S. & Sandberg, D. (2007). Numerical simulation of hot-pressed veneer products: moulding – spring-back – distortion. *Wood Material Science and Engineering, 2*(3/4):130-137.

Pizzi, A., Leban, J.-M., Zanetti, M., Pichelin, F., Wieland, S. & Properzi, M. (2005). Surface finishes by mechanically induced wood surface fusion. *Holz als Roh- und Werkstoff, 63*(4):251-255.

Rapp, A.O. & Sailer, M. (2001). Oil heat treatment of wood in Germany: state of the art. In: *Review on heat treatment of wood. Proceedings of special seminar of Cost Action E22: Environmental Optimization of Wood Protection,* Rapp, A. (ed.), Antibes, France, pp. 45-62.

Rapp, A.O., Brischke, C. & Welzbacher, C.R. (2006). Interrelationship between the severity of heat treatments and sieve fractions after impact ball milling: a mechanical test for quality control of thermally modified wood. *Holzforschung, 60*(1):64-70.

Schrepfer, V. & Schweingruber, F.H. (1998). Anatomical structures in reshaped press-dried wood. *Holzforschung, 52*(6):615-622.

Shigematsu, M., Fukaya, M., Sasaki, Y. & Tanahashi, M. (1998). Three dimensional image analysis on molding process of wood compressed by high-pressure steam technique. In: *Future fusion: Application realities for the virtual age. Proceedings of 4th International Conference on Virtual Systems and Multimedia, (VSMM'98),* Thwaites, H., (ed.), Ohmsha, Tokyo, vol.2, ISBN4-274-90268-4, pp. 372-379.

Stamm, A.J. (1964). *Wood and cellulose science*. The Ronald Press Company, New York.

Stamm, A.J. & Seborg, R.M. (1941). Resin-treated, compressed wood. *Transaction of the American Institute of Chemical Engineers, 37*:385-397.

Stamm, A.J, Burr, H.K. & Kline, A.A. (1946). Staywood. Heat stabilized wood. *Industrial and Engineering Chemistry, 38*(6):630-634.

Stamm, B. (2006). *Development of friction welding of wood – physical, mechanical and chemical studies*. PhD. Thesis No. 3396, EPFL, Lausanne, Switzerland.

Stamm, B., Natterer, J. & Navi, P. (2005). Joining wood by friction welding. *Holz als Roh- und Werkstoff, 63*(5):313-320.

Stevens, W.C. & Turner, N. (1970). *Wood bending handbook*. Woodcraft supply Corp., Woburn, USA.

Suthoff, B., Schaaf, A., Hentschel, H. & Franz, U. (1996). *Method for joining wood*. Offenlegungsschrift DE19620273 A1. Deutsches Patenamt.

Tanahashi, M. (1990). Characterization and degradation mechanisms of wood components by steam explosion and utilization of exploded wood. *Wood Research, 77*:49-117.

TANAHASHI, M., KYOMORI, K., SHIGEMATSU, M. & ONWONA-AGYEMAN, S. (2001). Development of compressive molding process of wood by high-pressure steam and mechanism of permanent fixation for transformed shape. In: *First International Conference of the European Society for Wood Mechanics*, Navi, P., (ed.). Lausanne, Switzerland, pp. 523-524.

VORREITER, L. (1949). *Holztechnologisches Handbuch, Band I.* (Wood technology handbook, volume 1.) Verlag George Fromme & Co., Wien.

WELZBACHER, C.R., WEHSENER, J., RAPP. A.O. & HALLER, P. (2008). Thermo-mechanical densification combined with thermal modification of Norway spruce (*Picea abies* Karst.) in industrial scale: Dimensional stability and durability aspects. *Holz als Roh- und Werkstoff, 66*(1):39-49.

YAO, Y., YOSHIOKA, M. & SHIRAISHI, N. (1994). Soluble properties of liquefied biomass prepared in organic solvents I. The soluble behavior of liquefied biomass in various diluents. *Mokuzai Gakkaishi, 40*(2):176-184.

CHAPTER 2

ANCIENT USES OF THERMO-HYDRO-MECHANICAL PROCESSES

2.1 INTRODUCTION

Our prehistory is usually divided into eras having received their names from the particular material that characterized the period, e.g., stone, bronze and iron. This division is based on archaeological finds, and since wood is rare in such contexts, it is easy to forget that there was also a "wooden age". This age began when humans came to our country and it continues yet today.

For thousands of years, wood has been used by man for the manufacture of tools, weapons, utility equipment and instruments, in addition to as a building material. With the collected experience of numerous generations, consideration has consciously been given to wood properties in many areas of processing and usage. A very early example of a product made of wood that has survived until today is the Kalvträsk ski which can be seen in Figure 2.1. Such skis were used ca 5200 years ago and have been preserved by lying embedded in a swamp in northern Sweden (Åström, 1993). The wood for the ski was cut with vertical annual rings and the front tip of the ski was probably bent into a curved shape with the help of heat from an open fire in order to improve the performance of the ski. The use of heat and water to give skis the right shape was practised by the Saami (Insulander, 1998) and may be the oldest evidence for the use of THM processes.

Fig. 2.1 5200 year old ski made of wood with vertical annual rings and formed by a THM process (photo by Sune Jonsson, Västerbotten Museum).

The simplest and in most cases also the strongest woodwork takes advantage of the natural shape of the timber. Bends, forks and holes can all be exploited in many ways. Even on a large scale, such as the building of houses or wooden ships, greater structural integrity can be achieved by using a single piece of timber having grown into the right shape rather than by joining several smaller pieces together. The building of wooden ships and stave churches in old times as well as of utility goods in the old farmer community exemplifies that it has traditionally been common to use naturally shaped timber, see for instance Halldin (1963); Holan, (1990), p.179.

One of the most graceful chair designs is the horseshoe armchair, which originated more than 500 years ago in the time of the Ming Dynasty. The back rail and arms of the chair form a continuous semicircle, gently descending toward the front with the terminals of the arms bent slightly back in a rhythmic yet reserved curve. Often coupled with tapered, S-shaped side posts, the effect is of a spherical void being gently cradled, simultaneously giving a sense of emptiness and wholeness. These inspired forms grew out of the rich tradition of Chinese craftsmanship, and integrating of aesthetic preferences with the science of joinery.

Of the existing horseshoe armchairs, most of the curved armrests are made of three to five segments with joints that fit so snugly that no nails or glue are required. In making the curved arm-rest of the horseshoe chair, craftsman selected wood with a naturally curved grain that would follow the curvature of the rail, ensuring that the grain would be straight and strong at the joint. This way of joining naturally grown curved sections has for example been practised in traditional Japanese wooden buildings. During the fourteenth century, the carpenter's square became popular in Japan. Its skilful use made possible the precise construction of complicated joints used in house building (Bramwell, 1976). In Norwegian stave churches, the longer construction elements, such as the nave's quadrant arches, consisted of curved wood elements made out of several pieces so precisely joined that even today it is difficult to detect the connection points (Bugge, 1953; Holan, 1990).

Egyptian workmen, whether they built vast temples or manufactured furniture, inspire both awe and admiration. But who were these people? How did they practise their different craftsmanships? Did they use techniques based on THM-processes or were such processes unknown? Were they well rewarded for their skilled work, or were they exploited? In order to answer such questions we need contemporary written texts, preferably written by and for the ancient workers themselves. So far, such documents have been found in very few archaeological sites (Lesko, 1994). However, the works settlement of Deir el Medina is among the most thoroughly documented and best-known communities from the ancient world.

The earliest evidence of true wooden furniture is found in the Egyptian society that existed some five thousand years ago. The exceptional conditions for survival in royal tombs have given us famous examples of furniture and other utility equipment. It is apparent that different pieces of furniture had been created by 3000 B.C., and there is no doubt that a skilled work force existed in Egypt. The origins of the technique of wood bending have been the subject of some dispute among experts, but it is probable that wood bending was known in Egypt around 1000 B.C. (Rivers & Umney, 2005). Unsubstantiated evidence suggests that some furnishing may have been fabricated by wood bending as early as 2500 B.C.

Egyptian tomb-drawings of Nebamen at Thebes (ca. 1400 B.C.) provide evidence of manufacturing processes where it is probable that the Egyptian carpenters used heat and moisture to bend wood, as depicted in Figure 2.2. The reliefs in the tomb of the scribe Hesire at Saqqara, dated ca. 2800 B.C., contain representations of furniture, including a chair whose seat has been described by Aldred (1954) in Singer (1954) as "strengthened by bent wood supports, in hard- and softwoods" and a stool "with bent wood reinforcements".

More immediate evidence that the process of bending wood was known to the Egyptians is provided by the paintings in the tomb of the nomarch Amenemhat at Beni Hasan, ca. 1971-1928 B.C. It contains an illustration, cf. Figure 2.3, of a bowmaker who holds a rod over a receptacle that possibly contains hot water, while straight rods

Fig. 2.2 Section of tomb-drawing of Nebamen at Thebes showing various operations on different parts of a chariot. In the upper picture, the carpenters are bending poles, and the drawing at the bottom to the left shows a wooden chariot-yoke that may have been THM-processed (from Latham, 1957).

Fig. 2.3 Bending of bows from a facsimile of a wall painting in the tomb of Amenemhat at Beni Hasan, ca. 1971-1928 B.C. (from Newberry, 1893).

and completed bows are displayed nearby (Ostergard, 1987). Only green and unseasoned wood could be bent by this method; seasoned wood of large cross-sectional dimension can only be bent by softening with the use of a steam chest and there is no evidence that this apparatus was used in ancient Egypt (Killen, 2000). Montet (1946) construed Egyptian paintings showing bowmakers as meaning that the heating and bending of the wood was carried out before debarking and that it may thus be understood that the wood was green.

The technique of laminating thin sheets of wood, with the grain of one sheet perpendicular to that of the next, i.e., plywood, was known to Egyptian carpenters at least from the period of the Third Dynasty, 2686-2613 B.C. (Killen, 1994).

Nearly two thousand years after some furnishing may have been fabricated by wood bending in Egypt, the classic Greek Klismos chair was introduced as a form of furniture. Some scholars believe that the graceful shape of the chair's sabre legs was achieved through a bending process (Ostergard, 1987). The Klismos chair is a light and elegant chair developed by the ancient Greeks, and presented in Figure 2.4. Perfected by the 5th century B.C. and popular throughout the 4th century B.C., the chair had four curving, splayed "saber" legs and curved back rails with a narrow concave backrest between them. In spite of the fact that the chair is familiar in different variants from frequent representations in vase paintings and reliefs, no original specimens have come down to us. In the pictures, the Klismos chair always has the same primary shape, but there is no unequivocal design of the construction of the chair and how it has been manufactured. The dramatic, concave sweep of its distinctive legs has led to the conclusion that the form was probably achieved by steam bending (Richter, 1966).

The outward legs of the chair, which at the back are transformed into a comfortable inclined back-support, make very high demands on both material and design. That fact that the shape must bear the extremely large loads to which the legs are

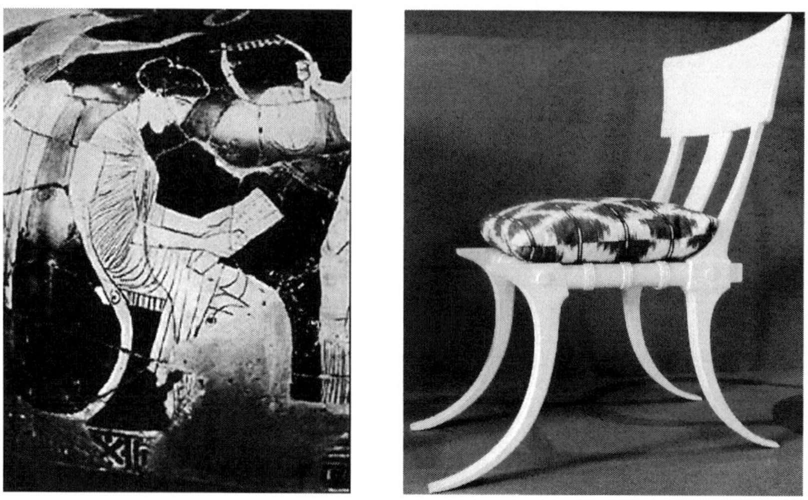

Fig. 2.4 Detail from a hydria found at Vári, Greece, showing the poet Sappho seated on a Klismos chair, ca. 420-440 B.C., in the National Archaeological Museum, Athens (left). A contemporary interpretation of the Klismos chair made by architect Åke Axelsson (right).

subjected supports the theory that the Greeks, already in the 5th century B.C., must have understood the technology of bending solid wood through heating and steaming. This hypothesis is corroborated by the Greek philosopher and scientist Theophrastus (372-287 B.C.). In his book, *Enquiry into Plants*, he states "In general those woods which are tough are easy to bend. The mulberry and the wild fig seem to be specially so; wherefore they make of these theatre-seats, the hoops of garlands, and, in general, things for ornament" and later "The work of bentwood for vessels is made of mulberry manna-ash elm or plane; for it must be tough and strong (Hort, 1916).

Another characteristic part of the chair is the broad upper rail at the back. In the 5th century B.C., it had a fairly weakly rounded flat shape, which can have been sawn and processed out of a single piece of solid wood. Later, in the 4th-3rd centuries B.C., the shape becomes strongly rounded and this development was probably achieved either by using a steamed and bent piece of solid wood or alternatively by the lamination of thin glued veneers.

2.2 WOODLAND CRAFTS

In older times, craftsmen in general had a profound, empirical insight into the subtle and diversified qualities of their materials. Knowledge handed down from generation to generation meant that they had a detailed understanding of how the properties of different kinds of wood could be used in an optimal way in various articles, whether for everyday use, for transport by means of carriages and boats, or for building purposes. An illustrative example within the furniture sphere is the so-called Windsor chair, depicted in Figure 2.5. The seat was traditionally made of durable hard-to-split elm (*Ulmus glabra*), the rear pins of tough and elastic yew (*Taxus baccata*) and the round-shaped back-piece of flexible ash (*Fraxinus excelsior*). The types of wood used in such old, solid wood furniture have in many cases probably been chosen on the basis of the physical properties of the particular wood species. In simple utility goods,

Fig. 2.5 A classic Windsor chair.

provincial furniture and interior decorations, the carpenter's choice has been limited to the species of wood growing locally. The species have been consciously used for various purposes depending on properties such as hardness, toughness and rot-resistance. An illustrative example of the provincial choice of different wood species is again the Windsor chair. When the tradition of manufacturing this type of chair spread to the USA, it was often made from other species of wood than those mentioned above (see for instance Dunbar, 1984; Abbott, 1989).

2.2.1 Bending of solid wood in woodland craft

Techniques for bending wood have, as already indicated, existed since prehistoric times, although early man used heat to help straighten rather than bend his arrows (Sentance, 2003).

Rigid pieces of wood may be rendered pliant by heating, steaming, or boiling, and these methods are used in such varied crafts as basketry, chair-making, and the shaping of tool handles and wooden hoops (Edlin, 1949). Thin rods will often bend satisfactorily without treatment, if the wood is handled whilst it is still green and unseasoned, and in other instances water alone may suffice to restore its pliancy. For walking sticks, craftsmen have often bent and bound wood while it is still alive and green, so that it will grow into the desired shape without being weakened by the stress of stretching and compression (Edlin, 1949).

Sometimes the material is warmed over a naked fire or in damp sand. The softening treatment may be used for straightening curved rods as well as for bending straight ones. In either case, the wood retains the shape in which it is dried and "set", as shown in Figure 2.6.

Ash wood is a strong, tough and pliant material that is also light in weight, readily cloven and easily bent either in the green state or after steaming. Consequently, ash is a common material in woodland crafts in regions where this species is commonly grown. Edlin (1949) describes in detail how split ash and beech can be bent into

Fig. 2.6 Bending a length of steamed ash to form the back frame of a Windsor chair (from Edlin, 1949).

circular shapes to form hoops and rims. This craft was widespread in the countryside of Great Britain until about the 1950's, and it has also been practised in many other parts of the world.

When making a hoop, a straight ash pole is first split into three or four segments with a froe (sort of knife), while being held in the angle of a stout three-legged break. The hoop-shaver then trims each piece to shape with his draw-knife, sitting astride his wooden shaving horse, which holds the wood in its vice-like grip. Each thin band is then warmed over a fire to render it pliable and then bent between two rollers or over a wooden frame. A more "modern", and simpler method is to steam the band in a steam chest over a water boiler. This renders the band so pliant that it can be coiled without further treatment. The final coiling is done on a rigid wooden frame with four, six, or eight spokes. The hoop is formed and held in its circular shape by being forced against adjustable pegs set on the spokes, until its two ends can be nailed together. Hoops are made in sets of six, known as coils, and each is coiled in turn within the pegs or within the stronger shive or "master hoop" that forms the pattern. The strongest and largest ash hoops are those used by coopers to set up their barrels, but ash hoops have been used in a wide range of applications, such as runners holding the sails to the mast, sieve rims, rims of military drums, bushel measures for grain, seed-lips for sowing grain in open fields, and old-fashioned spinning-wheels. Bent ash wood has also traditionally been used for the handles of various tools employed by farmers, foresters, craftsman, carpenters etc.

Before the modern composite materials were developed, ash was the sportsman's wood. Ash was found in the ordinary tennis racket, the frame of which may consist of a single piece of wood bent, after steaming, into a circular form with it ends extending down into the handle. Cleft ash is used, because its grain is unbroken throughout, which is not always the case with sawn material. Later, tennis rackets were made of laminated frames.

2.3 THM-PROCESSING IN THE CONSTRUCTION OF WOODEN VESSELS

Vessels of wood have been in use on water for thousands of years. Before ships or even boats, there must have been a long succession of floats and rafts used as aids to swimming. However, details of drawings, carvings, and models of Egyptian, Greek, Assyrian, and Minoan times, and even the work of Scandinavian Neolithic artists, representing early vessels do not suffice to show how they were constructed (Digby, 1954). It is even less possible to determine whether any part of the vessel construction was the result of a thermo-hydro-mechanical (THM) process.

In the introduction, the fabrication of a canoe from a log is presented. A logboat is carved out of a single log and then expanded over an open fire. Such a boat is light and has much more room than a regular logboat. Extra boards can be added to increase its size. This type of vessel belongs to the group of boats that are called primitive boat-types, i.e., boats built of skin, bark, reeds or hollowed-out tree trunks, that have been used by man for thousands of years. Dug-out canoes have been found wherever there are suitable forests. They vary in shape depending on local requirements and on the available material. The limiting factor is the tree diameter. The size of the hull was often increased during manufacture by filling the hull with wet and perhaps even heated sand, the weight of which forced the sides apart (Digby, 1954). Today such boats are used by different aboriginal groups, e.g., Greenlanders, Indians in Canada and the Punan Bah people in Borneo. The Punan Bah people use dugouts or longboats for transport on the many rivers that flow through the landscape of their great island. A special feature of these longboats is that the tree-trunk, after being hollowed out with an axe, is widened over a fire. This technique is also known from prehistoric times with regard to longboats in northern Europe. After widening, a plank is sewn to each side of the boat and we thus have the first step towards the development of the clinker-built ships with which we are familiar from the Viking period (Crumlin-Pedersen, 1970; Nicolaisen & Damgård-Sørensen, 1991).

Nicolaisen & Damgård-Sørensen (1991) have studied the technique for fabricating longboats in Borneo and they summarize the main steps of the boat-building process as

- selection and felling of the tree;
- preparation and cleaving of the trunk;
- preliminary shaping of the base;
- trimming of the base;
- expanding the hull by heat; and
- sewing.

The Punan Bah build their longboats of the same kind of species. Ideally they use meranti (Shorea gen.), but to save time during building softer species can also be employed. The Punan Bah select their wood on the basis of a variety of criteria, all associated with their beliefs in a shared destiny for the boat-builder and his boat, as well as taking into account the quality of the wood, its hardness, accessibility and size.

The boat-building process begins in the rain forest, where the tree has been felled. Here, the branches are cut off, and the bark removed. The length of the boat is then measured out on the bare trunk and the selected piece of the trunk is cut. Thereafter, a cleavage-line is drawn lengthwise, exactly following the fiber direction of the trunk. The trunk is cleaved into two halves with the help of wooden wedges. Both halves of the trunk can be used for longboats if small boats are to be built, whereas for large boats, only one part of the trunk is utilized.

The preliminary shaping of the base naturally involves following sub-phases: removing the sapwood, hollowing out the trunk, and shaping the base. The preliminary boat is then transported to the place were it will be finished.

After the preliminary shaping, the base has to be trimmed and smoothened, both externally and internally, to give the final shape and thickness. This is a time-consuming precision work done by hand with the help of adzes and axes.

The widening of the base takes place over a fire: the entire hull is placed over a large bonfire until the wood reaches a temperature at which it becomes flexible. The sides are then twisted outwards and downwards with the help of large forked branches so that the base is flattened and widened, while the ends lift themselves, and the sheer in the base increases. The widening is first and foremost designed to improve the stability of the boat by increasing the sheer in the sides. In addition, the expanded boat is hardened so that it can better withstand the attacks of rot and insects.

The hull is first placed over the fire with the underside downwards and, after about five minutes, it is turned over so that the charred underside is faced upwards. After about an hour, the hot wood is stretched outwards and downwards in the central region. When the wood is sufficiently stretched, the sides are fixed with the help of a wooden bar between them. The widening has not been completed at this position but, in order not to split the wood, it is necessary to stretch the boat over the whole of its length before returning to the center for the final widening. The total increase in width at the middle of the hull is about 14% based on the original width. Before expanding, the hull was thus about 0.7 m in width and 7 m in length. Figure 2.7 shows a drawing of such a hull before and after the expansion. It can be worthwhile to notice that the expansion process used by the Punan Bah people in Borneo differs from that used by the North American Indians, and which is presented in the introduction.

The sewing involves a number of processes that gradually transform the expanded boat-base into a proper boat with lashed-on strakes, stem and stern. The sewing process involves the following phases: the preparation of the lashing material, the forming and fastening of the thwarts, the dressing and fastening of the strakes, and finally the dressing and fastening of the stem- and stern-transoms.

The tradition of expanding the hull by heat from an open fire can also be found as far away from Asia as Northern Europe. The Hjortspring boat, built around 350-300 BC and found in the 1920s in Denmark, is a late link with the millennial development of the log boat. Some part of the construction can however be linked to the tradition of skin boats (Crumlin-Pedersen, 1970). It had an external length of 21 meters and was a 2-meter wide, clinker-built wooden boat with space for a crew of 22-23 men who propelled the boat with paddles (Rosenberg, 1937). The boat is one of the oldest

(a)

(b)

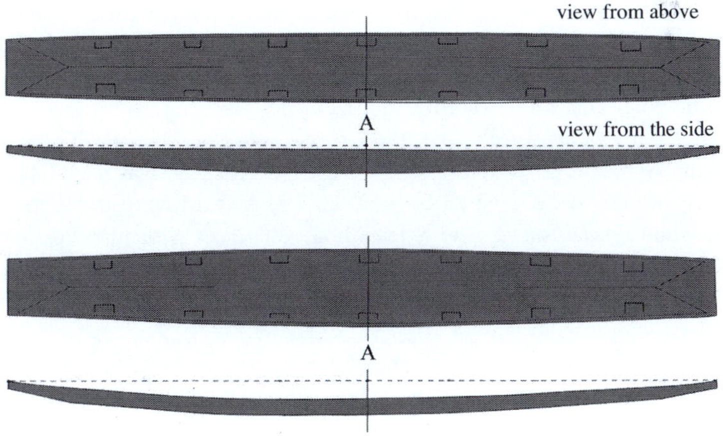

Fig. 2.7 A longboat under construction (a) before and (b) after expansion of the hull. The width at point A is 660 mm before and 750 mm after the widening. The distance between a line between the aft and fore end of the boat (the dotted line in the figure) and the rail at point A is 30 mm before and 150 mm after the widening. The length of the boat is 7 m (after Nicolaisen & Damgård-Sørensen, 1991).

find of a wooden plank ship in Scandinavia and its closest parallels are the thousands of petroglyph images of Nordic Bronze Age ships.

Five large lime-wood planks made up most of the boat. According to the reconstruction, there was no keel, but a 15.3-m long – up to 50-cm wide – lime plank with carved out cleats for fastening the frames. At each end, it had a V-shaped section on which flanged stem and stern pieces were sewn. It continued some distance fore and aft of these to form the bases of the lower of the characteristic double protrusions. Whether the bottom plank has been expanded from a hollowed-out trunk or manufactured directly to shape cannot be established. Figure 2.8 depicts a reconstruction of the Hjortspring boat.

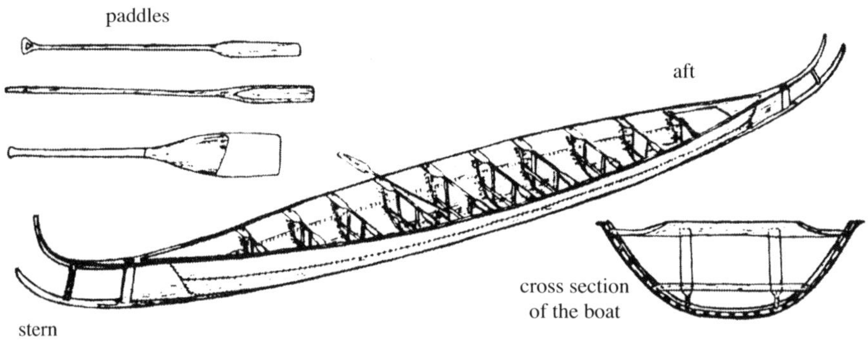

Fig. 2.8 Reconstruction of the Hjortspring boat (from Crumlin-Pedersen, 1970).

The reconstruction of the Hjortspring boat indicates a plausible line of development from the simple log boat to one with sides extended by planks. This evolved into boats of the Hjortspring type, which is essentially a log with two added strakes of planking. In time, the planks grew in number and the log turnes into a proper keel. This corresponds well with the shell-based way of designing boats according to Nordic tradition, where the frames are inserted after the planking.

Expanded log boats were in use at the same time as clinker-built boats, right up to the Viking Age. The so-called Björke boat, cf. Figure 2.9, was found by the shore of a lake in Sweden and can be dated to about 450 A.D. Most of the wood was well preserved when it was found. The Björke boat was built with many different techniques: the bottom was an expanded lime tree with added boards of pine wood, the frame was made of spruce lashed to the hull with juniper withes, and the added boards were fastened with iron nails.

The construction process for the expanding looks fairly like that still used by the Punan Bah people. It was practised in Finland until at least the 1930s for small boats of aspen (Nikkilä, 1947).

In Egypt, where all habitable and cultivated land was irrigated by the Nile, we should expect a very early start in navigation and wooden shipbuilding. The first datable boat representations are models founded in graves. The very oldest appears to be from the Neolithic (ca. 6000 B.C.) site of Merimda Beni Salaam and it seems most

Fig. 2.9 Reconstruction of the Björke boat from 450 A.D. at the Viking ship Museum in Roskilde, Denmark.

likely that these boats were made of papyrus bundles, not wood (Vinson, 1994). It is impossible to say when the first wooden ships were built in Egypt, but finds reveal wooden boats as old as 5000 years. However, the early Egyptian boats and ships were of quite a simple construction without any keel, and the boat-building technology evolved independently within Egypt in response to local conditions (Petrie, 1933). The way in which the boats and ships are built reflects aspects both of the legitimization of power and of participation in a regional trade network at least occasionally accessing the Red Sea before the third millennium B.C. (Ward, 2006). A contemporary and detailed description of Egyptian boats and boat-building techniques is given by the Greek historian Herodotus (ca. 484-425 B.C.) in his history of the Greco-Persian Wars (Godley, 1920). It is difficult to find evidence that their shipwrights used any THM processes.

An odd water vessel is the ***quffa*** that was probably invented in the regions of the Tigris and Euphrates and has been used in this region since at least the seventh century B.C. and probably goes back to the beginning of civilization (Digby, 1954). This vessel is a circular skin boat in which the flexible leather or hide cover is stretched over an internal frame of bent wood. Craft of this sort are not directly in the line of evolution, which developed into the later standard types of ships with keel and ribbed hull and which derived ultimately from the dug-out canoe.

With regard to design, Scandinavian wood treatment perhaps reached its peak during the Viking Age. Both the Viking ships with their inventories and the early Norwegian stave churches are artistic masterpieces that are difficult to surpass even today. The ships are especially fascinating since they were built to a superb standard with a high prestige value and a refined technology from which we still have things to learn when designing modern yachts, cf. Figure 2.10.

The term Viking ship does not describe a single type of ship but is a collective designation for the vessels built from the end of the 8th century to the beginning of the 12th century in accordance with common Nordic boat-building traditions. Viking ships were open, clinker-built vessels with a side rudder and a single mast. They were powered by oars or driven by the wind, using a square sail. Their hulls, which had a

Fig. 2.10 A reconstruction of a small boat from the Viking Age at the Viking ship Museum in Roskilde, Denmark.

light and flexible design, were held in shape by lateral timbers (floor timbers, beams and knees) and by longitudinal reinforcements (stringers and keelson). The longships were primarily warships, manned with a maximum of warriors who also acted as oarsmen. After a defeat in battle, the ship might still be handled by a crew numbering only a third or a quarter of the original number. In the 17 m-long Skuldelev 5-replica Helge Ask, the full crew was 26 oarsmen (Crumlin-Pedersen & Olsen, 2002). In longer ships, the number of men was increased by two for each additional meter, as the average spacing between the oars was one meter or slightly less. The largest Viking longships were probably manned by as many as 80 men. In broad ships like the Norwegian Gokstad ship, there would be room for several more men than the number of oarsmen. This was also the case in the broad and deep cargo ships. They could be sailed by a crew of only 4 to 5 men, but they could transport a much larger number of people if necessary, for instance for the transfer of settlers from Norway to Iceland.

The Vikings were skilled shipwrights who, on the basis of experience and tradition, could build ships that were adapted for particular purposes and specific seas. Their beautiful longships with an extensively curved stern and stem put great demands on the professional skills of shipbuilders as well as on the quality of the wood material. Viking shipbuilders preferred to build their ships from oak in southern Scandinavia and from pine in northern Scandinavia. Other species of wood were, however, also used for special purposes, such as ash for the upper planks in the longships. An interesting observation is that the Vikings probably did not use heat to render the wood pliable during the construction of the ships, and it is notable that they did not use heat even for the bending of the highly curved boards for the hull.

The building method is called "by eye", i.e., without drawings and from the outside inwards: the keel was first laid out, usually in one piece, after which the stern and stem were raised. The planking was then assembled with iron rivets in clinker fashion, creating the lines of the hull by varying the shape of the planks and the angle from one plank to the next. After the bottom part had been built up with the planking, the bottom frames, usually of naturally bent wood, were cut to shape and inserted. The sequence could then be repeated with the sides of the ship. The same working process can also be seen in the Bayeux Tapestry from ca. 1070 A.D., which from a Norman perspective reflects the circumstances surrounding the invasion and battle of Hastings in 1066.

The secret enabling the Vikings to bend oak wood in the hull to a comparatively great extent is the technique of splitting the log radially with the help of axes and wedges and bending the wood in the green state. For detailed studies of the conversion method, see for example Olsen and Crumlin-Pedersen (1967), McGrail (1982) or Indruszewski (2004).

In the manufacture of boards for the hull, the log was first split into two halves, then into 1/4 and 1/8-parts and sometimes down to 1/32-parts, cf. Figure 2.11. The number of boards was largely dependent on the diameter of the log and 32 pieces were considered to be the upper practical limit from a single log (Olsen & Crumlin-Pedersen, 1967). When a log is split into halves or more, the split follows the grain orientation in the longitudinal direction, so that fewer fibers are cut than in the sawing or shaping of a log with an axe. The sapwood was usually removed at this stage. The board width was, therefore, determined by the radius of the log minus sapwood and the inner part of the heartwood. For example, a log with a diameter of 1 m is required to obtain an approximately 30-cm wide board.

McGrail and McKee (1974) mentions that some of the major advantages of radially split boards include: the longitudinal fibers (grain) being parallel to the board sides contributing considerably to the mechanical strength; and dimensional movements due to changing humidity being greatest in the smallest dimension of the board, i.e., its thickness. Godal (1995) states that radially split wood is "watertight" (the rays are parallel to the sides of the boards), resistant to cracking and generally more durable.

Viking ships were built of fresh timber. Fresh or green wood is tougher than its dry counterpart and is also more pliable. By splitting the log into smaller and smaller pieces, boards that were light and thin but, at the same time, unprecedentedly tough were obtained. If we are careless and try to build a copy of a Viking ship today with sawn and less carefully selected wood, the boat will either break or be too heavy. The great advantage of green wood is that the ship retains its shape when launched. Since the boat is built in its wet form, no swelling or shrinkage and thereto-related problems occur. Experiments show that the radial splitting of large oak logs can be done without undue difficulty because the rays assist the cleaving.

With a combination of high knowledge of the wood material and skilled shipwrights, the Vikings used radially split green wood in their fabrication. Their ability to produce complicated bent wood constructions without the use of heat is unique in shipwrighting. The heating of wood either from an open fire, boiled water or steam has been used for at least a thousand years before and after the Viking Age.

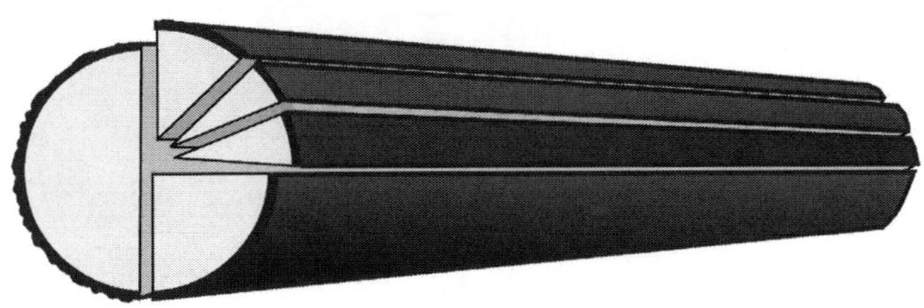

Fig. 2.11 Radial splitting of a log into sectors used for boards for the hull of a Viking ship.

The medieval shipwrighting for cargo ships in the Scandinavian countries was a direct continuation of the shipwrighting of the Viking Age when it comes both to the shape of the hull and the technology (Halldin, 1963). It was based on the clinker-building technique with rather thin boards throughout the Middle Ages, and the carvel-building technique probably did not come into use until the late thirteenth century. It was vital that the timber used for building the different parts of the ships was selected according to its natural shape, e.g., natural bent wood to produce frames, knees, etc., as illustrated in Figure 2.12. The fiber structure of the wood is thereby maintained, which gives the ships strength and stability.

Fig. 2.12 Natural bent wood for the building of ships. A drawing used for teaching purposes at the Maritime Academy in Stockholm (Halldin, 1963).

As in the Viking Age, the timber was produced through radial splitting of oak logs with the help of an axe and wedges, a technique that was used until the 16th century. From these wedge-shaped pieces of timber, the boards were axed. These were heated over an open fire and doused by water to make them pliable, so that they could be given the appropriate shape for their place in the hull. Figure 2.13 presents this technique.

Fig. 2.13 Softening of boards over an open fire (top) and the installation of pliable boards on the hull (Claesson Rålamb, 1691).

The ancient method, when shipbuilders made the boards pliable by heating over an open fire, gave way only very slowly to the gentler treatment by boiling in water or steaming in a steaming vessel. For example, steaming vessels were not used in Swedish shipyards until the mid-eighteenth century (Halldin, 1963).

2.4 THE MANUFACTURE OF WOODEN CASKS

2.4.1 Traditional coopering

Wooden casks deserve to be remembered for the important role that they have played in history. Casks have arguably been the most significant shipping container in history over at least 2000 years of civilization (Twede, 2005). The cask-maker is a specialist craftsman with a set of purpose-made tools. He is known in Britain as a ***cooper*** and this is now a common British family name. In the United Kingdom, ***cask*** is the general term and the word ***barrel*** technically refers to a cask with a capacity of 36 gallons, but the techniques are the same for a ***firkin*** which holds nine gallons, a ***hogshead*** holding 54 gallons or a ***butt*** holding 108. The coopering can be divided into three main branches; wet, dry and white, were wet coopering is the most advanced skill (Kilby, 1971). The wet cooper made casks with a bulge to hold liquids. The dry cooper fabricated casks with a bulge to hold a wide variety of dry commodities. The white cooper made straight-sided, splayed vessels. There were thus two basic types of casks: tight casks mainly designed to hold liquids, heavy products and explosives, and slack casks used for dry goods.

Casks probably evolved from open-top wooden tubs with straight sides, wider at the top than at the bottom (Twede, 2005). One of the earliest representations of a wooden tub has been found in the Egyptian tomb of Hesy (2690 B.C.) showing a wooden cask for measuring corn (Kilby, 1971). A wall painting in a tomb at Beni Hasan shows a cooper and tubs with staves being used in the grape harvest (Newberry, 1893).

In the first book of Kings in the Bible, (Ch. 18, Ver. 33), it is told that, at the time of King Ahab of Israel, in the ninth century B.C., the prophet Elijah defeated the priest of Baal. Three times he filled four barrels of water and poured them over the burnt offering. In the same book, a widow had a "handful of meal in a barrel".

Liquid-tight wooden casks with a bulge and made from staves were invented during the time of the Romans (Twede, 2005). Pliny the elder mentioned that casks were used for wine in the Northern part of the Empire, where wood was a more abundant natural resource than clay (Rackham, 1945). Cask-making may have developed concurrently with the ship-building technology. The materials, modes of construction and tools are similar: thin slabs of wood being bent into curved shapes and bound together. Figure 2.14 shows a reconstruction of a cask from the Viking Age made of spruce. As already mentioned, the Vikings were skilled shipbuilders and also good coopers. For further reading about the ancient history and development of coopering, see for instance the work of Kilby (1971).

Casks are also made and used for different purposes in the Far East, but they are shaped differently from European casks in that they are wider at the top than at the bottom and have straight sides with no bulge.

Fig. 2.14 A reconstruction of a wooden cask of Norway spruce from the Viking Age.

The manufacture of casks is, as has already clearly emerged, a very traditional craft. The techniques to produce casks have varied over the centuries and have also shown local variations, but the traditional manufacturing of casks with hand tools did not change much until the end of the nineteenth century when coopering started to be industrialized and various new production methods were invented. From about the 1950s, the competition from metal casks, bulk beer tanks, bottles, cardboard boxes, fiber-board drums, cans and large metal shipping containers incorporating new methods, together with rising costs, caused the bottom to drop out of the market for wooden casks.

The traditional production of the casks can be divided into eight sequences (Kilby, 1971) which are

1. selection and preparing of the raw material for staves;
2. shaping of the staves;
3. raising-up the cask;
4. firing and bending of the cask;
5. chiming the cask;
6. shaving the inside of the cask;
7. placing the heads into the ends of the cask and placing permanent hoops, and boring of tap holes; and
8. pickling of the casks.

Cask work starts with the cooper selecting the wood, called blanks, to be used to make staves. Casks have been made out of all kinds of timber. The earliest casks were made of palm; later the Celts used Pyrenean silver fir (*Abies alba* Mill.), and no doubt every kind of timber has been experimented with at various times, until the most suitable species, depending upon availability, quality and competitive costs, became the prerogative of particular branches of the trade (Kilby, 1971).

The white cooper worked largely with oak, but beech was also used quite extensively, and chestnut, ash, elm, teak, sycamore, pine and fir and even yew were employed when taste was not the main consideration. The wet cooper exclusively used oak because of demands relating to the taste and smell that can be transferred from the wood to the contents of the cask, the strength of the cask, the bending properties of the wood during the manufacture of the cask, and the need for the wood to be tight so that the contents would not leak (Kilby, 1971).

The most important properties of the wooden staves for tight casks is that they are free from knots and have vertical annual rings, i.e., that the logs are quarter sawn (Kilby, 1971). The vertical annual rings help to prevent seepage and to control warping and shrinkage. In early times, the logs were split with wedges which resulted in stronger and more straight-grained staves than their sawn counterparts. The staves are dried before further handling. They are then cut into appropriate lengths, widths and thicknesses, and inspected to make sure that each stave is properly quartered, and that it has no defects in structure, cracks or sapwood. A defect in structure on the convex side of a stave may signify that it will crack when being bent, whereas a defect on the concave side could possibly be hollowed out and therefore might not matter.

The shaping of the staves in their straight state is called ***dressing*** and includes four stages: tapering the edges of the stave ***(listing)***, shaping the outside (the back) of the stave ***(backing)***, shaping the inside of the stave ***(hollowing out)***, and finally preparing the angle on the edge of the stave corresponding to the radius of the cask on the end and in the belly ***(jointing)***, cf. Figure 2.15. After the staves have been dressed they are ***raised up***. This means that the staves are put together tightly at one end with the help of a raising-up hoop prior to the cask being bent. The staves must be matched according to toughness. If a soft stave is placed between two hard ones it will be forced outwards and crack either in the firing or later by wear. The staves are loosely assembled with an iron hoop around the top and another hoop is hammered over them, pulling them into a round shape. In early times, these hoops were usually made of cleft wood from ash or willow. When all the necessary staves are raised up in the hoop, it is hammered tight. The cask is now ready for firing.

During firing, the cooper placed the cask on a cresset, a small burning brazier, until the staves were hot right through, which he assessed by holding the back of his hand against the outside of the stave. The so-called truss hoops were then driven over the cask. During heating, the wood can be moistened by adding water on the outside and inside of the cask to make the wood more pliable. In about 1920, stout (ca. 31 mm in thickness) or extra stout (ca. 38 mm in thickness) casks became more common and it then became necessary to prepare the wood more thoroughly before firing. This was done either by steeping the cask in a tank of water which was kept boiling for half an hour, or by putting it over a jet of steam and lowering a bell, or a large cask, over the top of it to keep the steam around the cask. This softens the wood and makes the staves more pliable for bending. At this point, some casks for wine or whisky are slightly charred to give a distinctive taste to the contents. When the staves are pliable enough, they are bent together and secured by hammering hoops of different sizes into position so that the shape of the cask is created.

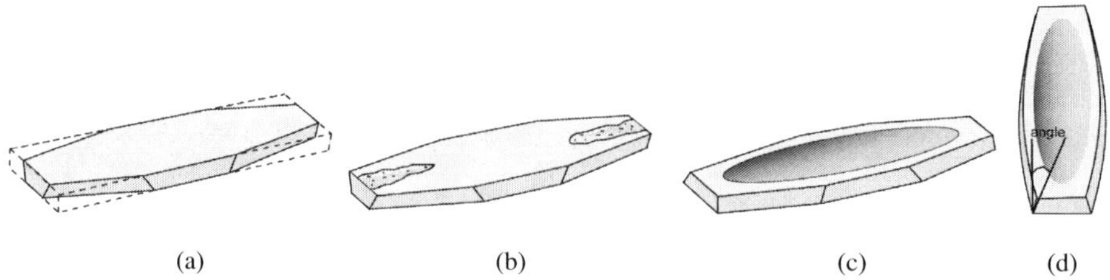

Fig. 2.15 Shaping of staves for a cask: (a) listing, (b) backing, (c) hollowing out, and (d) jointing.

The firing step is essential in order for the cask to acquire *set*, i.e., so that, when the hoops are taken off, the staves will remain bent and not spring back into a straight position as they would if the cask was subjected to less than half an hour's firing. This spring-back is said to be more common when the casks are bent by machinery than when they are bent by hand, but this depends strongly upon the quality and toughness of the timber used and on the exactness of the joints. A slightly rounded joint will cause a stave to crack in the pitch.

The next task for the cooper is to put a bevel on each end of the cask and create the grooves for the heads. This is called ***chiming*** the cask. It is always best to do this while the cask is warm from the firing, as the wood then seems to be easier to cut. The cut must be straight so that the head fits flat, at a uniform depth, instead of being formed as a sloping cylinder. The chiming includes adjusting the length of the staves and creating a chime with an adze, making the end grain perfectly straight with a topping plane, leveling the wood where the groove will be placed with help of a ***chiv***, a small plane hung at a fixed distance from a piece of wood which slides around the top of the cask, and finally cutting the groove into which the head will fit with the help of a tool similar to the chiv.

The next job is to shave the inside of the cask so that it is perfectly smooth. This is done by means of an inside-shave, a small rounded plane with wooden handles on either side, or with a stoup-plane, simply a small rounded plane.

Finally, the heads of the cask are fitted into the ends, and the rings, which will have become loose, are replaced with permanent ones that are heated so that they bind the staves tightly together as they shrink. The circular head, made from straight quarter-sawn boards joined together with dowels, is cut with a beveled edge that is inserted into the groove. A bung-hole is then drilled into the cask through which the cask will be filled and thereafter closed with a cork or a bung.

In order to prepare an oak cask to hold beer, wines or spirits for the first time, it was customary to pickle the cask, i.e., to fill it with a solution of brine and sodium carbonate in order to neutralize the acid-tasting tannin in the wood. The cask was allowed to soak for three days and was then filled with clear water for a day before it was washed out and filled with the liquid in question.

2.4.2 Machine coopering

As early as 1811, a patent was granted to the firm of Plasket and Brown Ltd. in England for a coopering machine designed to cut the heads for casks (Hebert, 1848). Most of the early machinery was crude and inefficient and coopers did not take this machinery seriously. Before such a machine could be installed, it was necessary to fit a steam engine, and this was an expensive proposition. It was not until the end of the nineteenth century that machinery became more common in coopering. There are numerous patents and other inventions that can be related to the industrialization of the coopering process but it is beyond the scope of this book to describe all these improvements. However, a few examples are given here to illustrate the machinery for the softening of wood and the bending process.

To soften the wood, machinery for steaming was used. For about twenty minutes before being bent, the casks were steamed in bells, large upturned tubs, which were dropped over the cask to trap the steam that poured out from jets under the cask. These steaming bells were used by many manufacturers, but there are many inventions that probably never reached practical use. Figure 2.16 shows an illustration from a patent of an invention related to improvements in heaters for cooperage purposes (Reed, 1890). The staves were assembled and held in place by truss-hoops, preferably with one head in place. The cask was entirely enclosed in the heating chamber which the inventor considered to give an effective and quick heating of the cask and thereby a faster shrinkage and set of the staves in the bent position. There was no additional steaming during the heating of the barrel.

For bending of staves, a so-called bell-press can be used (Diener & Roth, 1941; Kollmann, 1955; p. 808). After the staves are put together tightly at one end with the

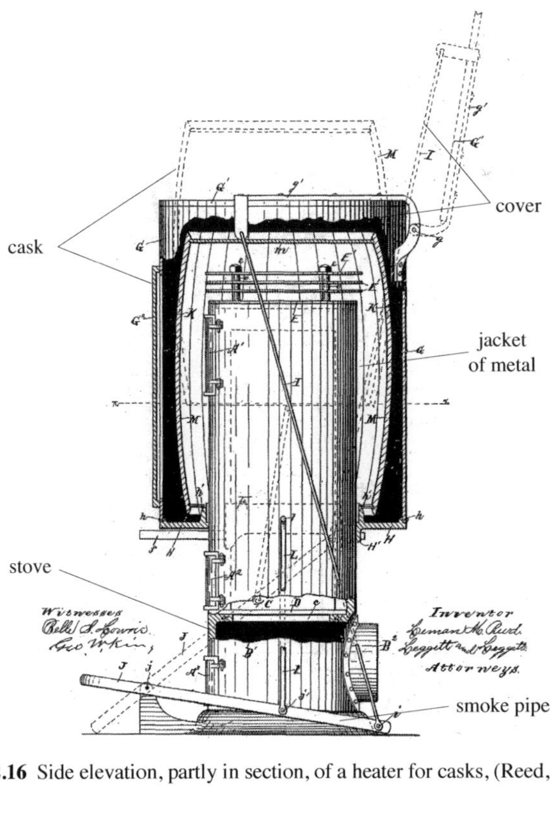

Fig. 2.16 Side elevation, partly in section, of a heater for casks, (Reed, 1890).

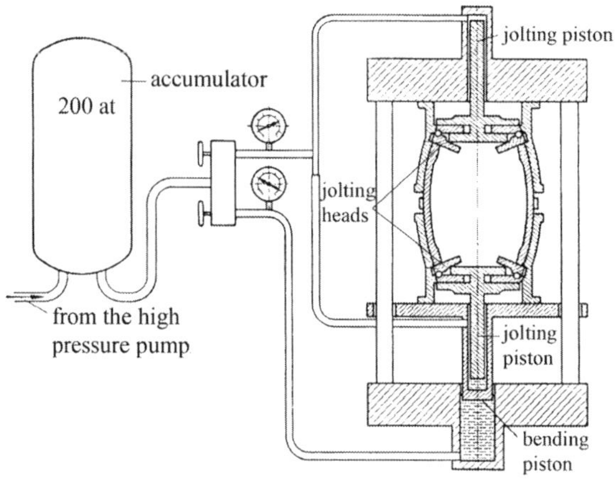

Fig. 2.17 Hydraulic bell-press for the bending of staves for a wooden cask (Kollmann, 1955).

help of a raising-up hoop, the cask is placed in the press in an upright position with the open end upwards. The cask is held with the help of a mechanical device and its open end is moved towards the inside of a conical end-stop (the bell). The staves then come together and the bulge of the cask is formed. The staves are subsequently fixed by a second hoop and the two-part clock is opened. The cask can then be taken out of the press.

When a cask is made of stout or extra stout staves, a pressure of 3-10 MPa is needed and it is normally obtained hydrautically. A schematic presentation of a hydraulic bell-press is shown in Figure 2.17.

2.5 THE INDUSTRIALIZATION OF SOLID WOOD BENDING

Historically, the industrial bending process not only enhanced the structural stability of a design but also allowed for the most economic use of materials and manpower. Commercially, the new process lowered the cost of construction and gave bent wood furnishing a competitive edge over products made by more conventional laborious means with elements that had been cut or carved, and then joined. The ultimate success of the bending process occurred after the middle of the nineteenth century when, through mass production, bent-wood furnishing flooded the market (Ostergaard, 1987).

The development of the technology of wood bending began when the first patent was granted in England in the first quarter of the eighteenth century and continued until the time when, through the work of Michael Thonét 140 years later, it was prepared for introduction into industrial manufacture (Šimoníková, 1992).

The three oldest patents, granted between the years 1720 and 1794 to J. Cumberland, H. Jacksson and J. Vidler, used different approaches to the process of softening the wood, including boiling, steaming and heating by conduction (Šimoníková, 1992). These operational procedures were intended for use primarily in the area of ship building. During the first decade of the nineteenth century, attempts to apply this process to the bending of laminated wood for the constructions of chairs by the Belgian Jean Joseph Chapuis (1755-1864) and the American cabinet-maker Samuel Gragg (1772-1855) met with considerable success. The remarkable feature of Gragg's work was his confident use of bent laminated wood; a goal not attained by Thonét – despite repeated experiments – until nearly fifty years later (Ostergaard, 1987).

Experiments with bent wood for purposes other than furniture making are also known from the early nineteenth century. In 1821, the Austrian wagon maker Melchior Fink was granted a patent for producing rims for bicycle wheels in one piece, although he never used this method on an industrial scale (Vegesack, 1987). The English craftsman Isaac Sargent became famous for his spiral staircases and Michael Thonét built cartwheels for the Prussian army in the 1840s (Vegesack, 1996).

2.5.1 The Thonét process

No other industrial method involving the thermo-hydro-mechanical treatment of wood has had such a great and global success as the Thonét process, an industrial

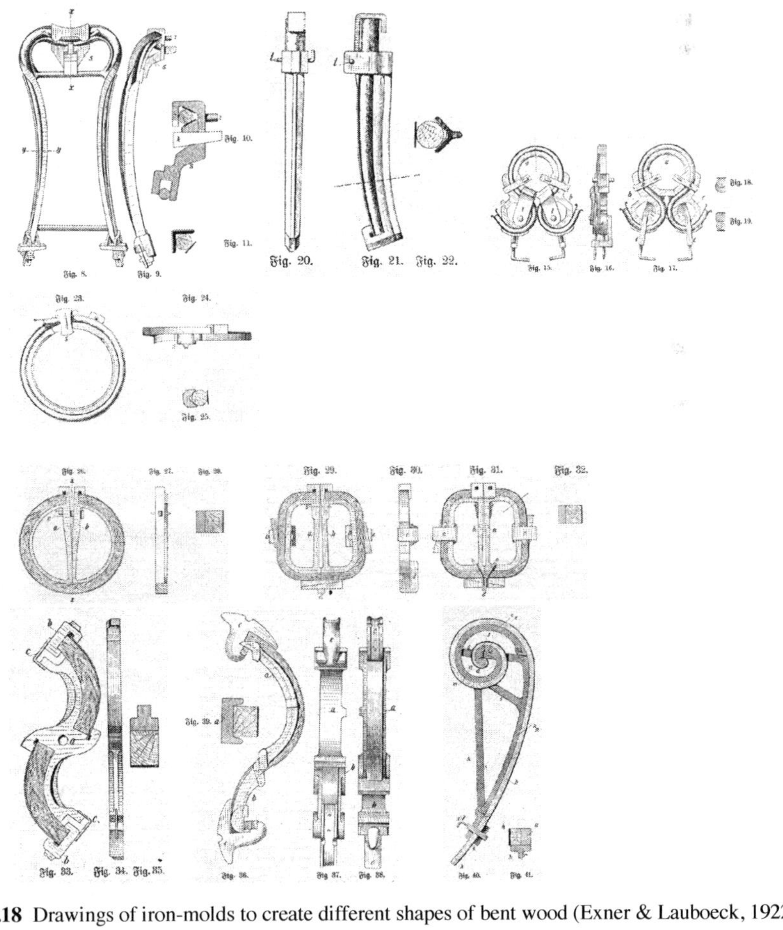

Fig. 2.18 Drawings of iron-molds to create different shapes of bent wood (Exner & Lauboeck, 1922).

process for bending solid wood invented by the German furniture maker Michael Thonét. The technique includes softening the wood by moisture and heat, and bending it in such a way that elongation of the wood on the convex side during bending is minimized. Consequently so is the risk of breaking the wood during bending. Before being dried, the bent wood is fixed in a specially designed iron-mold to form the final shape as depicted in Figure 2.18.

2.5.2 The Vienna Chair that conquered the world

The inventor of the Vienna Chair came from Germany. Michael Thonét was born in 1796 in Bopard-am-Rhein, 20 km south of Koblenz, where he established himself as an independent furniture maker in 1819. Michael Thonét was the first person to adapt the art of bending solid wood from the shipbuilding industry to furniture manufacture.

In 1830, he began his first experiments with bent wood. Sheets of veneer were laid parallel to the grain and were cut into regular strips of uniform size. These were then boiled in a solution of glue. Once bundled together and attached to wooden, or later to metal, frames they would be bent into the desired form. Once curved, this pliable material lost in thickness and thin strips of veneer were thus added to compensate for the loss. An additional disadvantage of this process was that the hygroscopic glue adsorbed water from the surroundings causing its strength to be reduced. In 1836, Thonét developed his special bent wood method to produce what became known as the Boppard layerwood chair, his first major breakthrough. However, his attempts to obtain patents for this technique in Germany failed. In 1841, he was granted a patent for the method in France, England and Belgium, and as a result, his company gained (or conquered) the world market for bent wood furniture.

In the spring of 1842, Thonét moved his activity to Vienna on the advice of the Austrian chancellor Klemens Metternich. In the years until 1849, he cooperated with Carl Leistler in the interior decoration of the Palais Liechtenstein in Vienna, where he laid a parquet floor and designed separate pieces of furniture.

Later that year, Michael Thonét was granted a patent for "Holz in beliebige Formen und Schweifungen zu biegen" (Wood bending, in any shape and form) by the *K.k. allgemeinen Hofkammer* in Vienna. This represented an imperial patent for the bent wood procedure, enabling beechwood slats to be pressed into a curved shape under steam. After the slats are heated to over 100 °C in steam chambers, two specially trained workers press them by hand, simultaneously and evenly, into curved cast-iron molds in which they are dried at about 70 °C for at least 20 hours. In 1853, the patent was renewed and it remained valid until 1869.

In 1849, together with his sons, Thonét started a workshop for the production of bent solid wood furniture, especially simple chairs intended for mass production. The first of these chairs, known as No. 4, was exhibited the next year at Nieder-Österreichischen Gewerbeverein (Lower Austrian business association) and thus became known to a wider public. The first order came from Café Daum in Vienna, where it was in use for more than a quarter of a century. The next delivery of 400 chairs went to a hotel in Budapest. In the 1850s, a method was developed for bending thick, solid pieces of beech wood, so-called "bugholz", on a large scale. This technique consisted of plasticizing the wood by treating it with hot steam, performing the bending in a special mold and allowing it to dry in a fixed position. At the same time, Thonét and sons went on to working almost exclusively with beech. In 1856, they started a new factory in Koritschan in Mähren in the middle of a rich beech forest district, leaving only a warehouse in Vienna.

In 1856, the simple chair (No. 14) which came to be the main product of the Thonét company was designed in Koritschan. During the most richly flowering eclectic period of the history of furniture, a chair was thus created which could not be more simple. It was light, aesthetically neutral and ideal in a transport perspective, since it could be delivered in parts which could rapidly be screwed together. Before 1900, the Thonét Bros factory had already produced over 50 million copies of their famous café chair No. 14 and it was being delivered in six parts as the world's first "knock-down" piece of furniture, shown in Figure 2.19.

In 1860, the first rocking-chair was manufactured, and this was produced in more than 100 000 items per year in the 1890s in Austrio-Hungary alone. In the 1860s, three more factories were started in Mähren and Hungary, including one in Bystřice, which is still one of the largest suppliers of bent furniture in the Thonét tradition.

When the first patents expired in 1869, competing furniture manufacturers were able to start production of the famous bent wood furniture using the same technology as that developed by Thonét; a technique still in use at many bent wood factories even today. The bent wood industry rapidly developed, so much so, that by 1893, 51 companies (25 in Austrio-Hungary) were in production. Jacob & Josef Kohn in Vienna became Thonét's largest competitor. While Thonét required one to two hours to make the wooden rods flexible using steam, his rival Kohn installed a machine that could produce identical parts within 3-5 minutes. This rendered it possible for 5500 pieces of furniture to be produced daily in the four factories belonging to the Kohn brothers.

Michael Thonét died in 1871. At the time of his death, the company had 4000 employees and an annual production of 450 000 items of furniture. Twenty years after Michael Thonét's death, there were 60 factories, mostly within Austrio-Hungary. The

Fig. 2.19 The famous Thonét café chair No. 14, ready for delivery in parts from a factory, and in its assembled form.

old Thonét models are still being manufactured all over the world. The Thonét family's present main factory was started in 1869 in Frankenberg in Hessen and new products are now also being manufactured together with the old models.

2.5.3 What was it that made the Thonét method so successful?

The art of bending wood by heating and moistening the wood material is, as mentioned before, several thousand years old and one may wonder what kind of innovation Michael Thonét added to the process to make it so successful.

Thonét used mainly the wood species beech, which through practical tests or from his experience as a furniture maker he knew to be better than any other species of wood for the purpose. Softening the square-cut or turned material by exposure to hot steam was well-known. Thonét's invention was such that, during the actual bending operation, the convex side of the wood was held with a steel band which prevented the wood from being stretched, thereby reducing the risk that the wood would break or that micro-fractures would develop during the bending process, particularly when tight curves were being produced. The Thonét method thus utilized the ability of the steamed wood to be locally considerably compressed in the fiber direction (20-30%) in a way which had not been done before. An additional idea that Thonét added to the bending process was that the wood pieces were bent over a fixture which gave it its final shape, as shown in Figure 2.20. Thereafter, the wood pieces were locked in the mold and dried with warm air. After the drying, the wood not only retained its bent shape but had also become stronger due to its increased density in the bent region.

The Thonét workshops and others later developed innumerable tools and machines in order to be able to produce the various furniture models that were designed and put on the market. An extensive description of the various machines that

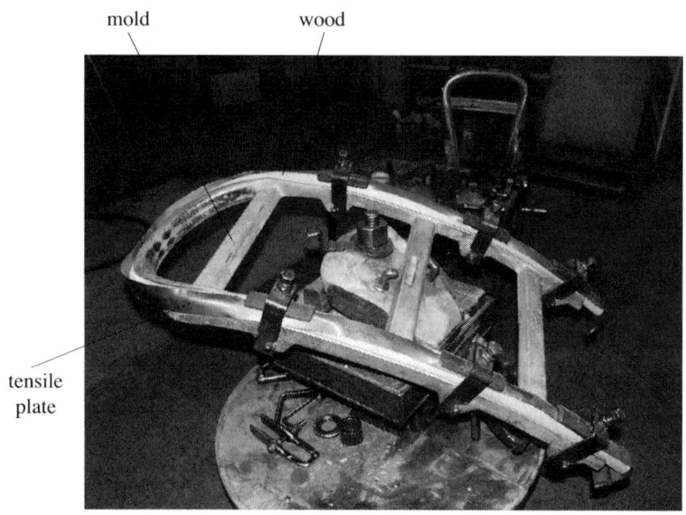

Fig. 2.20 A fixture for the bending and locking of steamed solid wood in the geometric shape required in the final product; in this case, the back of a chair. Note that the strap (tensile plate) on the convex side of the material remains in position until the wood piece has been dried.

were used for the bending of solid wood around 1900 has been presented by Exner and Lauboeck (1922).

In the rich design world of the nineteenth century, Thonét's bent wood furniture is a paradoxical phenomenon. With a matter-of-fact graceful design, completely conditioned by the manufacturing process, the furniture calmly broke the traditions of form which were characteristic of the latter half of the century. These items of bent wood furniture are the first to be mass-produced. They were developed for the new market created by the industrial society, and represent furniture for a new age. In furniture-history presentations, they play an obscure role, perhaps due to their lack of a specific time-connection. Some of these chair types are nevertheless still the world's most frequently produced furniture.

Thonét, with his initial experiments on laminates and the subsequent bending of solid wood, began the first large-scale production system based on interchangeable parts in furniture-making. He introduced a simple functional design which enabled him to achieve a commercial as well as a technical success.

2.6 THE USE OF VENEER FOR CURVED SHAPES

Veneer was used either constructionally or as a decorative layer to disguise poor quality wood. The earliest use of veneer has been shown to be largely for the purpose of display and ornamentation rather than for creating a curved shape or for increased strength (Knight & Wulpi, 1927).

The earliest evidence for the use of veneer comes from Egyptian tomb findings and reliefs. Little is known, however, of the methods used, e.g., how the wood was sawn into sheets or how a glue, which 3500 years later still holds thin layers of face wood to the heavier base or core, was prepared, but the work has endured and speaks for itself. Figure 2.21 shows a mural record found in Thebes that describes the production of an intarsia or a plywood construction from about 1490-1436 B.C. Nevertheless, there is as of yet no proof that the Egyptians used heat and moisture in

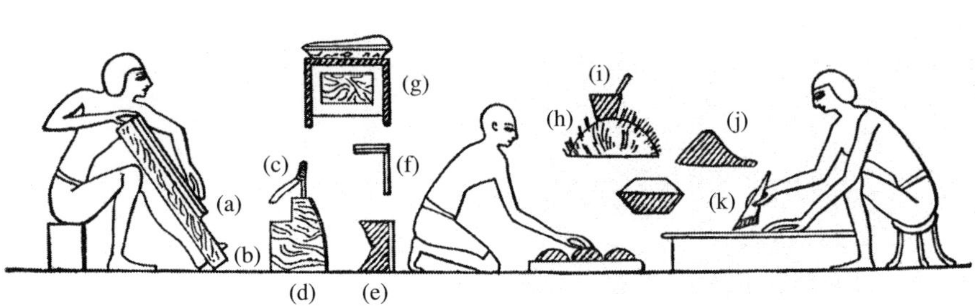

Fig. 2.21 Mural record of veneering, discovered in the Sculpture of Thebes and dated as early as the time of the Thutmosis III (1490–1436 B.C.). The man to the left is applying veneer on a core of ordinary wood, the man in the middle is grinding something and the man to the right is applying glue with a brush. A piece of dark wood (a) applied to one of ordinary quality (b) and an adze (c) fixed into a block of wood (d). Some tools and equipment: a ruler (e), a right angle (f), a box (g), a glue pot (i) on the fire (h), a piece of glue (j) and a brush (k). (From Knight & Wulpi, 1927).

combination with mechanical force, i.e., a true THM process, when they shaped their veneer constructions.

Killen (2000) says that Egyptian carpenters began to laminate thin sheets of wood as early as the Third Dynasty (2686–2613 B.C.) in an attempt to fabricate a large sheet of material that was dimensionally stable and equally strong in all directions. An example of six-ply wood, where the grain of one sheet is at right-angles to the next, was discovered in a sarcophagus within the Third-Dynasty step pyramid complex of Djoser at Saqquara (Lauer, 1933; Lucas, 1936, Makkonen, 1969). These sheets of plywood formed the sides, ends and base of a coffin. The thickness of each veneer was approximately four millimeters and their widths were found to range between 40 and 300 millimeters.

Gaius Plinius Cecilius Secundus (23-79 A.D.), better known as Pliny the Elder, dedicates in his *Natural History* a whole chapter to the art of veneering (see for instance Rackham, 1945). Pliny gives a complete description of the early use of veneers and also describes a number of species that are suitable for veneer making, but he does not mention any products or manufacturing method that can be linked to a THM process.

In his historical work embracing 40 volumes, Polybios (ca. 203-120 B.C.) gave a good description of the armament of the Roman army. One type of soldier carried a semi-cylindrical shield, a so-called *scutum*. The shield was 65 cm wide and 1.2 m in length and made of two layers of wood glued together, covered with leather that was attached to the wood by iron fasteners (Tarn, 1948). Sentance (2003) says that the legionaries of the Roman army were equipped with shields constructed from laminated sheets of steam-bent birch wood held together with animal glue.

The bow is a weapon for hunting and warfare, used by humans for a very long time. The bow is normally of solid wood, where yew was a commonly used species in Europe for a long time. A technically more advanced weapon is the two-wood bow, where a thin layer of tough wood is fixed to a hard, compressible belly. The term "two-wood bow" is used by Insulander (1999) and is a translation of the Northern name of the Saami bow, *tvividr*. The oldest known two-wood bow is from Korekava in Japan, a find dated to 2600 B.C. (Rausing, 1967). This type of bow is considered to be typical of the Fenno-Urgian peoples and is particularly widespread in the western region of northern Asia (Adler, 1902).

Insulander (1997) has studied the construction of the Scandinavian Saami two-wood bows. The oldest has been dated to ca. 200 B.C. but most of the finds are from the Middle Ages. This type of bow was constructed from two strips of wood, one of birch or another deciduous tree and one of compression wood from pine, attached together with a glue made from the skin from the perch fish (see also Schefferus, 1673; Linné, 1732, 1737). The compression-resistant compression wood formed the belly of the bow, and the tenacious birch formed the back. The bow has static ears, which were more or less recurved and the bow was probably reflexed, cf. Figure 2.22.

The laminate of birch and pine compression wood has exceptional qualities that, together with static ears, resulted in bows with an excellent performance. It has also been shown that fire was used during the construction of the bow. To protect the bow against moisture and water that could destroy the gluing, the bow was covered with a thin layer of birch bark.

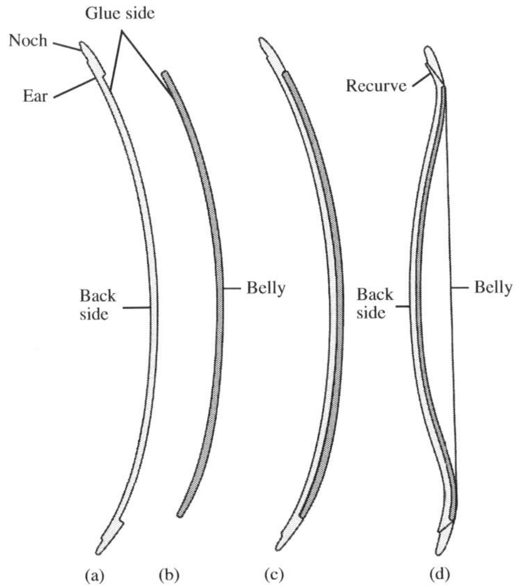

Fig. 2.22 The construction of the Fenno-Urgian two-wood bow. (a) Back strip of birch or other deciduous tree. (b) Belly strip of compression wood from pine. (c) Relaxed two-wood bow, slightly reflexed. (d) Two-wood bow braced, slightly recurved ears.

With a few exceptions, some of which have already been mentioned, it was not until the nineteenth century that veneer or thin wood sheets started being used for purposes other than covering flat surfaces or surfacings that were curved and of irregular shapes for embellishment purposes. As mentioned already, in the first half of the nineteenth century, Michael Thonét and others started experiments with multi-layered

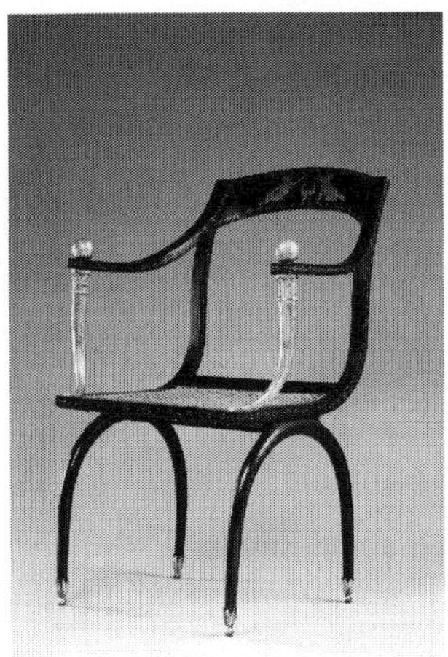

Fig. 2.23 A painted armchair by Jean Joseph Chapuis made of bent laminated wood, solid wood, cane, metal and gesso (Carnegie Museum of Art, Pittsburgh).

constructions of veneer, in some cases in combination with solid wood, to create curved shapes. Thonét developed his technique towards the bending of solid wood.

Jean Joseph Chapuis (1765-1864) built chairs out of laminated bent wood and used laminated wood for the legs and the continuous side rail, as shown in Figure 2.23. A series of four solid wooden rods, each about 6 mm thick, were glued together, bent, bound in a mold, dried, and then shaped to the appropriate form with a rasp. Whether or not Chapuis used steam to bend sections is unknown (Ostergaard, 1987).

Samuel Gragg (1772-1855) was an innovative and inventive Boston craftsman, who deposited a patent in 1808 for what he dubbed his "elastic chair" (Gragg, 1808; Podmaniczky, 2003). Although he began as an ordinary chairmaker manufacturing the Windsor chairs that were popular in the early nineteenth century, Gragg is revered for his elastic chair patent, which involved a new method of bending continuous strips of wood with steam to form the back, seat and front legs of the chair, cf. Figure 2.24. While the stylish product shares a profile with the classical Klismos chair that was popular at the time, the result transcends its own period and looks forward to later bent wood furniture and even the streamline modern designs of the twentieth century.

Gragg's innovative elastic chairs take two forms, both of them using bent wood to make the chairs exceptionally strong and light. One version suggests that it preceded the second, more developed variation. The first variation is constructed with bent strips of wood forming the back stiles (vertical back posts) and seat rails as continuous pieces. Forward-swept front legs are attached separately to the front corner of the seat. The carved "goat hoof" feet of these legs give the chair its name. Gragg's more highly developed second version of the elastic chair has the back stiles, seat rails, and front legs all formed with a single S-bent strip of wood. This chair is generally referred to as "fully elastic". The long parts of the frame were steamed in a box and then clamped in a pattern frame until they dried in the desired shape.

Fig. 2.24 The fully elastic chair made by Samuel Gragg about 1808. The chair is made of birch wood in the front rail, American white oak in the center seat strip and white beech in the front legs. The entire chair is painted yellow with dark and light brown outlining different elements. (Winterthur Museum).

John Henry (Heinrich) Belter (1804-1863) originated from the same traditions as Thonét but developed his ideas in the United States. His patents are related to the bending of laminates of wood in two directions around formers to shape such items as chair backs and bed frames (Schwartz *et al.*, 1981). It is however, a common misunderstanding that Belter patented the manufacture of laminated woods or plywood.

Belter immigrated to the United States from Germany in 1833 and is listed in New York City trade directories as a cabinet maker by 1844. Belter's name is associated with the finest examples of Rococo Revival furniture. He is known particularly for the work he did in laminated rosewood, and only recently has it been confirmed that he was not the only furniture maker at this time in the United States to use

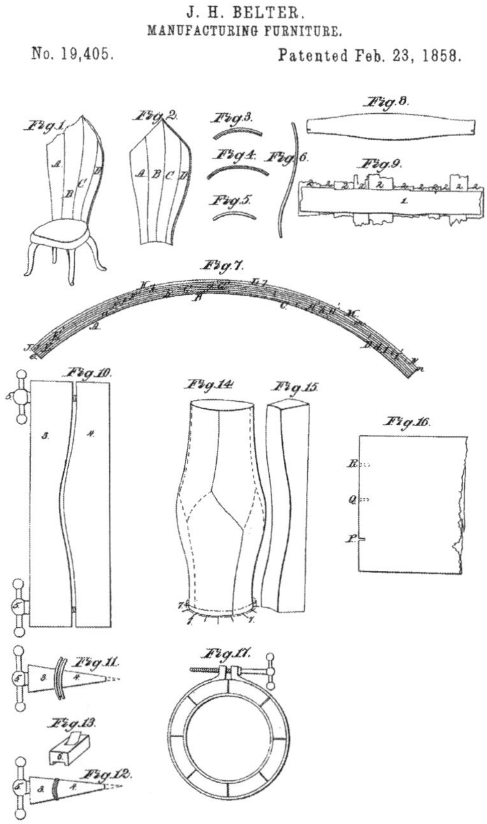

Fig. 2.25 A drawing from the patent by Belter illustrating a method for producing curved shapes of laminated veneer (Belter, 1858). Figures: 1-2. chair-back, 3-5. horizontal sections of chair-back, 6. vertical section of chair-back, 7. horizontal section of chair-back in a larger scale, 8. one of the compound staves of which the chair-back is composed (A-D in figure 1), 9. lamination of staves, 10-12. clamps between which the rough staves are held to be shaped, 13. tool for reducing the edges of the staves, 14-15. inside and outside of the dies (cawl) by which the chair-back is perfected (several chair-backs are manufactured at the same time and are after gluing sawn along the dotted lines), 16. scores in each end of the staves for fixing them in position, 17. horizontal section of the inside and all the outside dies with the hoop and screw for compressing them.

laminated wood (Schwartz *et al.*, 1981). Belter was the only New York craftsman to obtain a series of patents on his innovative and unique design and very high-quality construction. He obtained four patents during his career, thus protecting his creative furniture-making endeavours; a patent for machinery for sawing arabesque chairs in 1847, a patent for bending wood in 1858, a patent for a bedstead design in 1856, and a patent for a bureau design in 1860 (Belter, 1847, 1856, 1858, 1860).

Rosewood is a very dense wood that is also brittle, which can make solid rosewood hard to use in complicated constructions, e.g., curved shapes. The development of the technique of bending laminated wood into curved shapes obviates this problem and also enabled Belter to create designs that were more flamboyant than anything made of solid rosewood. The 1858 patent involved the production of laminated furniture that curves in two planes, rather than just one, to form the section of a sphere and for the method to accomplish its precise assembly. Stave construction is specifically mentioned for fashioning spherical-shape backs in this patent, cf. Figure 2.25.

The making of a grand piano rim is an interesting example of the process of laminated bending, and something that could not be accomplished in solid wood, cf. Figure 2.26. The rim is normally made by laminating flexible strips of hardwood to the desired shape, a system that was developed by Theodore Steinway in 1880 (Steinway, 1880a,b), as shown in Figure 2.27.

The Steinway grand piano case is manufactured in one piece from eighteen layers of maple veneer, each the same length as the perimeter of the case. The layers are glued and pressed together in a special press designed by C. F. Theodore Steinway (Steinway, 1880a). The piano case had previously been manufactured from several pieces joined together, but lamination gives an exceptionally stable case and a very rigid fixture for the sounding board. The method was soon adopted by other grand piano manufactures and is now the most commonly used technique.

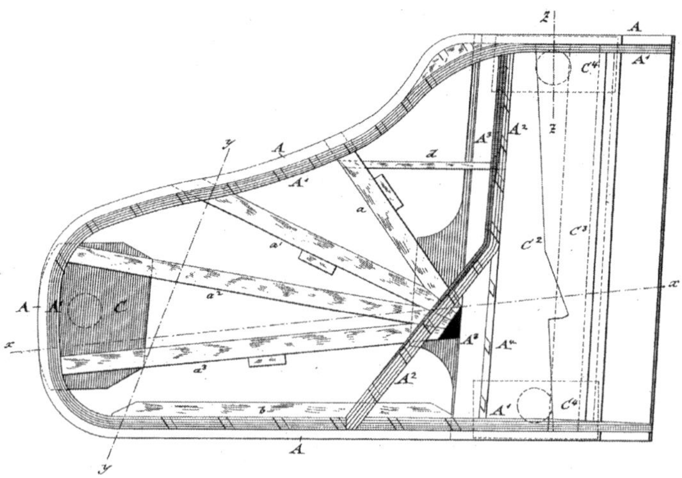

Fig. 2.26 A plan of the supporting-frame of a grand piano-forte, showing the bent rims (A), stiffening-braces (a, a′, a^2 and a^3), and consoles (C) (Steinway, 1885).

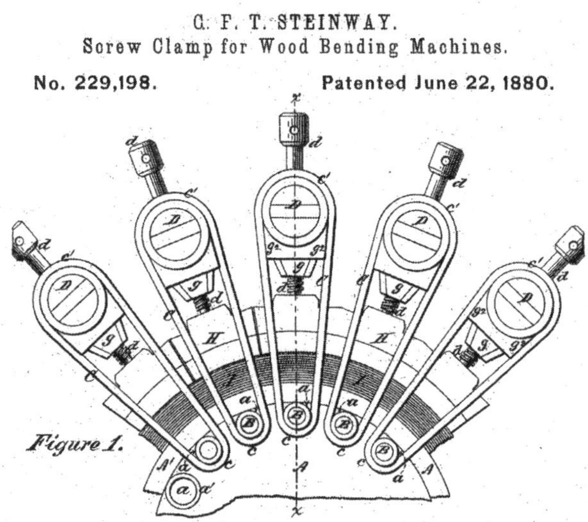

Fig. 2.27 Top view of a former or template provided with the improved taper-linked clamps designed by Theodore Steinway (Steinway, 1880a).

Stringed instruments, of the violin and banjo families, are made of laminated veneer, and some combine curved shapes with beautiful matching of figure and artistic inlays. Drum sides were in former times also made of veneer.

The economic aspects of using thin sheets of rare and beautifully figured veneers in order to secure the maximum surface of a matched and uniformly patterned figure, instead of utilizing the full timber thickness of rare woods, did not make their appearance until the middle of the nineteenth century (Knight & Wulpi, 1927). Possibly, in the older days when labor costs were low, and the transportion of bulky materials was extremely slow and expensive, the element of cost may have entered into the adoption of this technique. However, today most of the popular woods can be produced at a lower cost if made of a solid rather than a laminated construction. It is the superiority of laminated wood for beauty, durability and strength that has maintained its use in better furniture and cabinet-work, regardless of the additional cost.

2.7 THE DEVELOPMENT OF THE THM-PROCESSING TECHNIQUES

Wood technology has influenced the progression of furniture design as well as architectural design in the last centaury; it has helped define wood as a material of choice for modern furniture and building design and can continue to meet new challenges. An approach where the manufacturing process was subordinate the design dominated the furniture production until the 19th century. Generally, it was first in the 20th century that the production of laminated furniture began to exhibit a design characterized by technology. It can be said that the last one hundred years have seen the greatest leaps in areas such as technology and design. The acceptance of the machine as a positive and creative aesthetic force marked the beginning of the modern era. Resultant new techniques enabled creative designers to go beyond the imitation of historical forms.

Laminated veneer may be one of the most important engineered wood materials during the 20th century. Modern laminated bending of veneer origins go back to the Industrial Revolution in the 19th century. It was during this time that design began to develop into a modern discipline with technical innovation as a key driving force. The laminated wood bending process was thus a product of the new machinery made possible by the Industrial Revolution. Many furniture designers who used bent or molded veneer attributed the early ground work to Thonét, and these companies made numerous machinery developments in furniture production. This laid the way for advanced molded veneer technology and industrial furniture production. Developments such as the veneer rotary cutter, new adhesives, and the hydraulic hot-plate press made laminated products possible. Such developments in woodworking machinery took place very early in the twentieth century; concurrently, developments in adhesives were also taking place. Laminated veneer as a material choice was essential for wood to keep up with the new demands and expectations of the industrial world and fulfilled the requirements created using technological processes that customized, modified and extended the physical properties of wood as a raw material. Since it was cheap, durable and easily accessible, it became an important medium for experimentation by modernist designers from the 1920s onwards. Many famous designer have worked with laminated veneer in their furniture during the 20th century, e.g., Gerrit Rietveld (1888-1964), Gustaf Axel Berg (1891-1971), Alvar Aalto (1898-1976), Gerald Summers (1899-1967), Marcel Breuer (1902-1981), Bruno Mathsson (1907-1988), Charles Eames (1907-1978), Eero Saarinen (1910-1961).

One of the newer technologies developed in the area of wood materials is the wood welding process. This method may be interesting in the furniture-making industry as well as in building construction in the future. It has long been known that during frictional processes, besides the occurrence of wear and abrasion, some portion of mechanical energy is transformed into heat. In the case of friction welding, the thermal energy generated by friction, which in most cases is undesirable, is used to join materials by encouraging wear in a controlled way, leading to efficient welding. Bevington (1891) invented a rotary welding process for the formation of tubes and in order to weld the ends of two tubes together, cf. Figure 2.28. The first connections of plastics were made in the 1940s by Henning (1942). In the 1950s, the friction welding process was developed in the Soviet Union and Czechoslovakia. With a number of techniques, friction welding was developed and applied to numerous materials (Vill & Shternin, 1957; Vill, 1959). With these processes, even different materials, e.g., steel and aluminum, can be joined. It wasn't until the 1990s that the first attempts were made to join wood by means of friction (Suthoff *et al.*, 1996; Suthoff, B. & Kutzer, 1997). Methods for welding wood were presented in two German patents and it is stated that two pieces of wood can be welded by means of an oscillating, rotating or linear frictional movement in an open system or under an inert gas or vacuum atmosphere. The technique of friction welding is detailedly presented in Chapter 7.

Possible viable alternatives to solid wood construction in eco-design are the use of paper and cardboard composites. These materials have been developed in recent years to a level where important properties such as strength levels can deem the

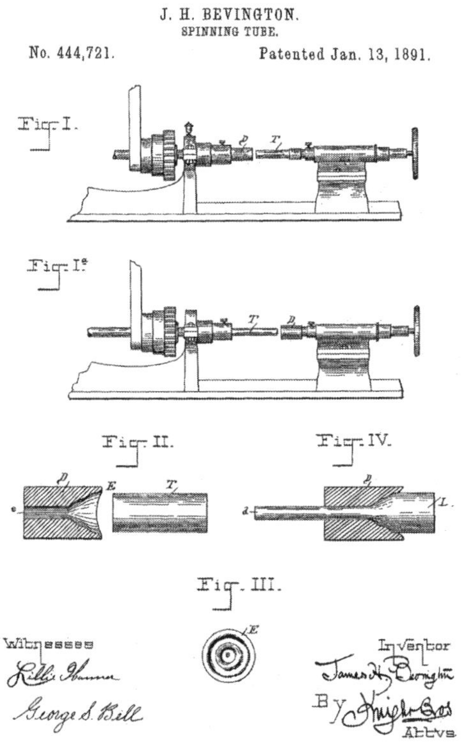

Fig. 2.28 A drawing from the Bevington patent from 1881 for rotational welding of tubes. Fig. I. Side elevation of a lathe for spinning tubes. Fig. Ia. Similar to I with the parts transported. Fig. II. Longitudinal section of a die to reduce a metal tube to one of smaller diameter. Fig. III. End elevation of a die. Fig. IV. The tube and die after the tube has been operated.

material useful for applications previously impossible. The material itself is interesting since the industrial process is very controllable, resulting in a product that performs consistently and to the required specifications.

Figure 2.29 shows a lighting fixture made in a material called DuraPulp and developed in close collaboration with the Swedish forestry industry company Södra together with the Swedish architects and designers Claesson Koivisto Rune. Durapulp is a combination of paper pulp and PLA (polylactic acid, a biodegradable plastic). The mix is heated to precisely 167 °C, which causes the plastic to encapsulate the paper fibers. The resultant material can be comparable to wood or a hard plastic. The bioplastic PLA is a recent development in the search for alternatives to petroleum-based polymers, and can be obtained by converting corn starch into a material with the characteristics of plastic. This composite of pulp and PLA results in a paper that can withstand weight, tension and humidity.

The technology uses paper as the basic material. The purpose of adding PLA is to alter the undesirable qualities of paper (such as its fragile nature, and generally short life span), converting them into strengths.

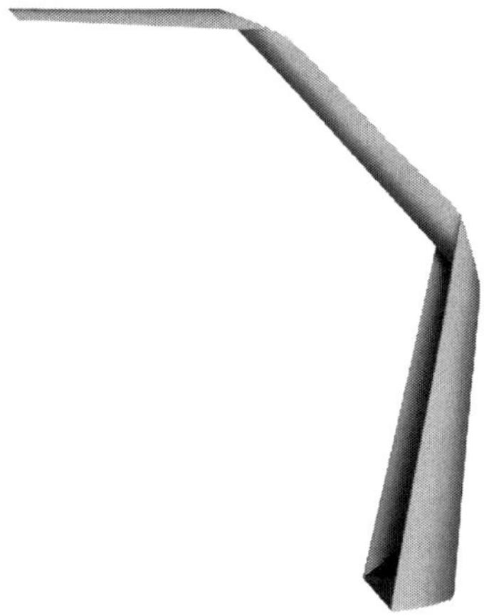

Fig. 2.29 Paper task light in DuraPulp by Södra and by the Swedish architecture and design studio Claesson Koivisto Rune. DuraPulp is made of 100 per cent renewable fibers and is fully biodegradable.

Wood is the ultimate renewable material. THM processing can improve the intrinsic properties of wood, rendering it possible to produce new materials and to acquire a form and functionality desired by engineers without changing its eco-friendly characteristics. There are numerous THM processing techniques and the number of these methods is growing continuously. Many THM wood treatments like wood molding, heat treatments, chip-less manufacturing, wood ageing with accelerating techniques, welding of wood, high thermo-hydro processing (steaming and boiling logs before slicing or rotary peeling), wood surface densification, wood folding, etc. have been developed recently or are in the developmental phase. The challenge for the future with these methods is that the wood often becomes brittle, and thus deteriorates with time. Furthermore, its strength decreases during THM treatment, again becoming worse over time, and there are also problems associated with up-scaling and during full-scale production. New processes are being developed towards closed systems and higher working temperatures (around 200 °C) to make the process more controllable and the material form-stable. The THM techniques are also being developed to produce wooden structural elements for building constructions.

In the future, we will see product improvement and development of new product-market combinations for industries with an interest in combined THM processing. Potential areas of application will include building construction, furniture manufacture, improving service lives of products through increased durability and stability, wood-finish compatibility and the identification of new market areas through peer-to-peer interaction.

2.8 REFERENCES

Abbott, M. (1989). *Green Woodwork. Working with wood the natural way*. Guild of Master Craftsman Publications Ltd., East Sussex, ISBN 0-946819-18-1.

Aldred, C. (1954). Fine wood-work. In: Singer, C., Holmyard, E.J., & Hall, A.R. (eds.). *A history of technology. Volume 1: From early times to fall of ancient empires*. Oxford University Press, New York, London.

ADLER, B. (1902). Die Bogen Nordasiens. (The bow from north Asia.) *Internationales Archive für Ethnographie, 15*:1-27.

BELTER, J.H. (1847). *Machinery for sawing arabesque chairs*. U.S. Patent No. 5208.

BELTER, J.H. (1856). *Bedstaed*. U.S. Patent No. 15552.

BELTER, J.H. (1858). *Improvement in the method of manufacturing furniture*. U.S. Patent No. 19405.

BELTER, J.H. (1860). *Bureau*. U.S. Patent No. 26881.

BEVINGTON, J.H. (1891). *Spinning tubes*. U.S. Patent No. 444 721.

BRAMWELL, M. (ed.) (1976). *The international book of wood*. Mitchell Beazley Publisher Ltd., London, ISBN 0-85533-081-3.

BUGGE, A. (1953). *Norwegian stave churches*. Dreyers Forlag, Oslo.

CLAESSON RÅLAMB, Å. (1691). *Skepsbyggerij eller adlig öfning tionde tom*. (Shipwrightry or noble practice.), Stockholm.

CRUMLIN-PEDERSEN, O. (1970). Skind eller Træ. (Skin or wood.) In: Hasslöf, O., Henningsen, H. & Christensen, A.E. (eds.). *Sømand, fisker, skib og værft: introduktion til maritim etnologi*. (Sailor, fisherman, ship and shipyard: introduction to maritime ethnology.) Rosenkilde og Bagger, Copenhagen, ISBN 87-505-0072-4.

CRUMLIN-PEDERSEN, O. & OLSEN, O. (eds.) (2002). *The Skuldlev ships I. Topography, archaeology, history, conservation and display*. Volume 4.1 in the series Ships and Boats of the North, The Viking Ship Museum, Roskilde, ISBN 87-85180-467.

DIENER & ROTH (1941). *Ein Faß*. (One barrel.) Diener & Roth Fassfabrik in Stuttgart.

DIGBY, A. (1954). Boats and ships. In: Singer, C., Holmyard, E.J. & Hall, A.R. (eds.) (1954). *A history of technology. Volume 1: From early times to fall of ancient empires*. Oxford University Press, New York, London.

DUNBAR, M. (1984). *Make a Windsor chair with Michael Dunbar*. The Taunton Press, Newtown, Connecticut, ISBN 0-918804-21-3.

EDLIN, H.L. (1949). *Woodland crafts in Britain. An account of the traditional uses of trees and timbers in the British countryside*. B. T. Batsford LTD, London.

EXNER, W.F. & Lauboeck, G. (eds.) (1922). *Das Biegen des Holzes. Ein für Möbelfabrikanten, Wagen- und Shiffbauer, Böttcher und anderen wichtiges Verfahren*. (Wood bending, an important process for furniture makers, wagon- and shipbuilder, coopers and others.) 4th ed. Verlag von Vernh. Friedrich Voigt, Leipzig.

GRAGG, S. (1808). *Elastic chair*. U.S. Patent, Boston MA 31 august, 1808. (The Patent Office records concerning this patent were destroyed in the 1836 fire. The full patent is not now available in the Patent Office records, but can be found at James Madison University, Carrier Library.)

GODAL, J.B. (1995). The use of wood in boatbuilding. In: Olsen, O., Skamby Madsen, J., & Rieck, F. (eds.) (1995). *Shipshape – essay for Ole Crumlin-Pedersen on the occasion of his 60th anniversary, February 24th 1995*. pp. 271-283, Roskilde, Denmark.

GODLEY, A.D. (tr.) (1920-1925). *Herodotus. The Persian wars (books 1-9)*. Loeb classical library, volumes 1-4.

HALLDIN, G. (ed.) (1963). *Svenskt skeppsbyggeri – en översikt av utvecklingen genom tiderna*. (Swedish shipwrightry – an overview of the development through the ages.), Allhems Förlag, Malmö, Sweden.

HEBERT, L. (1848). *The Engineers and mechanics Encyclopædia: Comprehending practical illustrations of the machinery and processes employed in every description of manufacture of the British Empire*. Thomas Kelly, Ltd, London.

HENNING, A. (1942). Das Schweissen thermoplastischer Kunststoffe. Deutsche Reich Patent 739 340, 1938. (The welding of thermoplastic materials.) *Kunststoffe, 32*:103-109.

HOLAN, J. (1990). *Norwegian wood – a tradition of building*. Rizzoli International Publications, Inc., New York, ISBN 0-8478-0954-4.

HORT, A. (tr.) (1916). *Theophrastus. Enquiry into plants (book 5). Loeb classical library, volume 1*. 5th ed. 1990, Harvard University Press, London, ISBN 0-674-99077-3.

INDRUSZEWSKI, G. (2004). *Man, ship, landscape. Ships and seafaring in Oder mouth area AD 400-1400. A case study of an ideological context*. PNM, Publications from the National Museum, Studies in Archaeology & History, Vol. 9, Copenhagen.

INSULANDER, R. (1997). The Fenno-Urgian two-wood bow – a missing link. *Bulletin of primitive technology, 14*(2):35-39.

INSULANDER, R. (1998). Tjurved – ett historiskt material med framtida potential. (Compression wood – a historical material with potential.) *Flora och Fauna, 93*(4):175-178, ISSN 0014-8903.

INSULANDER, R. (1999). Den samiska pilbågen rekonstruerad: en jämförande analys av fynd från Sverige, Norge och Finland. (Reconstruction of the saami bow: a comparsion between finding in Sweden, Norway and Finland.) *Fornvännen, 94*(2):73-87, ISSN 0015-7813.

KILBY, K. (1971). *The cooper and his trade*. John Baker Ltd, London, ISBN 0-212-98399-7.

KILLEN, G. (1994). *Egyptian woodworking and furniture*. Shire Egyptology Series No. 21, Shire Publication Ltd., Buckinghamshire, ISBN 0-7478-0239-4.

KILLEN, G. (2000). Wood – Procurement and primary processing. In: Nicholson, P.T. & Shaw, I. (eds.). *Ancient Egyptian materials and technology*. Cambridge University Press, Cambridge, pp. 353-368, ISBN 0-521-45257-0.

KNIGHT, E.V. & WULPI, M. (eds.) (1927). *Veneers and plywood. Their craftmanship and artistry, modern production methods and present-day utility*. The Ronald Press Company, New York.

KOLLMANN, F. (1955). *Technologie des Holzes und der Holzwerkstoffe. Zweiter Band*. (Technology of wood and wood products. Second volume.) Springer-Verlag, Berlin.

LATHAM, B. (1957). *Timber – Its development and distribution. A history survey*. George G. Harrap & Co, Ltd, London.

LAUER, J.-P. (1933). Fouilles du service des antiquités à saqqarah (secteur nord, Novembre 1932-mai 1933). (Excavations of the Department of Antiquities at Saqqara (north sector, November 1932-May 1933)) *Annales du service des antiquités de L'Égypte, 33*:155-166.

LESKO, L.H. (1994). *Pharaoh's workers: The villagers of Deir el Medina*. Cornell University Press, Ithaca, London, ISBN 0-8014-2915-3.

LINNÉ, C.v. (1732). Iter Lapponicum. Dei gratia institutum 1732. In: Smith, J.E. (1811). *Lachesis Lapponica or a Tour in Lapland, now first published from the original manuscript journal of the celebrated Linnaeus*. London.

LINNÉ, C.v. (1737). *Flora Lapponica*. (The flora of Lapland.) Schouten, Amsterdam.

LUCAS, A. (1936). The wood of the third dynasty, ply-wood coffin from Saqqara. *Annales du service des antiquités de L'Égypte, 36*:1-4.

MAKKONEN, O. (1969). Ancient forestry: an historical study. Part II: the procurement and trade of forest products. *Acta Forestalia Fennica, 95*:1-46.

MCGRAIL, S. & MCKEE, E. (1974). *Building and trials of the replica of an ancient boat: the Gokstad færing. Part 1. Building the boat*. National Maritime Museum, Maritime Monographs and reports, No. 11.

MCGRAIL, S. (ed.) (1982). *Woodworking techniques before A.D. 1500. Papers presented to a symposium at Greenwich in September, 1980, together with edited discussion*. National Maritime Museum, Greenwich, Archaeological Series No. 7, BAR International Series 129, ISBN 0-8605-4159-2.

MONTET, P. (1946). *La vie quotidienne en Egypte au temps des Ramsès*. (Daily life in Egypt in the time of Ramses.) Librairie Hachette.

NEWBERRY, P.E. (1893). *Archaeological survey of Egypt, Part I. Beni Hasan*. Griffith F.L. (ed.), Kegan Paul, Trench, Trübner & Co Ltd., London.

NICOLAISEN, I. & DAMGÅRD-SØRENSEN, T. (1991). *Building a longboat. An essay on the culture and history of a Bornean people*. The Viking Ship Museum, Roskilde, Denmark, ISBN 87-85180-16-5.

NIKKILÄ, E. (1947). *En satakundensisk Äsping och dess europeiska motsvarigheter*. (A satakundensisk Äsping (a type of boat) and its European analogues.) *Folk-liv: Acta Ethnologica Europaea, 11*:33-46.

OLSEN, O. & CRUMLIN-PEDERSEN, O. (1967). A report of the final underwater excavation in 1959 and the salvaging operation in 1962. *Acta Archaeologica, 38*:73-174.

OSTERGARD, D.E. (1987). *Bent wood and metal furniture 1850-1946*. The American Federation of Arts, New York, ISBN 0-295-96409-X.

PETRIE, W.M.F. (1933). Egyptian Shipping. *Ancient Egypt and the east, I*:1-14.

PODMANICZKY, M. (2003). The incredible elastic chairs of Samuel Gragg. *Magazine Antiques, 165*(May):138-146.

RACKHAM, H. (tr.) (1945). *Pliny. Natural History (books 12-16). Loeb classical library, volume 4*. 5th revised ed. 2005, Harvard University Press, London, ISBN 0-674-99408-6.

RAUSING, G. (1967). *The bow: Some notes on its origin and development*. Acta Archaeologica Lundensia, No. 6, Lund, Sweden.

REED, L.M. (1890). *Heater for cooperage purposes*. U.S. Patent No. 426325.

Richter, G.M.A. (1966). *The furniture of the Ancient Greeks, Etruscans and Romans*. Phaidon, London.

Rivers, S. & Umney, N. (2005) *Conservation of furniture*. Elsevier, Butterworth-Heinemann, Oxford, ISBN 0-7506-09583.

Rosenberg, G. (1937). *Hjortspringfundet*. (The Hjortspring find), Nordiske Fortidsminder 3(1), Gyldendal, Copenhagen.

Schefferus, J. (1673). *Lapponia*. ex officina Christiani Wolffii, Frankfurt am Main. (See also translation by John Scheffer (1674). *The history of Lapland*. Oxford.)

Schwartz, M.D., Stanek, E.J. & True, D.K. (1981). *The furniture of John Henry Belter and the rococo revival*. E. P. Dutton, New York, ISBN 0-525-93170-8.

Sentance, B. (2003). *Wood – the world of woodwork and carving*. Thames & Hudson Ltd., London, ISBN 0-500-51120-9.

Šimoníková, J. (1992). *The furniture of Bystřice pod Hostýnem*. Ton, Bent wood furniture factories, Bystřice pod Hostýnem, Czech Republic.

Singer, C., Holmyard, E.J. & Hall, A.R. (eds.) (1954). *A history of technology. Volume 1: From early times to fall of ancient empires*. Oxford University Press, New York, London.

Steinway, C.F.T. (1880a). *Screw-clamp for wood-bending machines*. U.S. Patent No. 229198.

Steinway, C.F.T. (1880b). *Upright piano-forte*. U.S. Patent No. 230354.

Steinway, C.F.T. (1885). *Piano-frame*. U.S. Patent No. 314742.

Suthoff, B., Schaaf, A., Hentschel, H. & Franz, U. (1996). *Verfahren zum reibschweißartigen Fügen von Holz*. (Method for joining wood.) Offenlegungungsschrift DE 196 20 273 A1. Deutsches Patent und Markenamt.

Suthoff, B. & Kutzer, H.-J. (1997). *Verfahren zum reibschweißartigen Verbinden von Holz*. (Method for joining wood.) Offenlegungungsschrift DE 19746 782 A1. Deutsches Patent und Markenamt.

Tarn, W.W. (1948). *Alexander the great, Volume II. Sources and studies*. (Appendix 24), Cambridge University Press, Cambridge.

Twede, D. (2005). The cask age: the technology and history of wooden barrels. *Packing Technology and Science, 18*(5):253-264.

Vegesack, A.v. (1987). *Das Thonet Buch*. (The Thonet book.) Bangert Verlag, München, ISBN 3-925560-09-2.

Vegesack, A.v. (1996). *Thonet – Classic furniture in bent wood and tubular steel*. Hazar Publishing Ltd., London, ISBN 1-874371-26-1.

Vill, V.I. & Shternin, L.A. (1957). The basic technology of friction welding. (translated title) *Glavelektroadard 4*.

Vill, V.I. (1959). Friction welding of metals. (translated title) *Mashgiz*, Moskow-Leningrad.

Vinson, S. (1994). *Egyptian boats and ships*. Shire Egyptology series No. 20, Shire Publication Ltd., Buckinghamshire, UK, ISBN 0-7478-0222-X.

Ward, C. (1976). Boat-building and its social context in early Egypt: interpretations from the First Dynasty boat-grave cemetery at Abydos. *Antiquity, 80*(307):118-129.

Åström, K. (1993). Skidan från Kalvträsk. (The ski from Kalvträsk.) *Västerbotten, 74*(3):129-131, ISSN 0346-4938.

CHAPTER 3

THE STRUCTURE AND COMPOSITION OF WOOD

3.1 WOOD

Wood is an important resource, and one of the few that are renewable. It is prevalent in our everyday lives and the economy. Wood and wood products are also a store for carbon and thus help to minimize carbon dioxide in the atmosphere.

The structure of wood is a result of the requirements of the living tree. Wood (also called xylem) has several roles, e.g., support, conduction and storage, and the cells performing these roles make up 60-90 percent of the volume (Wangaard, 1981). Within the living tree these cells are dead, i.e., the protoplasm is absent, leaving hollow cells with rigid walls. The support role enables a tree stem to remain erect despite the height to which the tree grows. Because of this height, the wood must also perform the role of conduction, which consists in transporting water and minerals from the ground to the upper parts of the tree. Finally, nutrients are stored in certain parts of the wood until required by the living tree. The only living cells in the tree are those storing nutrients.

The bark of a tree carries out the roles of protection, conduction, and nutrient storage. The bark protects the xylem from extreme temperatures, mechanical injury, etc. The conduction role in the bark, as opposed to in the wood, is carried out by living cells. These cells conduct the nutrients created by photosynthesis in the leaves or needles down to the tree stem to be used for producing new cells and for storage in the nutrient-storing cells of the wood and bark. The study of wood for scientific or engineering purposes is often simplified if the role of the cells in the living tree is kept constantly in view.

Wood is a bio-polymeric and composite material. Its structure is cellular and porous, and its excellent properties as a material are to a large extent due to its highly organized composition in several hierarchic levels from parts of the tree down to the molecular level. Differences in the hierarchic levels are very important for the properties of wood materials. Although the wood structure is diverse, it always consists mainly of cellulose, hemicelluloses and lignin, together with small amounts of extractives, such as terpenoids, resin acids, fatty acids, pectin, proteins, and inorganic matter.

The polymeric components of wood and its porous structure allow its properties and structure to be modified under the combined effects of temperature, humidity and mechanical action (THM action). For a better understanding of the fundamentals of THM processing, it is important to have a solid knowledge of the structure and chemical composition of wood. It will be noted that the structure of wood is variable, and that this variation is mainly due to the hereditary feature of its structure, its biological origin and also to the environmental changes and cyclic conditions under which the tree grows.

3.1.1 Classification of woody plants

In practice, woods and the trees that produce them are divided into two categories: softwoods and hardwoods. The woody plant of the softwood tree is relatively simple and uniform compared to that of the hardwoods. The terms softwood and hardwood have no direct relation to the softness or hardness of the wood. In fact, hardwood trees such as aspen and balsa have softer and less dense wood than many softwood trees.

Species that are taller than six meters when mature and which have a dominant single stem are called ***trees***; those species that do not reach six meters are called ***shrubs*** (Hoadley, 1990). Seed plants, to which all trees and shrubs belong, are divided into two categories; ***gymnosperms***, i.e., plants with naked seeds, and ***angiosperms***, i.e., plants with covered seeds. The gymnosperms consist of softwoods, yew trees and their relatives, and ginko; whereas the angiosperms include all other flowering plants. The angiosperms are further divided into two groups; ***monocotyledons*** and ***dicotyledons***. The former means that these plants have only one heart-leaf in the seed, such as grasses and bulbous plants. The second group, dicotyledons plants, are hardwood trees and most herbaceous plants other than smooth broad-leaved plants. Figure 3.1 shows how the softwoods and hardwoods are divided according to this system.

Needle-like leaves characterize softwood trees, and such trees are commonly known as evergreens because most of them remain green all the year round and annually lose only a portion of their needles. Most softwoods also bear scaly cones within which seeds are produced, and are therefore often referred to as conifers. In contrast to softwoods, hardwoods bear broad leaves which generally change color and fall in the autumn in temperate zones and produce seeds within acorns, pods, or other fruit bodies. Moreover, hardwood trees are therefore often referred to as broad-leaved trees.

The woods of the softwoods and hardwoods differ greatly. The relatively simple cellular structure of the softwood, consisting primarily of tracheids and ray cells, makes the wood easy to use, to modify or to make more uniform for the final product. In hardwoods, the fibers, rays, and vessels all contribute significantly to the wood quality, but their differences, importance and interaction make manipulation of the wood for a high-quality final product difficult. The different distribution of vessels between ring-porous and diffuse-porous woods, as shown in Figure 3.13, also complicates the refinement of the hardwood species.

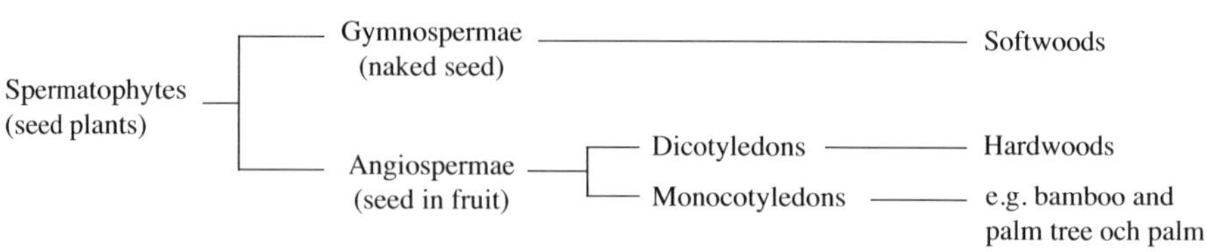

Fig. 3.1 Softwoods and hardwoods are included in the botanical division spermatophytes.

3.1.2 Scientific names, local names and trade marks

The scientific names of plants and animals create a clear and internationally comprehensible system. The system of classifying and naming living organisms is called taxonomy, where, for example, a timber is grouped in: kingdom, divisions, classes, orders, families, genera and species. The example in Figure 3.2 shows the classification of Scots pine (*Pinus sylvestris* L.).

Carl Linnaeus (1735) created the system where each species is given a Latin name in the second degree (binomial) which is unique to that species. The first leg of the Latin name indicates the genus which embraces many other species. The second stage is the name which, for example, can point to something of the nature of the features or the name of a botanist. Together they create a name which is unique for that species. The full botanical name is followed by the Latin binomial name, often abbreviated, of the scientist who first described the species under that name, such as L. for Linnaeus, seen in Figure 3.2. A species may then in turn be divided into sub-species. In botany, subspecies is one of many ranks below that of species, such as variety, subvariety, form, and subform. The subspecific name is preceded by "ssp." or "subsp.". Any botanical name including a subspecies, variety, etc., is called a ternary name.

Kingdom: Viridiplantae
└── Division: Pinophyta
└── Class: Coniferopsida
└── Order: Coniferales
└── Family: Pinaceae
└── Genus: Pinus
└── Species: Pinus sylvestris

Binomial name: *Pinus sylvestris* L.

Fig. 3.2 Scientific classification of Scots pine (*Pinus sylvestris* L.).

The system of giving a timber a botanical name is based on the entire tree's characteristic features and not only on the wood properties. This means that the higher levels of classification can include woods with very different properties, while the lower levels are generally composed of woods with relatively similar characteristics. For example, the relatively different species oak, beech and chestnut belong to the same family, Fagaceae, but to different genera, Quercus, Fagus and Castanea. Various species of oak, for example, European oak, American oak, Turkey oak, which have similar properties, belong to the same genus.

Trade names of wood do not always follow the botanical nomenclature. The same species may have different local names depending on where you are in the world, and may also have one or several trade names. The Northern Europe ordinary species pine (*Pinus sylvestris* L.) and spruce (*Picea abies* L. Karst.) go under common names such as Scots pine and Norway spruce. It also happens sometimes that several species of pine and spruce, but from the same family, are sold under one and the same trade name; e.g., redwood or white wood.

Another example is the name "mahogany" that has been linked to over 300 different botanical species. The trade name mahogany was originally Spanish mahogany.

The well known Honduras mahogany is closely related to Spanish mahogany and signifies several species of the same genus. African mahogany refers to several species from two distinct but connected families, which in turn are part of the same family as the American mahogany species. In addition, the Philippine mahogany consists of several species from different families and genus. This means that wood referred to as mahogany can exhibit great differences in appearance and behavior. A similar confusion can occur with other types of wood such as ash, oak, teak and walnut.

3.2 THE THREE PRINCIPAL SECTIONS OF WOOD

When discussing the structural features of wood, it is important to indicate which surface is being referred to. Three distinctive planes exposing different views of the wood structure can be noted. A cut perpendicular to the longitudinal direction of stem is called a ***transverse*** or ***cross section***, a cut in the radial plan is called a ***radial section*** and a longitudinal cut tangent to the annual rings is termed a ***tangential section***. It is also necessary to distinguish between the three local principal directions of wood; axial (*l*), radial (*r*) and tangential (*t*), which are the principal local directions responsible for the wood anisotropy, as well as for its mechanical, physical and technological properties. Figure 3.3 schematically illustrates the local principal axes and three corresponding sections: the radial, tangential and cross sections.

A microscopic examination of the wood on these three sections gives us the overall information of the structure and the general character of the wood, portraying

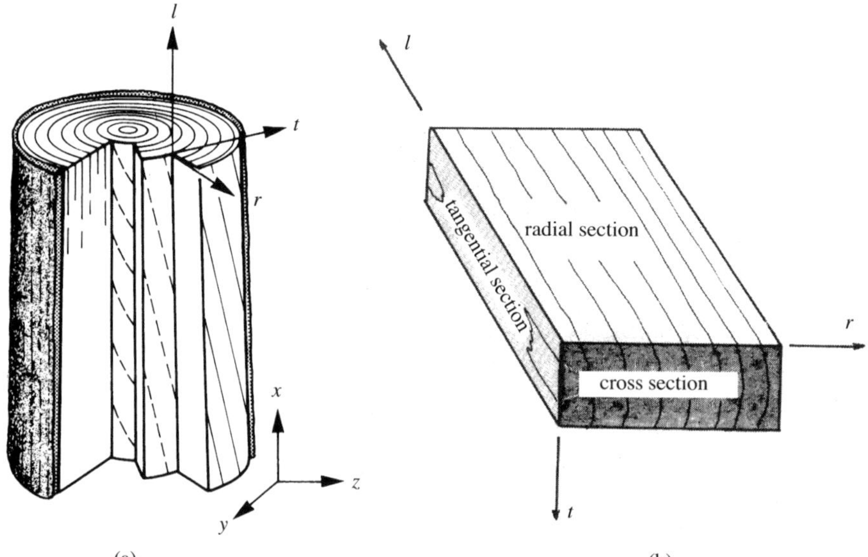

Fig. 3.3 A schematic representation of the three principal axes and sections; (a) local principal axes of a wood stem; longitudinal (*l*), radial (*r*) and tangential (*t*), (b) definition of sections and principal directions for a rectangular specimen.

the morphological characteristics of the woody plant. This is the most reliable method of wood identification, and microscopic examination is most important with softwoods, where visual features are characterized more by similarities than by differences among woods.

3.3 FORMATION OF THE WOOD AND THE BARK

3.3.1 Wood cells

The basic unit of the structure of a tree is the ***cell***. A set of cells of similar origin, or with a similar function, is called a ***tissue*** and various tissues may be classified as for instance conducting tissue or storage tissue. Thus, wood is a tissue, i.e., a collection of various kinds of cells that are produced by division in the cambium. The cells are typically elongated, consisting of an outer cell wall encompassing a cell cavity.

All wood cells have specific functions to perform in the living tree. As already mentioned, the cells can be divided into three groups: conducting cells, supporting cells, and storage cells. Conducting and supporting cells are elongated, axial cells. In hardwoods, they consist of vessel elements, and the supporting cells consist of

(a) (b)

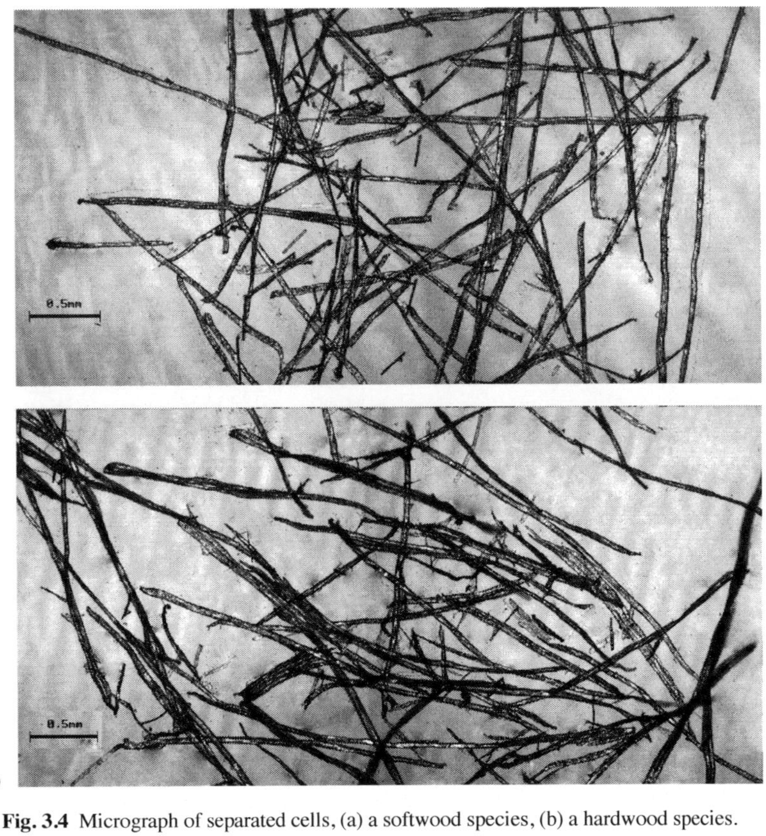

Fig. 3.4 Micrograph of separated cells, (a) a softwood species, (b) a hardwood species.

fibers. In softwoods, tracheids perform both these functions. Conducting and supporting cells are dead cells whose cell cavities are filled with water and air. Storage cells consist of short, thin-walled parenchyma cells with living cell contents. They are involved in the storage and distribution of food, and remain living as long as they belong to the sapwood, i.e., the water-transporting part of the stem.

Wood is formed by various types of cells oriented in the longitudinal or radial direction of the stem. These cells are bonded to others and form various tissues. To study these cells, one can separate them from each other, either mechanically or by chemical means. For comparative purposes, maceration is essential to obtain for example quantitative data on the length of a vessel element and imperforate tracheary elements (Carlqvist, 1988). One of the chemical means used in the laboratory is known as the Franklin maceration process (Franklin, 1945). Small pieces of wood are first saturated with water and then plunged into a mixture of hydrogen peroxide and acetic acid in equal proportions. The whole is then placed in an oven at 60 °C for 48 hours. During this maceration process, the cells dissociate from each other due to the intercellular lignin layer becoming dissolved. Figure 3.4 presents micrographs of chemically macerated cells of a softwood and a hardwood. During the mechanical separation of cells, on the other hand, it is necessary to first saturate a small piece of wood with distilled water to facilitate the cutting of a very thin layer of approximately 100 μm by a microtome under a microscope. Subsequently, with the help of tweezers, it is possible to remove a cell rather easily from the wet layer.

3.3.2 Formation of wood

All plants, including trees, can grow larger by primary growth from their branch tips. Trees are also capable of secondary growth, i.e., an increase in diameter of existing tissue, resulting in secondary tissue. Wood tissue is formed through the division of cells in the cambial zone, i.e., the zone located between the wood (xylem) and the inner bark (phloem). The cambium tissues, i.e., the cork cambium and the vascular cambium, are the only parts of a woody stem where cell division occurs; undifferentiated cells in the vascular cambium divide rapidly to produce secondary xylem tissue on the inside and secondary phloem tissue on the outside.

The cambium tissues contain living material called protoplasm in their cell cavities. As the newly divided cells develop, they assume different sizes and shapes and perform various functions in the tree. Certain cells called parenchyma remain alive for years, but most of the cells, e.g., tracheids, fibers and vessel elements, develop into cell types that lose their protoplasm within days after being created. These cells are named sclerenchyma cells.

Longitudinal growth proceeds at the tips of the stem, branches, and roots, whereas radial growth takes place in the vascular cambium. The annual growth in a tree can be schematized by the superposition of layers with a conical shape, where each layer represents the wood formed in one year. Such a scheme is given in Figure 3.5. It shows the growth of a portion of a tree stem during six consecutive years.

Tissues are also distinguished as being meristematic, i.e., involved in the new cell formation, and permanent. There are regions within a plant where growth has ceased at least temporarily and in which cells and tissues have become fully differentiated and

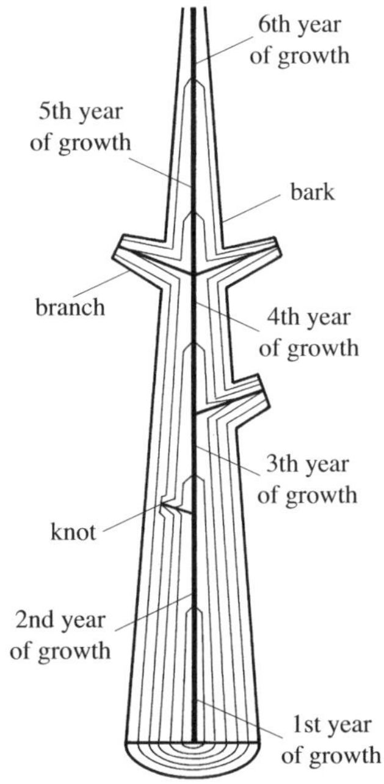

Fig. 3.5 Diagrammatic representation of the annual growth of a tree stem by the successive superposition of wood layers.

mature. Parts of and sometimes the entire permanent tissue may again become meristematic and then be involved in further cell formation. An example of this is the formation of the cork cambium, which arises in the permanent tissue (Panshin *et al.*, 1964).

The regions in which plant tissue are formed are called ***meristems***, and this is the tissue in all plants consisting of undifferentiated cells (meristematic cells). The development of the woody stem, as well as the roots and the branches is the combination of an elongation and a growth in thickness resulting from two different processes provided by meristems. Differentiated plant cells cannot generally divide or produce cells of a different type. Therefore, cell division in the meristem is required to provide new cells for the expansion and differentiation of tissues and for the initiation of new organs, providing the basic structure of the plant body.

Apical meristems (primary meristems or growing tip) are found at the tips of stems and roots and increase the length of these sections. Their main function is to begin the growth of new cells in young seedlings at the tips of roots and shoots. Specifically, an active apical meristem lays down a growing root or shoot behind itself, pushing itself forward. The parenchyma cells include the apical meristematic cells of roots and shoots, and the green photosynthetic cells of leaf. Meristems are specialized where most new cells arise.

Lateral meristems (secondary meristems) are found just under the surface along the length of the stem or root and the lateral meristems surround the established stem of a plant and cause it to grow laterally (i.e., become larger in diameter). There are two types of lateral meristems:

- Vascular cambium, producing secondary xylem and secondary phloem, a process that may continue throughout the life of the plant. This is what gives rise to wood in plants.
- Cork cambium, giving rise to the outer bark of a tree, see sect. 3.4.1.

The wood cells can also be divided into groups according to their orientation: cells oriented in the longitudinal direction of the stem, the branches, and roots; and cells lying in the radial direction, perpendicular to the stem axis. The cambium consists of two types of cells: fusiform (spindle-shaped) initials and ray initials. The fusiform initials give rise to the longitudinal cells of the xylem and phloem, whereas the ray initial cells are at the origin of the radially oriented ray cells – made of parenchyma – and the radial tracheids. Figure 3.6 shows a schematic of the fusiform and ray initial cells of softwoods and hardwoods as well as their various longitudinal and radial cellular products.

The addition of new cells in xylem and phloem is accomplished by the tangential division of the cambial initials, as shown schematically in Figure 3.7. At the first stage of division, the initial cells divide to generate two new cells: one of these is placed either on the xylem side (xylem mother cell) or on the cambium side (phloem mother

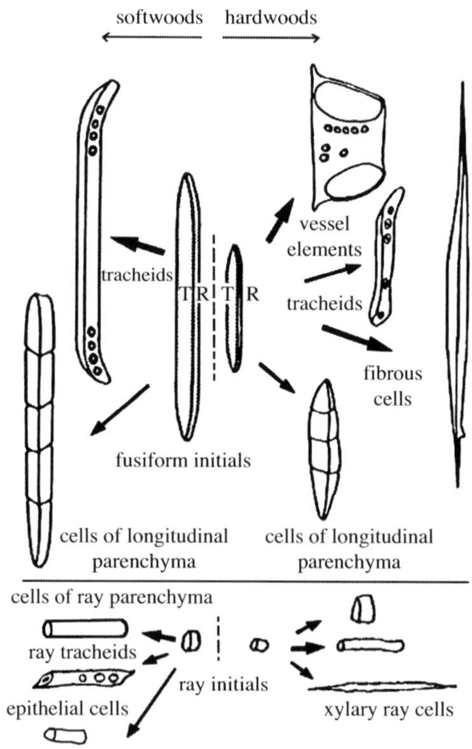

Fig. 3.6 A schematic representation indicating the manner in which longitudinal cells are derived from fusiform cambial initials (upper part of the figure) and how transverse cells are derived from ray initials in softwoods and in hardwoods.

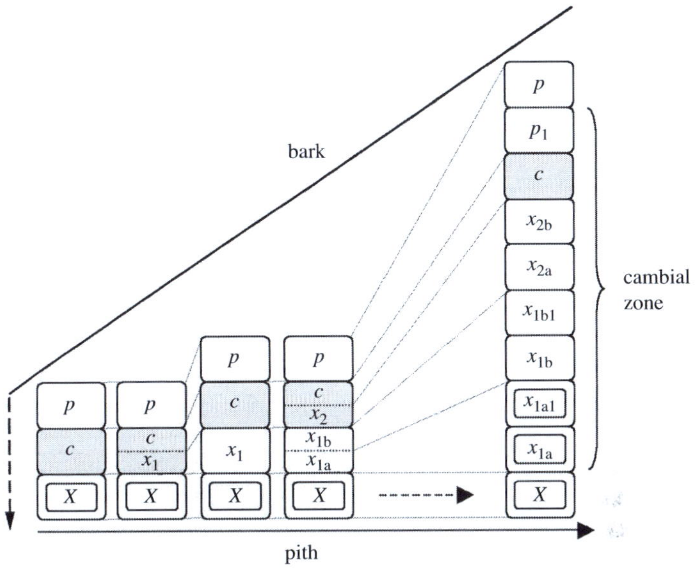

Fig. 3.7 A schematic transverse section view of the tangential division of the cambial longitudinal initials (after Panshin *et al.*, 1964). p – mature phloem cells, c – cambial initials, x – mature xylem cells, x_i – xylem mother cells capable of dividing several times or directly into mature cells, p_i – phloem mother cells capable of dividing several times or directly into mature cells.

cell), while the other increases to the original size of the cambial initial and continues to function as an initial cell. The cambial initials continue to divide in this manner throughout the growing season, with one cell always remaining a cambial initial. The newly formed xylem or phloem mother cells may either mature directly into permanent xylem or phloem cells or divide one or more times before they mature.

The formation of woody cells by division in the tangential plane is the origin of the diameter growth and the increase of the distance between the cambial zone and the center of the tree (pith). To preserve the sapwood continuity in the tangential direction as the stem increases in diameter, division of the initial cell in the radial plane gives rise to two new initial cells, both belonging to the cambial. Thus, the number of initial cells increases by increasing the stem diameter. It should be noted that the rhythm of division of the initial cells in the tangential plane is faster than of the radial plane.

The complete development of a wood cell consists of several successive phases. During the division of a meristem cell, a fine membrane of separation between the cells appears. This intercellular layer, consisting of pectic substances, joins together the wood cells. Lignification takes place in this layer during differentiation. The cell lignification always starts on the level of this joint membrane. The initial thickness of this layer is 0.5 to 1.5 μm. The third phase (maturation phase) consists of cell-wall thickening. In softwoods, the new cell widens in the radial direction giving a uniform pattern through the radial alignment. In a different way, the hardwood cells expand more or less in two directions: radial and tangential. Expansion of the cells in the tangential direction pushes the adjacent cells outward in the transverse plane, and this explains the rather irregular pattern of the cells in hardwood.

When the cell reaches its final dimensions, the cellular protoplasm deposits on the inner surface of the intercellular membrane prior to the development of layers with different structure. As a result, the interior of the cell contains the primary layer and three secondary layers S_1, S_2 and S_3. The primary layer consists of tangled celluloses in which lignin and hemicelluloses settle. Each secondary (S) layer is made up of several layers consisting of cellulose, hemicelluloses and lignin. The secondary layer contains a large proportion of cellulose in the form of microfibrils. The last stage in the development of the cell is the formation of a lignin matrix through the various layers of the cell starting from the intercellular membrane. A majority of the cells die shortly after the lignification phase is finished, except for some that remain alive and exert a reserve function. These cells die after a much longer period. It should be noted that the phases of development of a cell, including birth, thickening, maturation and lignification, occur almost simultaneously.

3.4 THE MACROSTRUCTURE OF WOOD

Almost all plants with which we are familiar have three main parts: roots, stems and leaves. The characteristic that separates trees from other woody plants is that they have a single main stem, the trunk or bole, and branches connected to the stem.

Wood is a material with a largely maintained biological structure composed of cellulose and lignin-based cells. The study of the structure of wood includes how the cells are built, and the grouping of cells into functional units and tissues of a characteristic size and shape. It is important to distinguish between macro-, micro- and ultrastructure. The structure that is visible to the naked eye, or with a magnifying glass up to 10 times magnification, is called the macrostructure. The structure visible only with a microscope is called the microstructure. In ultrastructure studies, it is the cell-wall composition that is in focus.

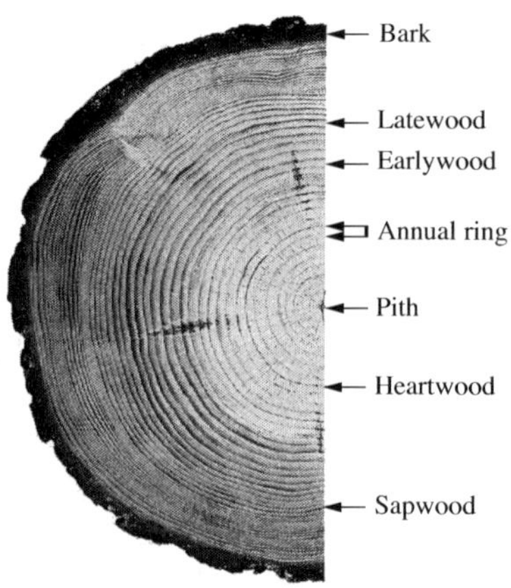

Fig. 3.8 The wood macrostructure visible in the cross section of Scots pine (*Pinus sylvestris* L.).

The main macroscopic parts of wood, seen in Figure 3.8, are
- bark;
- pith;
- vessels of hardwood;
- sapwood and heartwood;
- annual ring divided into early- and latewood;
- rays; and
- resin canals of softwood.

The texture, color and scent of wood are also often included in its macro-structure. Juvenile wood and reaction wood are two wood types that are of great importance and will therefore be dealt with here.

3.4.1 Bark

The trunk has an outer covering, called the ***bark***, that protects the wood from extremes of temperature, drought, and mechanical injury. The bark usually means tree tissues outside the cambium. What is commonly called bark includes a number of different tissues, cf. Figure 3.9. The bark can be divided into the inner bark and the outer bark. The former, which in older stems is living tissue, includes the innermost area of the periderm. The outer bark in older stems, includes the dead tissue on the surface of the stems together with parts of the innermost periderm and all the tissues on the outer side of the periderm.

Cork is an external, secondary tissue that is more or less impermeable to water and gases, and is also called the phellem. Cork cell walls contain suberin, a waxy substance that protects the stem against water loss and the invasion of insects into the stem, and that prevents infection by bacteria and fungal spores. The cork produced by

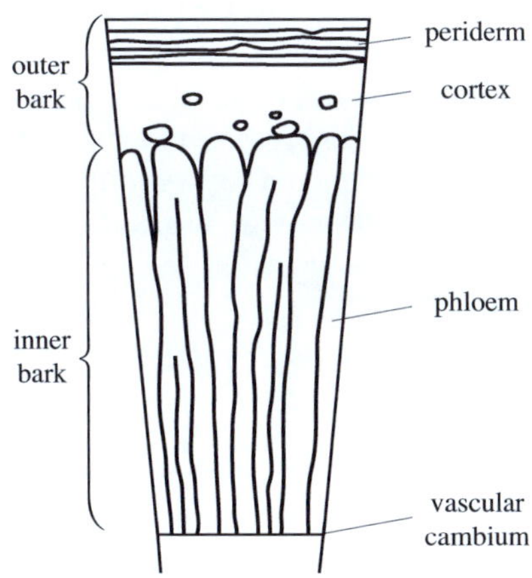

Fig. 3.9 A schematic representation of bark from the outside to the inside of a mature woody stem. The different layers are: cork (phellem), cork cambium (phellogen), phelloderm (the periderm includes these three layers), cortex, and phloem. Bark is composed of periderm and phloem, cortex and the cells that produce these tissues.

the cork cambium (phellogen) is normally only one cell layer thick and divides periclinally to the outside producing cork. The phelloderm, which is not always present in all barks, is a layer of cells formed by and interior to the cork cambium. Together, the phellem (cork), phellogen (cork cambium) and phelloderm constitute the ***periderm*** (Dickison, 2000; Pereira, 2007).

The cortex is the primary tissue of stems and roots. In stems, the cortex is between the periderm layers and the phloem, and is composed mostly of undifferentiated cells, usually large thin-walled parenchyma cells. Phloem is a nutrient-conducting tissue composed of sieve tubes or sieve cells mixed with parenchyma and fibers. Products derived from bark include spices such as cinnamon and other flavorings, tannin, resin, latex, medicines, poisons, various hallucinatory chemicals and cork.

3.4.2 Pith

Under the bark is the wood, in the center of which is the pith. The pith of a tree is formed during the first year of growth and becomes a storage area for impurities that are deposited from the active xylem during the growth of the tree. Pith consists of soft, spongy parenchyma cells, and is located in the center of the stem and in the center of the roots. In some plants, the pith is solid, but in most cases it is soft. A few plants, e.g., walnut, have a distinctive chambered pith with numerous short cavities. The pith varies in diameter from about 0.5 mm to 8 mm. Freshly grown pith in young new shoots is typically white or pale brown, but it usually darkens with age. It may be inconspicuous, but it is always present at the center of a trunk or branch. The roots have little or no pith and the anatomical structure is more variable. The shape of the pith varies between species and can be used to identify different varieties.

3.4.3 Vessel elements of hardwood

The fundamental anatomical difference between hardwood and softwood is that hardwoods contain specialized conducting cells called vessel elements. This cell type is found in virtually all hardwoods but rarely in softwoods. Vessel elements are generally much larger in diameter than other types of longitudinal cells and the vessels are in general shorter than both hardwood and softwood fibers. A number of vessel elements are linked end to end along the grain to form long tube-like structures. Both the size and arrangement of the vessels in the cross-section of a wood sample are used to classify hardwoods, as seen in Figure 3.13.

3.4.4 Sapwood and heartwood

As the tree grows older, it no longer needs the whole cross section of the xylem part of the stem for fluid transport. At this time, the life functions of living cells in the central part of the stem start to decline and heartwood develops. The new wood cells thus created are added to the sapwood, while the older cells adjacent to the heartwood gradually change to form new heartwood. The proportions of sapwood and heartwood vary according to species, the age of the tree, the position in the tree, the rate of growth, and the environment (Hillis, 1987).

The sapwood is the outer, newly generated part of the trunk that, in the living tree, contains living cells and reserve material (e.g., starch) and it also holds the water-transporting cells. Heartwood is the inner and central part of the trunk, which, in the living tree, comprises only dead and non-water-transporting cells, and in which the reserve materials have been removed or converted into extractives. Heartwood formation can also be found in the roots of many species, especially in the region near the stem (Hillis, 1987).

In some varieties, a zone, usually about 1-3 annual rings, can occur between the sapwood and the heartwood for a short period of time. This zone is called the transition zone and is defined by Hillis (1987) as a narrow, pale-colored area surrounding some heartwood or injured regions, often containing living cells, usually devoid of starch, often impermeable to liquids, with a moisture content lower than that of the sapwood and sometimes also below that of the heartwood.

The volume percentage of living cells (parenchyma cells) in the sapwood varies between 5 and 40% of the volume of the total tissue (Hillis, 1987). The death of these cells and the transition of the sapwood to heartwood are accompanied by secretion of oxidized phenols, which are generally the origin of the pigmentation of heartwood. In trees in which heartwood and sapwood have the same color, the death of these cells does not lead to pigmentation. The substances secreted by the trees are called

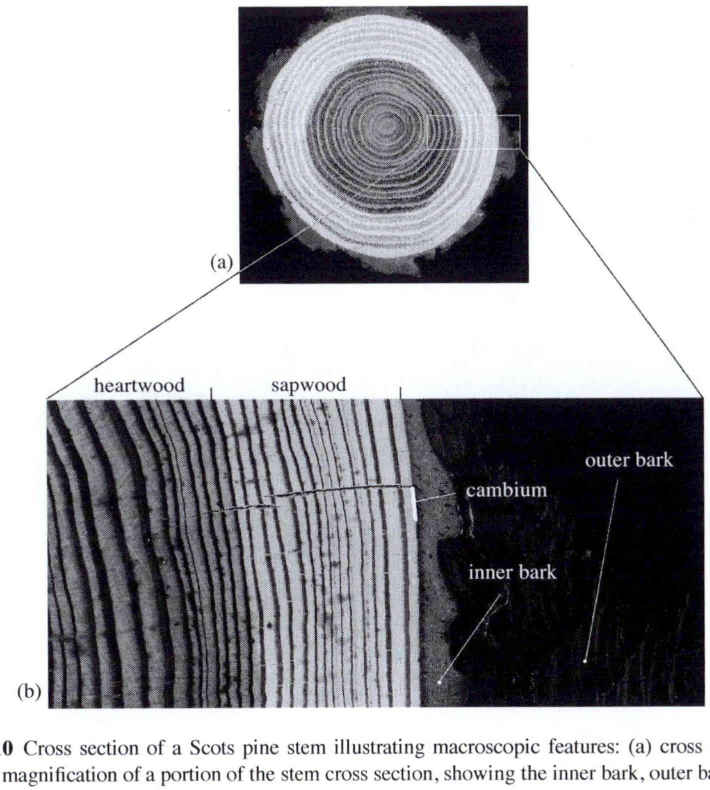

Fig. 3.10 Cross section of a Scots pine stem illustrating macroscopic features: (a) cross section of the log; (b) magnification of a portion of the stem cross section, showing the inner bark, outer bark, cambium, sapwood and heartwood.

extractives. They are more or less toxic for wood-decaying organisms and help the wood to resist fungi and insects.

Figure 3.10 shows an example of the difference in color between sapwood and heartwood. Sapwood is often clearer than the heartwood, but in many species this color distinction between the sapwood and heartwood does not exist. Rather, the difference involves function and moisture content.

In contrast to heartwood, sapwood in the living tree has a very high moisture content. In Scots pine there is also a large variation in moisture content between earlywood and latewood in the sapwood, where the moisture is mainly found in the earlywood. In the heartwood, there is no difference in moisture content between earlywood and latewood, cf. Figure 3.11.

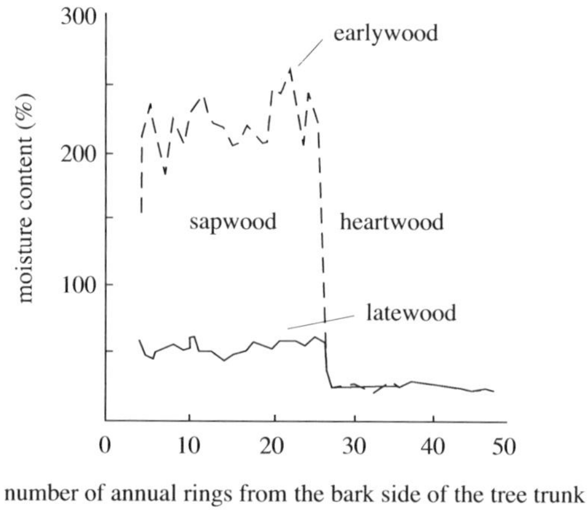

Fig. 3.11 Moisture content of earlywood and latewood of sapwood and heartwood of Scots pine (*Pinus sylvestris* L.) (Vintila, 1939).

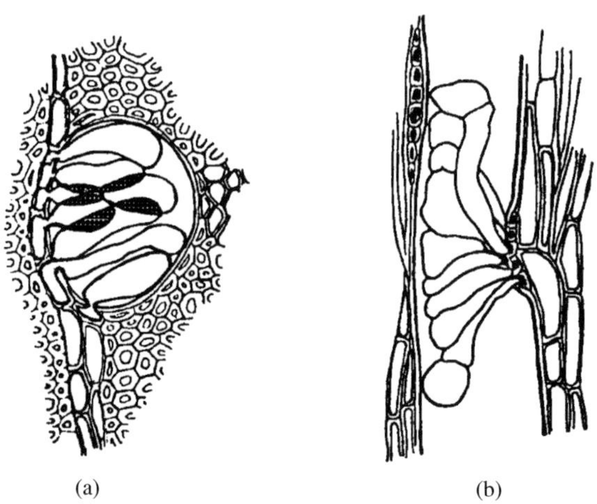

Fig. 3.12 Tyloses in a hardwood vessel in (a) transverse and (b) longitudinal sections. The tyloses effectively prevent water transportation in the vessel (Bosshard, 1974).

During heartwood formation in a number of species of hardwoods, the vessels are filled with so-called tyloses, which are outgrowths of parenchyma cells into the hollows of vessels, as seen in Figure 3.12. Tyloses are significant in that they partially, or often completely, block the vessels in which they are present, a situation that can be either detrimental or beneficial depending upon the use to which the wood is put. The existence of tyloses in the heartwood vessels of white oak, and the relative lack of them in red oak, is the reason why white oak is preferred in the manufacture of barrels, casks and tanks for the storage of liquids. In contrast to this beneficial feature of tyloses, wood in which they are well developed may be difficult to dry or impregnate with different chemicals (Bowyer *et al.*, 2007).

3.4.5 Annual growth

The growth rate and periodicity of trees growing in the tropical forests located around the equator are not the same as those of trees in temperate forests. In the former, the growth is orchestrated by the alternation of the dry and rainy seasons, while in temperate forests, it is the seasonal growth during the vegetation period which characterizes the annual ring. The growth starts at the beginning of spring, continues in summer and stops in the autumn. The part produced in the spring is called earlywood and that in the summer is called latewood.

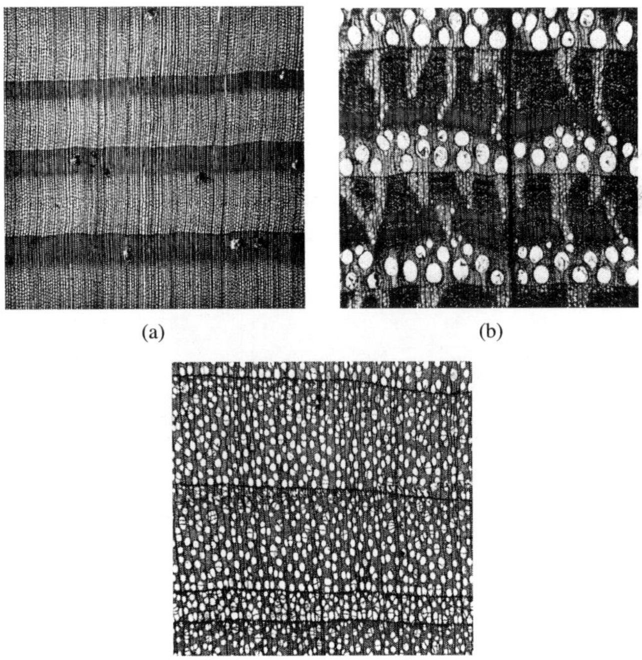

(a) (b) (c)

Fig. 3.13 Cross-section view of (a) a softwood species, Scots pine (*Pinus sylvestris* L.), (b) a ring-porous hardwood species, oak (*Quercus petraea* (Matt.) Liebl.) and (c) a diffuse-porous hardwood species, goat willow (*Salix caprea* L.).

At the beginning of the vegetative growth, trees form a new layer of wood between the existing wood and the bark all over the trunk, the branches and the roots. The annual rings can often be easily distinguished because of differences in structure and color between the earlywood and latewood. When discussing woods, it is customary to divide the annual rings into three classes: softwood, ring-porous hardwood and diffuse-porous hardwood, see Figure 3.13.

Earlywood and latewood in softwood

In temperate softwoods, there often is a marked difference between latewood and earlywood. The latewood is denser than that formed early in the season, see Figures 3.14 and 3.15. The cells of the earlywood have thin walls (approximately 2 μm) and an important lumen, whereas those of latewood have thicker walls (approximately 5 μm) but a narrower lumen. The strength is in the walls, not the cavities. Hence, the greater the proportion of latewood, the greater is the density and strength. The width of an annual ring is not nearly as important for the density of softwoods as the proportion and nature of the latewood in the ring.

If a heavy piece of pine is compared with a lightweight piece, it can seen at once that the heavier one contains a larger proportion of latewood than the other, and therefore shows more clearly demarcated annual rings. In white pines,[1] there is little contrast between the different parts of the ring, and as a result the wood is very uniform in texture. In hard pines,[1] on the other hand, the latewood is very dense and deep-colored, presenting a decided contrast to the soft, straw-colored earlywood.

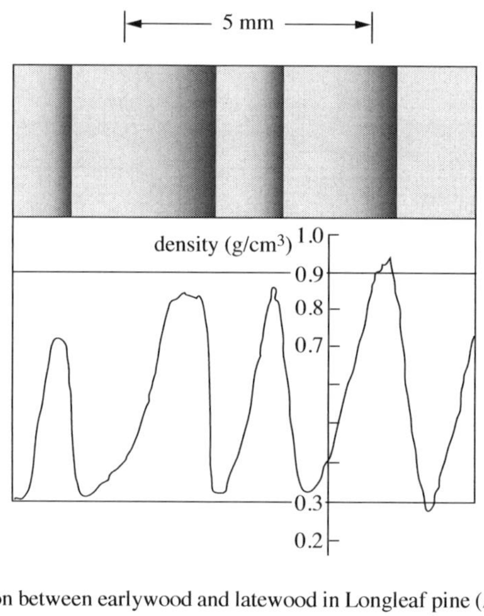

Fig. 3.14 Density variation between earlywood and latewood in Longleaf pine (*Pinus palustris* Mill.) after Phillips *et al.* (1962).

[1] There are three main subgenera of Pinus, the subgenus Strobus (White pines or soft pines), the subgenus Ducampopinus (Pinyon, Bristlecone and Lacebark pines), and the subgenus Pinus (yellow or hard pines). This classification is based on cone, seed and leaf characteristics.

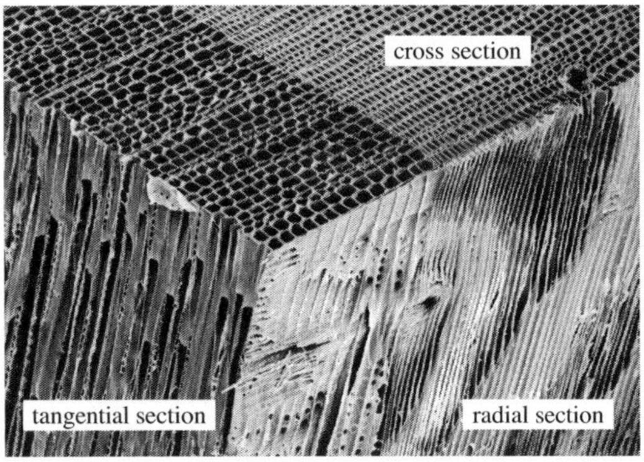

Fig. 3.15 3D scanning electronic micrograph of Norway spruce (*Picea abies* (L.) Karst.) with a uniform structure.

The border between earlywood and latewood is, in contrast to the annual ring border, diffuse and the change for instance density and cell-wall thickness is gradual through this transition zone. To distinguish between earlywood and latewood, a definition is needed and there exist numerous such definitions. The most universally accepted is the one by Mork (1928), that states that the cells are classified as latewood when the double wall thickness is greater than the lumen diameter. Since then, more accurate definitions have been proposed, e.g., by Phillips (1962), based on a beta ray method, and by Jagels and Dyer (1983), based on digital image analysis of the shape of the cell cross section.

Earlywood and latewood in ring-porous woods

In ring-porous woods, each season's growth is always well defined, since the large vessels formed early in the season are on the denser tissue of the year before, see Figure 3.13b. In the case of the ring-porous hardwoods, a definite relation seems to exist between the rate of growth of timber and its properties. This may be briefly summed up in the general statement that the more rapid the growth or the wider the annual rings, the heavier, harder, stronger, and stiffer is the wood. It should be kept in mind that this applies only to ring-porous woods such as oak, ash, hickory, and others of the same group, and it is, of course, subject to some exceptions and limitations.

In ring-porous woods of good growth, the thick-walled, strength-giving fibers are usually most abundant in the latewood. As the width of the ring diminishes, this latewood is reduced so that very slow growth produces comparatively light, porous wood composed of thin-walled vessels and wood parenchyma. In good oak, these large vessels of the earlywood occupy 6 to 10 per cent of the volume of the log, whereas in inferior material they may make up 25 per cent or more. The latewood of good oak is dark-colored and firm, and consists mostly of thick-walled fibers forming one half or more of the wood. In inferior oak, this latewood is much reduced both in quantity and quality. Such a variation is very largely due to the rate of growth.

Earlywood and latewood in diffuse-porous woods

In the diffuse-porous woods, the demarcation between annual rings is not always clear, and in some cases it is almost (if not completely) invisible to the naked eye.

Conversely, when there is a clear demarcation there may not be a noticeable difference in structure within the annual ring, see Figures 3.13c and 3.16.

In diffuse-porous woods, as has been stated, the vessels are even-sized, so that the water-conducting capability is scattered throughout the ring instead of being concentrated in the earlywood. The effect of rate of growth is not, therefore, the same as in the ring-porous woods, approaching more the conditions in the conifers. In general, it may be stated that such woods of medium growth afford stronger material than when they grow very rapidly or very slowly. In many uses of wood, the total strength is not the main consideration. If ease of working is prized, the wood should be chosen with regard to its uniformity of texture and straightness of grain. Such uniformity and straightness generally occur when there is little contrast between the latewood of one season's growth and the earlywood of the next.

Fig. 3.16 3D scanning electronic micrograph of beech (*Fagus sylvatica* L.), with a more differentiated structure.

3.4.6 Rays

All the transverse cells found in any given wood are included in the wood rays, a ribbon-like aggregate of horizontally oriented cells. The rays are formed by the cambium and extend in the radial direction in the xylem, see Figure 3.17. Rays provide an avenue by which sap can be transported horizontally either to or from the inner bark (phloem layer).

Rays may contain ray parenchyma, ray tracheids and ray epithelial cells. Not all of these three kinds of cells occur in every ray and rays are usually composed predominantly of ray parenchyma cells, with ray tracheids forming one or more marginal rows of cells and an occasional row of cells in the body of a ray. When transverse resin canals are present, rows of epithelial cells and the resin canal cavity are also included in the ray.

The size of the rays differs significantly between species. They can be anything from slightly visible to completely invisible to the naked eye. The variation between species is great, and rays are therefore very useful in identification. In softwoods, rays are usually one cell (or a maximum of two cells) wide in the tangential direction and 1 to 20 and sometimes up to 60 cells high. The rays in hardwoods vary between one and several cells, depending on the species. These rays can always be observed on the tangential, radial and transverse sections.

Wood rays consisting of nutrient-storing cells extend in the transverse direction from the bark toward the center of the tree at a right angle to the annual rings. The first formed rays extend from the bark to the pith and are called primary rays, the others extend from the bark to some later-formed annual ring outside the pith and are called secondary rays.

Rays have a major influence on wood properties, not least for the strength. Those who have chopped firewood have hardly failed to notice that this is easiest if the axe goes straight through the rays of a piece of wood. This is because the opening follows the direction of the rays, and in this direction, the rays are the weakest zones in the wood, causing the wood to split easily. The more the axe chop deviates from this perfect ray direction, the more rays will thus "reinforce" the wood, and prevent its splitting.

Rays are also one of the main causes of transverse hygroexpansion anisotropy, i.e., the rays restrain dimensional changes in the radial direction, and their presence is partially responsible for the fact that, upon drying, wood shrinks less radially than it does tangentially. The mechanism causing differential transverse hygroexpansion has been the subject of considerable controversy in the literature on wood science for many years, see for instance Skaar (1988).

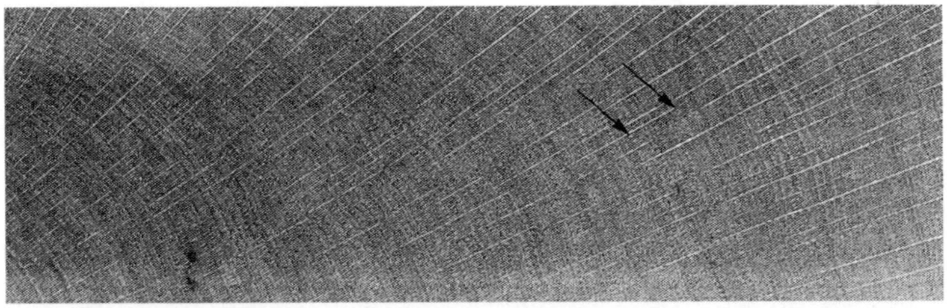

Fig. 3.17 Rays (indicated by the arrows) in the cross section of White oak (*Quercus alba* L.).

3.4.7 Resin canals of softwood

A characteristic feature of many softwoods is their resin content, which is often sufficient to give them a clear fragrance and make newly sawn timber sticky. Resin canals or resin ducts are tube-like intercellular spaces that transport resin in both the vertical and horizontal directions. These vertical and horizontal resin canals are interconnected, and form a uniform network in the tree (Ilvessalo-Pfäffli, 1995).

Resin is formed in epithelial parenchyma cells, and can in some species also be stored in special resin canal cavities called resin pockets. These cells supply resin to the channels and pockets. Resin canals are found only in some softwood species.

Longitudinal resin canals are observed as small holes in the cross section, whereas transverse resin canals, that are located inside the rays, are observed in the tangential section.

3.4.8 The texture, color and scent of wood

The texture of wood is material-dependent, i.e., it depends on the type of wood and on how the wood is built up. A piece of wood can show a great variation in hue depending on for instance the type of wood, the content of extractive substances, heartwood or sapwood, and age. For most types of wood, the annual ring orientation in the cross-section of the wood is very important for the texture. A tangential surface with horizontal annual rings becomes mottled whereas the radial surface with vertical annual rings has an even and harmonious pattern, see Figure 3.18. The word "figure" sometimes has a more specific meaning, denoting a wood surface with a distinctive or decorative appearance (Hoadley, 1990). The term "figured veneer", for example, implies more than just the common patterns caused by wood structure. Special figures may result from uneven heartwood pigmentation, irregular annual ring formation, deviation in cell and grain direction, or any combination of these.

Color is one of the most conspicuous characteristics of wood and, although quite variable, it is one of the important features used in identification as well as for adding aesthetic value. Basic wood substances, i.e., cellulose and lignin, have little color of their own, so any distinctive color is associated with heartwood (Hoadley, 1990). A dark color always indicates heartwood, but a light color can be either heartwood or sapwood. Some wood also undergoes considerable color changes with age or up on exposure to UV-radiation.

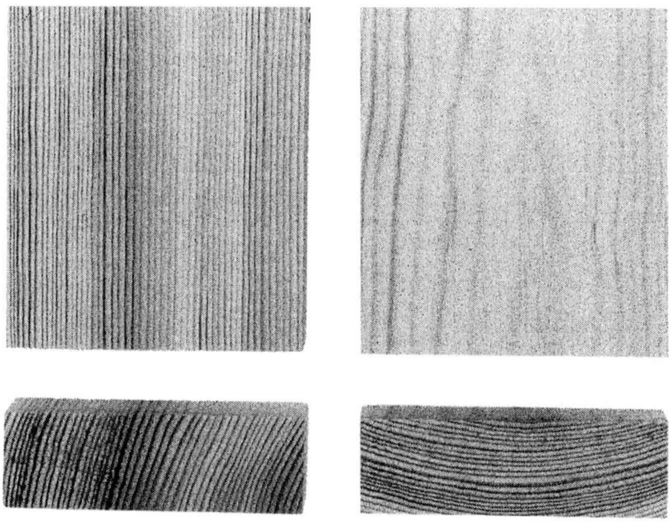

(a) (b)

Fig. 3.18 Influence of annual ring orientation on the texture of the flat side surface (upper part of the figure) of Scots pine (*Pinus sylvestris* L.); (a) radial section with vertical annual rings, (b) tangential section with horizontal annual rings.

Certain woods have distinctive odors. Many softwoods, as well as numerous tropical woods, are known for their aromatic character. The odor is due to volatile extractives or resins in the wood.

3.4.9 Juvenile wood

The wood first laid down by the cambium near the center of the tree has characteristics that differ from that formed at a large number of annual rings from the pith. It is called juvenile wood. Juvenile wood is sometimes referred to as core-wood or crown-formed wood and the mature wood as outer-wood. Although juvenile wood occurs in both softwoods and hardwoods, it is usually much less evident in hardwoods. The source of juvenile wood is primarily from young plantations, thinnings, top wood, plywood cores and the harvesting of young stands.

Juvenile wood is formed nearest the pith at all heights in the tree, cf. Figure 3.19. Many characteristics are used to assess the juvenile zone, the most common being density and cell length, although several other characteristics are used. Each has a different curve of development from the pith outward so that the definition of the juvenile zone depends on the characteristic used. Bendtsen (1978) showed for hard pines that density, strength, cell length, cell wall thickness, transverse shrinkage and late wood percentage increase from pith to bark. The general pattern for the fibril angle, longitudinal shrinkage and moisture content of the same species is that these properties decrease from pith to bark. Moreover, Zobel and Talbert (1984) concluded that the chemical composition of the juvenile wood also differs from that of mature wood. In most softwoods, the lignin content is higher and the cellulose content lower

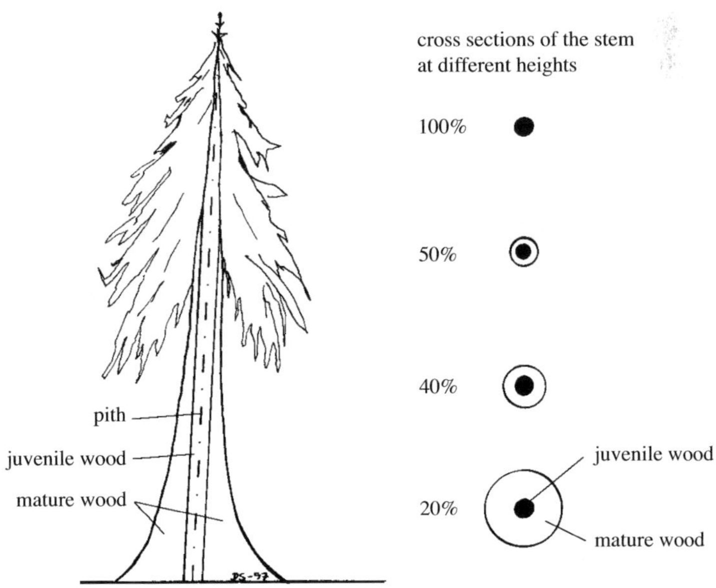

Fig. 3.19 Juvenile wood occurs around the pith and roughly forms a cylinder up the tree. The proportion of juvenile wood increases towards the top of the tree.

in the juvenile wood. In hardwoods, Zobel and Talbert (1984) say that the proportion of, and the chemical make up of, cellulose and lignin vary and that the holocellulose content is higher in juvenile wood than in mature wood.

Considerable work has been done to understand the relationship between juvenile wood and mature wood, or how many annual rings from the pith that are required to give a good estimate of the wood formed nearer the bark. A very thorough review of the area of juvenile wood has been presented by Zobel and Sprauge (1998). Zobel and Buijtenen (1989) state that the term "juvenile wood" is misleading as the wood characteristics within the juvenile zone are not uniform but rapidly changing throughout the zone from the pith outwards, cf. Figure 3.20a. Following this rapid change, there is a leveling off, creating a transition zone leading to the mature wood where changes in wood properties are much less rapid, i.e., there is no absolute shift from juvenile wood to mature wood in any given year. Figure 3.20b shows a schematic pattern of the density variation from pith to bark for ring-porous hardwoods, diffuse-porous hardwoods and softwoods.

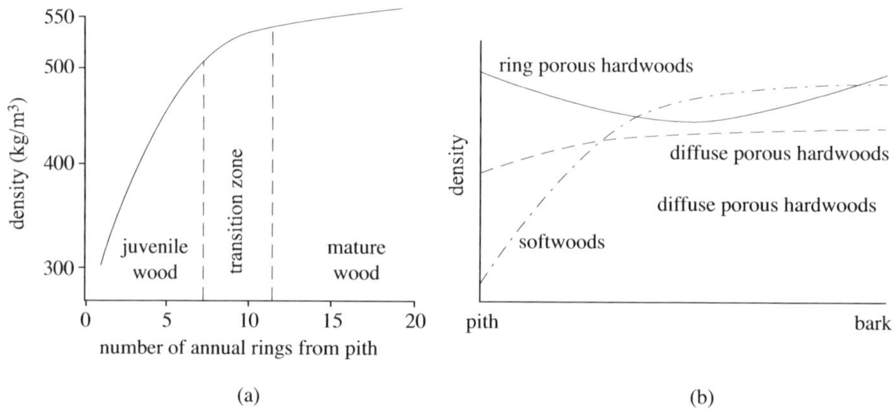

Fig. 3.20 (a) A schematic presentation of the change in density from pith and outwards in Loblolly pine (*Pinus taeda* L.). The change in density is rapid in the juvenile wood (Zobel & Buijtenen, 1989). (b) A schematic illustration of the radial change in density from pith to bark for ring-porous and diffuse-porous hardwoods as well as for softwoods (Zobel & Sprauge, 1998).

3.4.10 Reaction wood

When a tree is growing on a sloping land surface or is exposed to a dominant wind direction, the load on the stem is unbalanced. The tree then starts to produce modified wood, known as reaction wood, to compensate for the unbalanced load. The formation of reaction wood is related to the process of straightening of leaning stems and the same happens in the branches and in the area where the branches connect to the stem.

Softwoods and hardwoods have adopted different strategies for the formation of reaction wood. The reaction wood of softwoods is called compression wood, because it forms on the lower, or compression-stress side of leaning stems. The reaction wood in hardwoods is distinguished from compression wood because of obvious differences

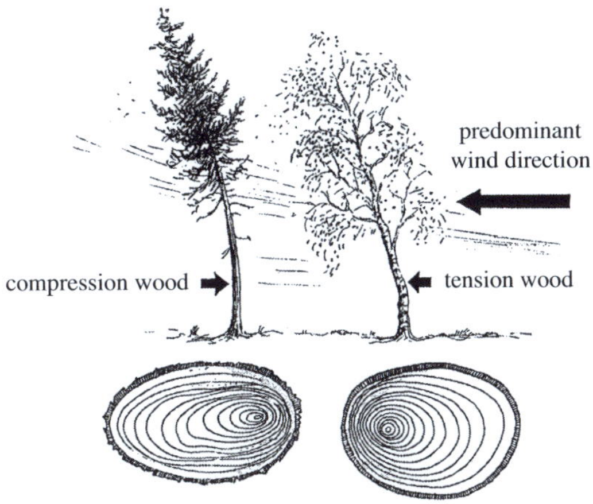

Fig. 3.21 Formation of reaction wood in softwoods and hardwoods as a result of a predominant wind direction over a long time.

in its physical properties, but the name "tension wood" arises from the fact that the increased growth takes place on the upper or tension-stress side of leaning tress, cf. Figure 3.21.

In both softwoods and hardwoods, the wood formed on the opposite side of the stem or branch to the reaction wood is known as opposite wood, while that on either side, lying between the reaction wood and the opposite wood, is referred to as lateral wood. In comparison with wood production in a vertically growing stem with almost perfectly circular annual rings, compression wood and tension wood are usually produced in larger quantities giving the stem a cam-shaped cross section with pronounced eccentricity with respect to the pith (Barnett & Jeronimidis, 2003).

Reaction woods have physical and mechanical properties that differ from those of normal wood and some of these characteristics are worth mentioning here. The compressive strength of compression wood is greater than that of normal wood, but as a consequence compression wood is also very brittle. This brittleness can be a problem during bending. The tensile strength and Young's modulus of tension wood are greater than those of normal wood. It also has a higher fracture toughness and impact resistance. The reaction woods have shrinkage characteristics differing from those of the adjacent normal wood, which can result in warping and cracking of the wood during drying.

Although it has not been generally recognized, reaction wood has many characteristics similar to those of juvenile wood (Zobel & Sprauge, 1998). In the softwoods, both juvenile wood and compression wood have short cells with flat micro-fibrillar angles and often a high lignin content; in hardwoods, juvenile fibers of both the diffuse- and ring-porous types are short and hardwoods have tension wood with a high cellulose content. When a tree is producing juvenile wood, it is especially susceptible to environmental forces that lead to the formation of reaction wood (Zobel & Sprauge, 1998).

3.5 THE MICROSTRUCTURE OF WOOD

The structure of wood is multi-leveled and hierarchical. In wood, it is possible to define more than ten different structural levels between the macroscopic (trunk) and the molecular level. In this section, the microstructure of the cellular wall is briefly explained. More detailed information on the structure of wood can be found in various specialized books, e.g., Jane (1956), Core *et al.* (1976), Butterfield & Meylan (1980), Dinwoodie (1981, 1989), Fengel and Wegener (1984), Schweingruber (1990).

The microstructure of wood can be observed by light and polarized microscopy up to a maximum magnification of about one thousand. On the other hand, the characteristics of the cell-wall ultrastructure can be investigated by indirect means that are capable of detecting the submicronic structures of wood. Different devices can be utilized for this purpose, e.g., high performance Scanning Electronic Microscopy (SEM), Transmission Electronic Microscopy (TEM), Atomic Force Microscopy (AFM) and X-Ray Diffraction (XRD). As the general structure of softwoods differs from that of hardwoods by the regularity of the cellular arrangement and the absence of vessels, and since the cells of softwoods have different shapes as opposed to those of hardwoods, the microstructures of the softwoods and hardwoods are here be described separately.

The following terminology misusage was pointed out by Zobel and Buijtenen (1989). The term "fiber" is commonly used for both the true wood fibers of hardwoods and the tracheids of softwoods. Although this is botanically incorrect, the general use of the term fiber must be recognized, since numerous publications refer to the fiber characteristics of the softwood as well as to the real fibers of hardwoods. Tables 3.1 and 3.2 give various general characteristics of some coniferous and hardwood species (Fengel & Wegener, 1984).

3.5.1 Microstructure of softwoods

The wood of coniferous trees consists of two types of cells; longitudinal tracheids and ray parenchyma, oriented both axially and horizontally. Most of the tracheids are longitudinal, while the parenchyma cells have a radial orientation. In addition to these two types of cells, other elements such as epithelial cells constitute longitudinal and horizontal resin canals. The transversal tracheids are not present in all species. The various types of softwood cells are presented in Table 3.1.

Table 3.1 Cell types and their functions in softwoods.

Cell type	Function
Longitudinal tracheids	Support, conduction
Parenchyma	
Ray parenchyma	Storage
Longitudinal parenchyma	Storage
Epithelial parenchyma	Secretion of resin
Short tracheids	
Ray tracheids	Conduction
Strand tracheids	Condustion

Longitudinal tracheids

The longitudinal or vertical tracheids constitute 90-95% of the volume of the softwoods (Ilvessalo-Pfäffli, 1995). Tracheids are long, narrow cells with closed ends and bordered pits, Figure 3.22. The length of a tracheid varies from 2 to 6 mm and the width of a tracheid varies between 14 and 60 μm (Ilvessalo-Pfäffli, 1995). One can note that the tracheids of latewood have a thick wall and a small lumen more suited to the function of mechanical support, as compared to tracheids of the earlywood, where their function is mainly to conduct sap.

The length of a longitudinal tracheid, which is a closed unit, is very small compared to the height of the tree. To ensure the conduction of sap within the tree, it is thus necessary for the tracheids to be functionally connected to other tracheids. The conduction between tracheids, in both the lateral and vertical directions, takes place through pits. Most of the pits are located in the radial walls. Figure 3.22 and the micrograph in Figure 3.23 illustrate the pits. Pits also exist on the tangential walls of the tracheids, but these are much less numerous.

The pits of earlywood tracheids are large and circular, averaging about 200 pits per tracheid, wheras latewood tracheids show rather small, slit-like pits, 10 to 50 per tracheid (Trendelenburg & Mayer-Wegelin, 1955).

Pits have two essential parts: the pit ***cavity*** and the pit ***membrane***. The cavity is open internally to the lumen of the cell, and it is closed by the pit membrane. Pits can be of many shapes and sizes and they are in general reduced to two basic types on the

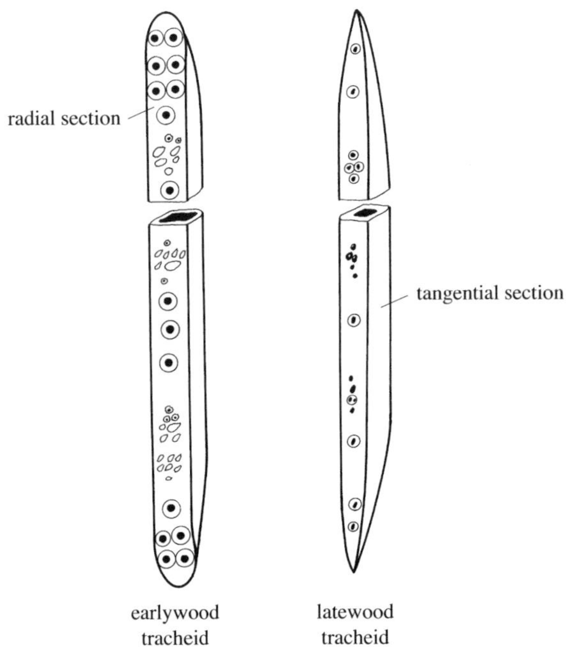

Fig. 3.22 Diagrammatic representation of an earlywood and a latewood tracheid. The length of these cells is approximately 100 times greater than their width.

basis of the shape of the cavity viz.: the ***simple pit*** and the ***bordered pit*** (Ilvessalo-Pfäffli, 1995).

In the simple pit, the cavity is almost straight-walled and only gradually widens or narrows toward the cell lumen. The lumen end of the cavity is known as the pit aperture. In the bordered pit, the cavity is constricted towards the lumen, forming a dome-shaped chamber, which is overarched by the pit border. In wood, the pits of adjacent cells are usually paired, forming three types, see Figure 3.24.

Table 3.2 Density, dimensions of and volume percentages of various cells in different tree species (Fengel & Wegener, 1984)[1].

	Coniferous Wood (temperate zones)			**Hardwoods** (temperate zones)		
	Fir *(Abies alba)*	Spruce *(Picea abies)*	Pine *(Pinus sylvestris)*	Beech *(Fagus silvatica)*	Oak *(Quercus robur)*	Poplar *(Populus ssp.)*
Density [kg/cm³]						
minimum	320	300	300	490	390	
average	410	430	490	680	650	400
maximum	710	640	860	880	930	
Fiber length (mm)						
minimum	3.4	1.7	1.4	0.6	0.6	0.7
average	4.3	2.9	3.1			
maximum	4.6	3.7	4.4	1.3	1.6	1.6
Fiber diameter (μm)						
minimum	25	20	10	15	10	20
average	50	30	30			
maximum	65	40	50	20	30	40
Vessels length (mm)				0.3-0.7	0.1-0.4	0.5
Vessels diameter (μm)				5-100	10-400	20-150
Cell percentage (average values on volume)						
Tracheids	90	95	93	38	44/58	62
Vessels				31	40	27
Parenchyma	scarce	1.4-5.8	1.4-5.8	4.6	4.9	scarce
Ray cells	9.6	4.7	5.5	27	16.2/29.3	11.3

(1) Fiber-tracheids or libriform fibers.

The ***simple pit pair*** is composed of two simple pits. It occurs between parenchyma cells and in hardwoods also between vessel elements and parenchyma cells, see Figure 3.24a. The ***bordered pit pair*** is composed of two bordered pits. It occurs between tracheids in softwoods and between vessel elements in hardwoods, see Figure 3.24b. The ***half-bordered pit pair*** is composed of a bordered pit and a simple pit. It is found in the contact zone between the longitudinal tracheids and the rays, and

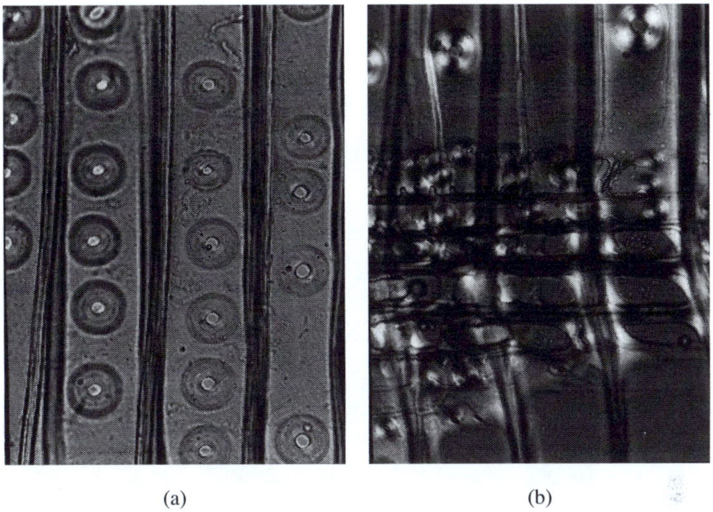

Fig. 3.23 Micrograph of the radial walls of tracheids presenting: (a) bordered pits, (b) simple pits between tracheids and ray parenchyma cells (in the center), and small bordered pits between tracheids and ray tracheids (edges).

is therefore also known as a cross-field pit, see Figures 3.23b and 3.24c. Their size, shape, and arrangement vary according to species, and cross-field pitting is the most important feature in the identification of softwood species at a micro-structural level. Half-bordered pit pairs are also found between vessel elements and parenchyma cells in hardwoods.

In the bordered pit pairs of most softwoods, the membrane has a thickening in the central zone called the ***torus***, which is somewhat larger in diameter than the aperture and is impermeable to water. The membrane around the torus, the ***margo***, is porous,

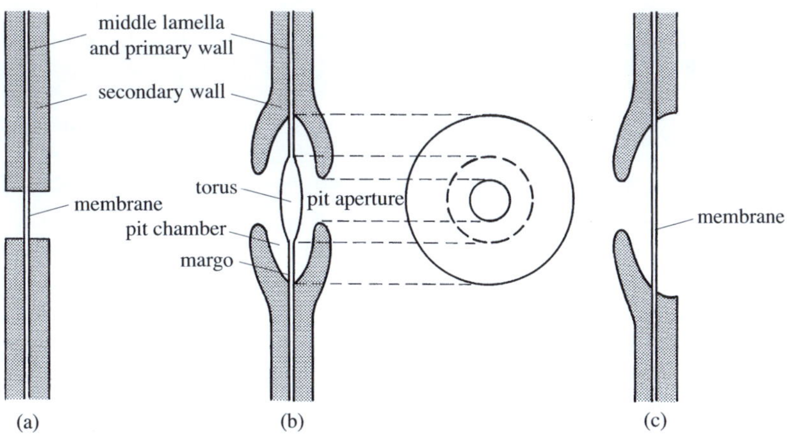

Fig. 3.24 Three types of pit pairs; (a) simple pit pair, (b) bordered pit pair, (c) semi-bordered pit pair.

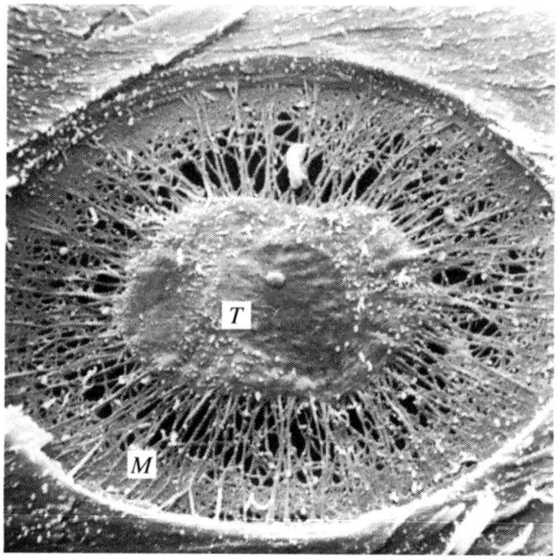

Fig. 3.25 Membrane of a bordered pit showing the torus (*T*) and the margo (*M*) through which water passes from one cell to the next.

see Figure 3.25. When the torus is pressed against one of the apertures, the passage of water is prevented. This phenomenon is called an aspirated pit and occurs when sapwood is transformed into heartwood or when the wood is dried. In heartwood, the pits are definitively blocked in this position.

The rays

The rays consist of radially oriented, brick-like and thin-walled parenchyma cells. The rays of softwoods are composed either of parenchyma cells alone or of parenchyma cells and ray tracheids. In coniferous trees, less than 25 parenchyma cells usually pile up to form a ray.

The ray tracheids are about the same size as the ray parenchyma cells. They are dead cells with small bordered pits leading to other ray cells and to longitudinal tracheids. The ray tracheids seem to be functionally limited to the occurrence of resin canals. Their functions include conduction, and the accumulation and storage of water and other substances in the radial direction.

3.5.2 Microstructure of hardwoods

The structure of hardwood timbers is more complex than that of softwoods. Moreover, during their evolution, hardwoods have developed special types of cells from the tracheid: vessel elements for conduction and fibers for support. Hardwoods are made up of various types of cells that are very variable in dimension and form. The different types of cells constituting the hardwood timbers are presented in Table 3.3 and in Figure 3.26. Practically all the hardwood timbers contain longitudinal vessels, longitudinal fibers, and longitudinal parenchyma cells as well as ray parenchyma cells. Table 3.3 provides certain cellular characteristics of hardwoods. It is interesting to observe that hardwood rays, unlike those in softwoods, are made exclusively of parenchyma cells.

Table 3.3 Different cell types and their functions in hardwoods.

Cell type	Function
Vessel elements	Conduction
Fibers:	
Libroform fibers	Support
Fiber tracheids	Support
Parenchyma:	
Ray parenchyma	Storage
Longitudinal parenchyma	Storage
Tracheids:	
Vascular tracheids	Conduction
Vasicentric tracheids	Condustion

Longitudinal cells

The longitudinal cells in hardwoods consist of vessels, tracheids, axial fibers, and axial parenchyma.

The vessels

A vessel is made up of a tube comprising successive cell elements. The connections of the vessels elements form very long continuous tubes in the tree. The volume of the vessels in hardwoods is variable, varying between 6 and 55%. The diameter of

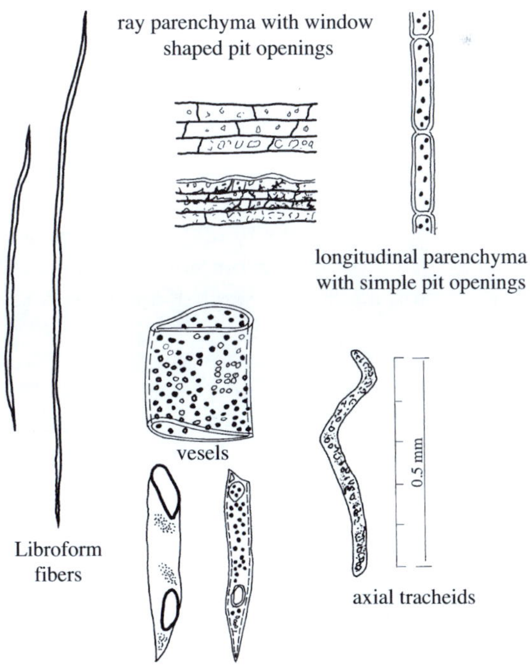

Fig. 3.26 Diagram showing the anatomical elements constituting hardwood.

the vessels varies between 20 and 300 μm. The passage of the sap in the longitudinal direction is made possible by wide openings (perforations) at each end of the vessel elements. In addition, water and sap can transfer to adjacent vessels laterally through small pits in the vessel wall. The pits connecting two laterally adjacent vessels are different from the bordered pits, since they are primarily simple pits without a "torus". This is shown in Figure 3.24a.

The fibers

The role of longitudinal cells is to provide mechanical support for the wood. They are long cells with thick and rigid walls, as shown in Figure 3.26. The volume percentage of the fibers varies between 25 and 75% in hardwood. The length of this type of cell varies between 0.8 and 2.3 mm. Figure 3.27 shows a micrograph of a hardwood microstructure with fibers, vessels and rays.

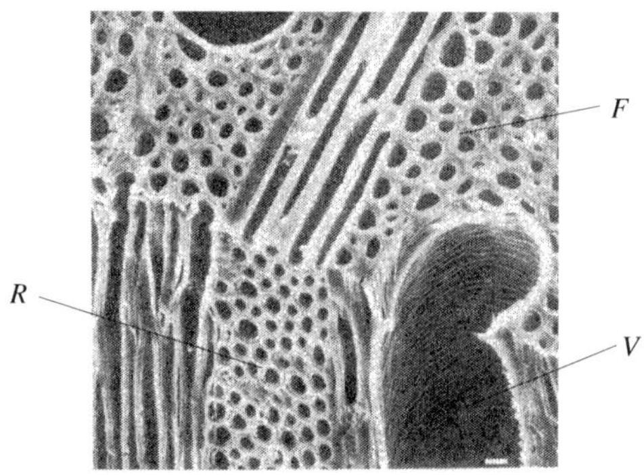

Fig. 3.27 3D SEM micrograph of a hardwood structure showing vessels (*V*), fibers (*F*) and rays (*R*).

The parenchyma cells and rays

Hardwood timber usually contains a greater volume percentage of longitudinal parenchyma than softwood timber. These cells fulfil a storage function for nutrients. The rays in hardwood timbers are formed by two or up to 40 radial cells in height, one to more than 20 in width, and sometimes in so great a number that some rays become visible to the naked eye.

In summary, hardwood timber is characterized by the presence of vessels, tracheids, fibers, longitudinal parenchyma and ray parenchyma. The vessels fulfil the role of conduction and the fibers, with their thick walls, ensure the flexible rigidity and mechanical support of the tree, often constituting most of the wood volume, up to 60%. The radial and longitudinal parenchyma cells ensure that there is a reserve of nutrient substances in the tree.

3.6 WOOD CELL-WALL STRUCTURE AND ULTRASTRUCTURE

With the help of a polarized, optical or electronic microscope, one can easily detect the various layers which form the cell wall. This wall consists of the primary wall

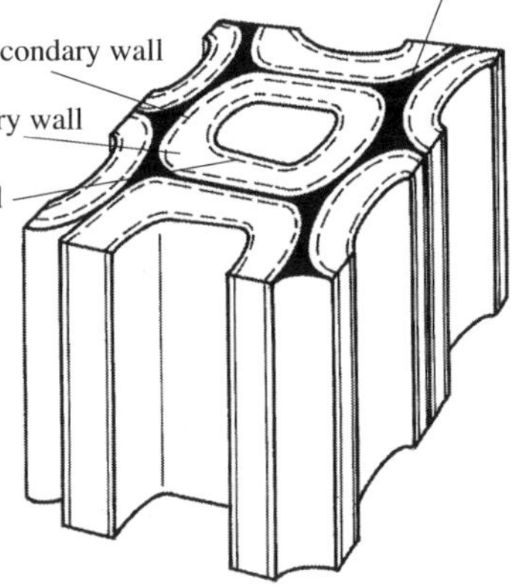

Fig. 3.28 Diagrammatic representation of a longitudinal cell (tracheid) surrounded by six other cells.

(P) and the secondary wall (S). The ***middle lamella*** (M), which is not an integral part of the cell wall, interconnects the cells. Frequently, in the literature, the middle lamella and primary wall are treated as a single entity and called the ***compound middle lamella***. Figure 3.28 shows a representation of a segment of a longitudinal cell (tracheid) surrounded by other cells. The secondary wall is made up of three distinct layers: S_1, S_2 and S_3.

3.6.1 Middle lamella

The middle lamella appears after the division of cambial cells. Its thickness varies between 0.5 μm and 1.5 μm. The optical microscope shows the existence of an important quantity of lignin in this layer, particularly if the lignin is dyed with a suitable stain. This layer joins the cells together. To separate the cells (for anatomical study or for the manufacture of paper pulp), techniques of maceration or chemical attack are used. The layer is destroyed and the cells separate.

3.6.2 Primary wall

The primary wall is very thin and measures approximately 0.1 μm in thickness. Like the middle lamella, it contains a large quantity of lignin, but very little cellulose. It is often difficult to distinguish the primary wall from the middle lamella.

3.6.3 Secondary wall

The view obtained under the polarizing microscope reveals that the secondary wall is made up of three layers (S_1, S_2 and S_3). In the latewood the S_2 layer is the thickest part of the tracheid wall and there is little difference in thickness between the S_1 and the S_3 layer.

The ultrastructure of the cell wall is made up of cellulose microfibrils, lignin and hemicelluloses which are not directly observable by optical microscopy. They can however be detected by AFM, TEM, SEM or by XRD.

Several models have been proposed for the ultrastucture of the secondary wall of the cells of coniferous trees. In all the models, it is suggested that the microfibrils have the shape of a spiral in Z form in the S_1, S_2 and S_3 layers. The angle of the spiral α is defined by the angle of the cell with axial direction and is called the microfibril angle, cf. Figure 3.29. According to Kollmann and Côté (1968), the three layers of the secondary wall are organized as composite multi-layers, see Figure 3.30. The microfibril angle of the S_2 layer ranges between 5-10° (latewood) and 20-30° (earlywood), that of the S_1 between 50 and 70° and that of the S_3 layer between 50 and 90°, Figure 3.31. These layers, on the other hand, are composed of concentric parallel laminae. The S_2 layer is composed of 30 to 40 laminae in the cells of earlywood and more than 150 laminae in those of latewood. The S_2 layer is definitely thicker than its neighbors and hence contributes in a dominant way to the mechanical and physical properties of the cell wall.

The studies made by Scallen (1974), Kerr and Goring (1975), Ruel *et al.* (1978) and Brändström (2001) have shown that the S_2 layer consists of disrupted laminae, where it is suggested that the cellulose microfibrils are a ribbon-like structure made up of several proto-fibrils banded together in the radial direction, see Figure 3.32.

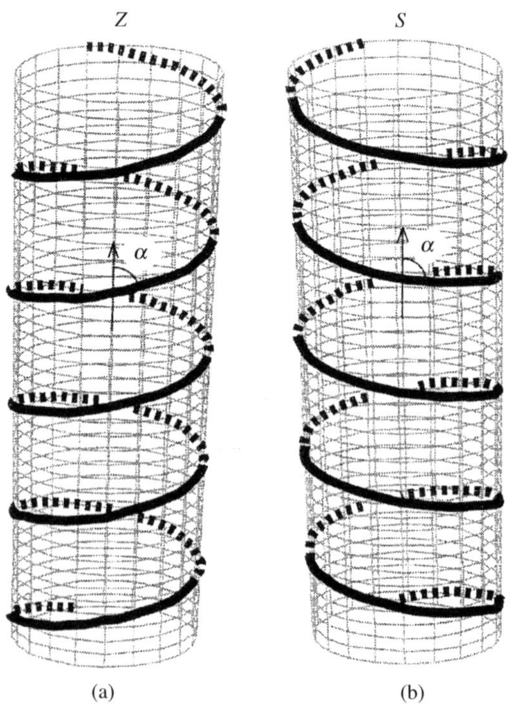

Fig. 3.29 Diagrammatic representation of the shape of a microfibril and the angle α in a concentric laminate of the S_2 layer: (a) spiral in Z, (b) spiral in S.

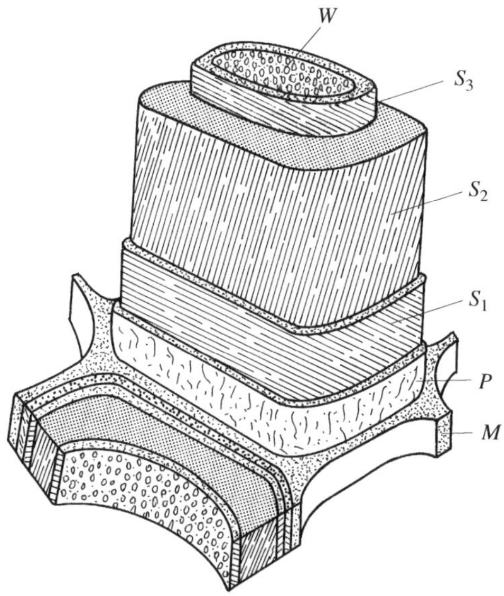

Fig. 3.30 Diagrammatic representation of the various layers of the cell wall. *M*: middle lamella, *P*: primary wall, *S*: secondary wall with its S_1, S_2 and S_3 layers, *W*: warts layer.

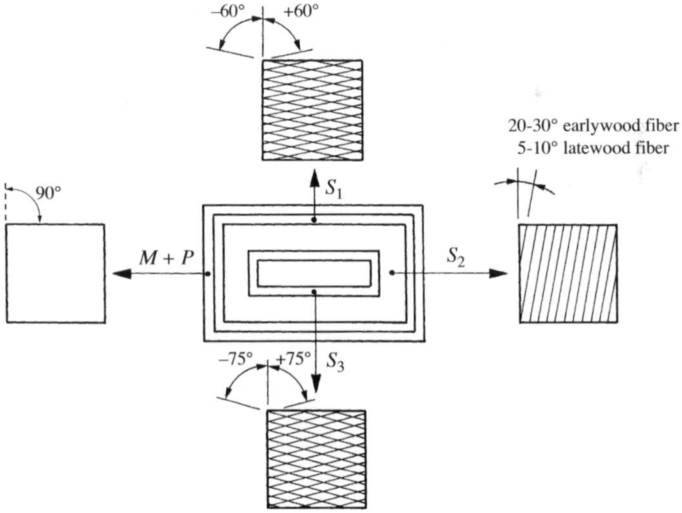

Fig. 3.31 Schematic diagram of the microfibril angle arrangements within the S_1, S_2 and S_3 layers.

Other models have been proposed, such as those of Sell and Zimmermann (1993), Sell (1994), Larsen *et al.* (1995), Singh (1997), Sell and Zimmermann (1998), Zimmermann and Sell (2000) who demonstrated that the microfibrils of the S_2 layers are organized (laminated) in the radial direction of the cell wall. The idea of a radial laminate structure is also supported by authors such as Larsen *et al.* (1995), Schwarze

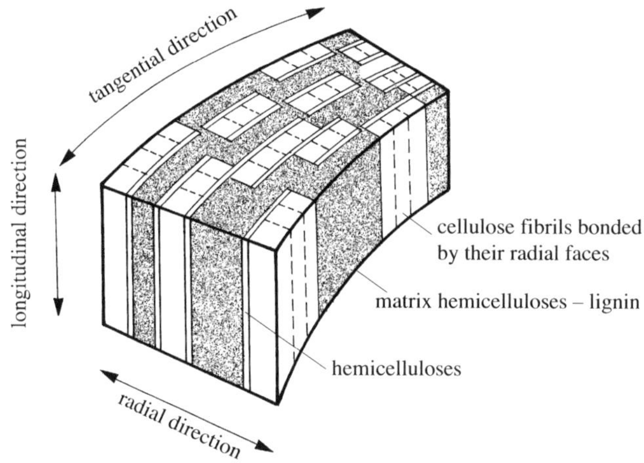

Fig. 3.32 Lamellar structure of the secondary layer S_2 suggested by Scallen (1974).

and Engels (1998) and Zimmermann and Sell (2001), who used the white wood rot fungus to determine the arrangement of microfibrils in the cell wall. Although these observations confirm the idea of a radial arrangement of the microfibrils in the S_2 layer, other studies such as those of Fahlén and Salmén (2001) using AFM or those of Nakashima *et al.* (1997) where high-resolution microscopy was applied to detect the ultrastructure of the cell wall, do not support the results obtained with wood rot fungi.

The microfibril angle varies from the pith to the bark (Fioravanti, 2001). It also depends on the activity of the cambium and the thickness of the annual rings, and is influenced by genetic and environmental factors. A complete list of the various techniques of measuring the microfibril angle and their advantages and disadvantages is given by Huang *et al.* (1998). Among these techniques, the X-ray diffraction proves to be fast, and it is with this method possible to measure the microfibril angle of a single fiber, due to the x-ray beam covering many fibers and indicating the average angles of the cells.

The results obtained by different methods are often contradictory. For example, the work of Herman *et al.* (1999) on single tracheids showed large variations in the microfibril angle within the annual rings with a strong diminution of microfibril angle from the earlywood to the latewood cells, whereas studies carried out by Reiterer *et al.* (1998) and Lichtenegger *et al.* (1999) using SAXS (small-angle X-ray scattering) on the same type of cell have pointed at a higher microfibrillar angle in the tracheids of latewood cells as opposed to in those of earlywood. It has been deemed important to understand the reasons for such differences in the results obtained by varying methods and to find that which gives accurate and reproducible results. A technique developed by Jang (1998) using confocal microscopy with polarization is called CLSM, which stands for Confocal Laser Scanning Microscopy. It is based on the dichromic fluorescence of the cell wall. When the cell wall is colored with specific fluorochromes having a strong affinity to cellulose, the direction of the cellulose microfibrils can be measured without interference from the other layers. This technique is however very laborious and the preparation of very fine specimens of single fiber is still an unresolved issue.

Based on the CLSM method, Sedighi-Gilani *et al.* (2005, 2006) studied the variation of microfibril angle in single tracheids. The microfibril angle was measured at different points in the S_2 of earlywood and latewood tracheids of compression wood and normal wood. Special zones like the border of the pits or the cross-field zone were investigated by focusing on arbitrary small areas along the wood tracheids. Figures 3.33 and 3.34 show the variation in microfibril angle in respectively the radial wall of an earlywood tracheid and in the tangential wall, radial wall and between two bordered pits. These results have demonstrated that the microfibril angle varied slightly within the radial wall.

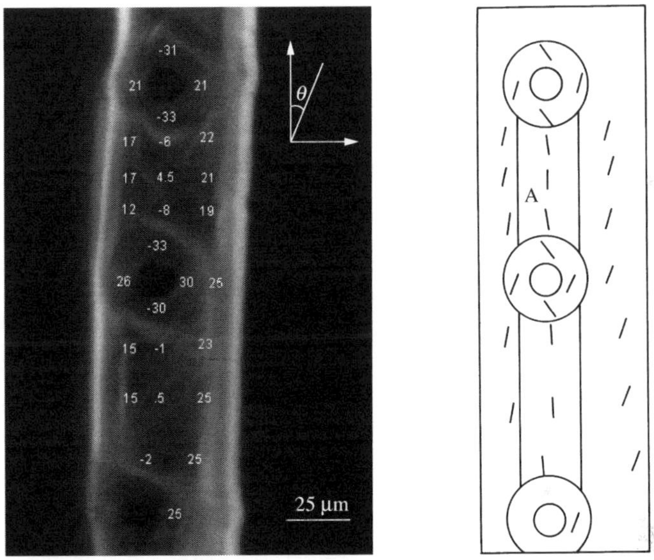

(a) (b)

Fig. 3.33 Varaiation of microfibril angles in the radial wall of an earlywood tracheid. (a) Measured microfibril angles and their location, (b) plotted microfibril angles as measured in (a) (Sedighi-Gilani *et al.*, 2006).

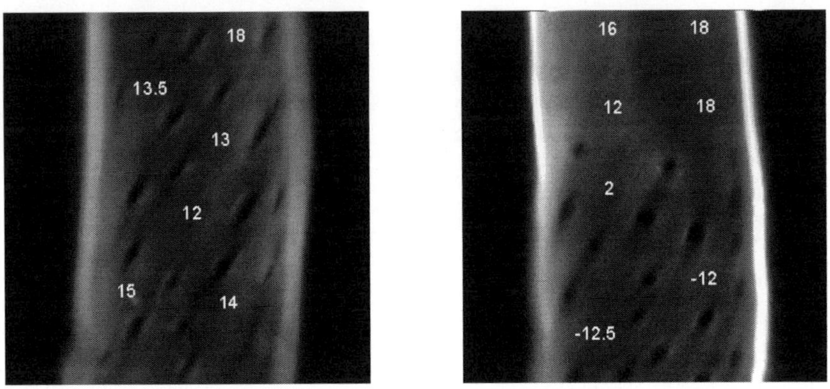

Fig. 3.34 Measured microfibril angles in the tangential wall, radial wall and between bordered pits (Sedighi-Gilani *et al.*, 2006).

3.7 CHEMICAL CONSTITUENTS OF WOOD

The main chemical components of wood are three polymers: cellulose, hemicelluloses and lignin. In addition there are other components called ***extractives***. These extractives are deposited in the cell walls during the formation of heartwood by secretion. This section briefly explains the structure and molecular arrangement of the polymeric

Table 3.4 Chemical components of wood, their polymeric natures and functions.

Component	Composition (%)	Polymeric nature	Degree of polymerisation	Basic monomer	Function
Cellulose	45-50	Linear molecule semi-crystalline	5000-10000	Glucose	Fiber
Hemicelluloses	20-25	Ramified amorphous molecule	150-200	Sugar essential non glucose	Matrix
Lignin	20-30	Three-dimensional amorphous-bonded	–	Phenolpropane	Matrix
Extractives	0-10	Polymerized molecule	–	Polyphenol	A protection element

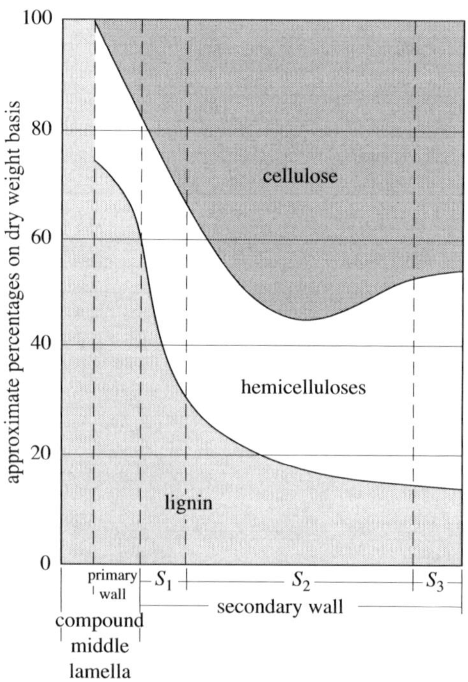

Fig. 3.35 Distribution in percentages of cellulose, lignin and hemicelluloses within the cell-wall layers (Panshin H *et al.*, 1964).

components of wood. A more detailed description on the molecular structure of these components can be found in specialized books (Fengel & Wegener, 1984; Lewin & Goldstein, 1991; Hon & Shiraish, 1991). Table 3.4 gives the volumetric percentage of each chemical component, together with its polymeric nature, degree of polymerization and function. It should be noted that the distribution of the chemical components of wood differ within the various layers of the cell, see Figure 3.35.

3.7.1 Cellulose

Cellulose is the most widespread material in nature. It is present in plants, algae, fungi and in certain animals. Cellulose comprises 40-50% of the dry substance in most wood species. Although the molecular structure of cellulose is known, there are nevertheless many unanswered questions regarding its molecular arrangement within the microfibrils.

Cellulose is a linear natural polymer (homopolysaccharide) made up of β-D-glucopyranose units which are linked together by $(1 \rightarrow 4)$-glycosidic bonds, see

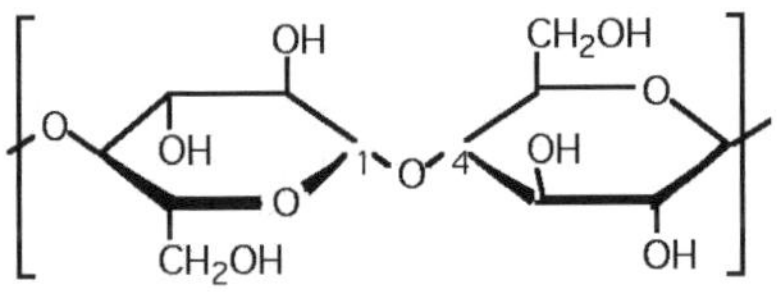

Cellobiose (1.03 nm)

Fig. 3.36 Model of cellobiose, the base unit of cellulose bonded by β- $(1 \rightarrow 4)$-glucosidic linkages.

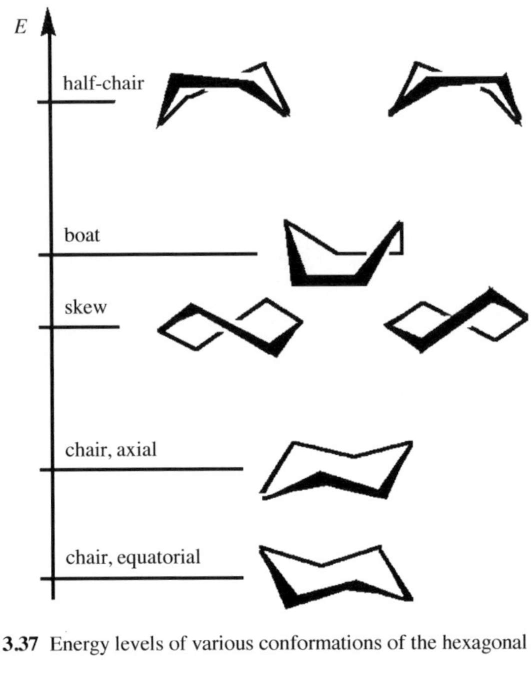

Fig. 3.37 Energy levels of various conformations of the hexagonal ring.

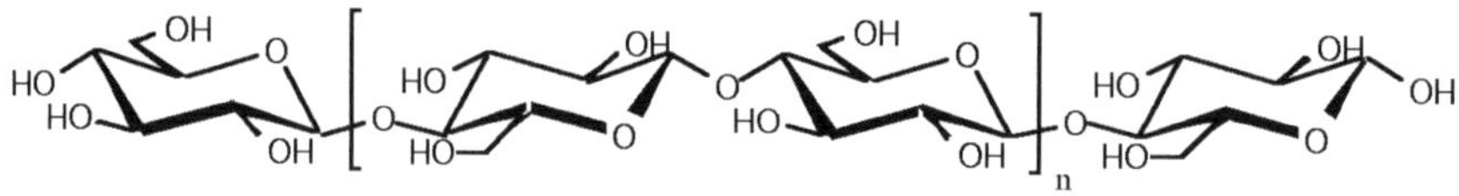

Fig. 3.38 Chemical formula of cellulose.

Figure 3.36. The chair conformation (equatorial) of the hexagonal ring of the β-D-glucopyranose unit is favored energetically, as shown in Figure 3.37. The α-D-glucopyranose polymerizes in a helicoid molecule (starch), whereas the β-D-glucopyranose polymerized in a linear molecule (cellulose), see Figure 3.38. The molecular weight of cellulose is high. For example, cotton has a degree of polymerization (DP) of approximately 10000 and that of wood is around 8000 (Fengel & Wegener, 1984).

Natural cellulose crystallizes in two polymorphic forms: I_α and I_β (Hardy & Sarko, 1996). The I_α form can be indexed by a triclinic mesh with a unit chain with $a = 0.674$ nm, $b = 0.593$ nm, c (axis of the chain) = 1.036 nm, $\alpha = 117°$, $\beta = 113°$ and $\gamma = 81°$. The I_β form is characterized by a monoclinic mesh with two chains with, $a = 0.801$ nm, $b = 0.817$ nm, c (axis of the chain) = 1.036 nm, $\alpha = \beta = 90°$ and $\gamma = 97.3°$, see Figure 3.39.

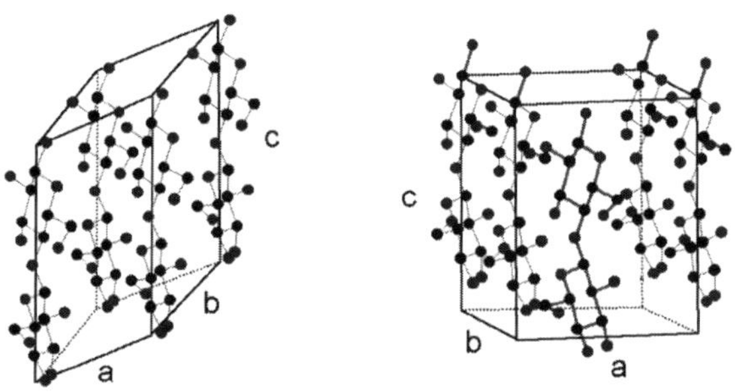

Fig. 3.39 Crystalline structure of cellulose I_α (triclinic) and I_β (monoclinic). (a) Monoclinic mesh with a unit chain of cellulose I_α, (b) monoclinic mesh with two chains of cellulose I_β.

The densities of the triclinic and monoclinic meshes are respectively 1.582 and 1.599 g/cm^3. The significantly higher density of the monoclinic mesh suggests a greater thermodynamic stability; due to the electrostatic interactions, the intra-plans are slightly higher in the monoclinic mesh. The crystallization of cellulose is made possible by the linear character of its chain, by the hydrogen bridges between the hydroxyl groups of its macromolecules and by the hydrogen bridges between the OH-groups of adjacent cellulose molecules, rendering the cellulose chain rigid, see Figure 3.40. Thus the rotation of the chain around the glycosidic bonds is very limited, as shown in Figure 3.41.

In wood, the cellulose chains crystallize partially to form the elementary fibrils. The width of such fibrils has been estimated by electron microscopy to about 2.5 nm and by X-ray diffraction to between 2 and 3.5 nm (bundles of 19 to 36 cellulose chains) based on Newman (1997). Figure 3.42 also shows two models of a group of

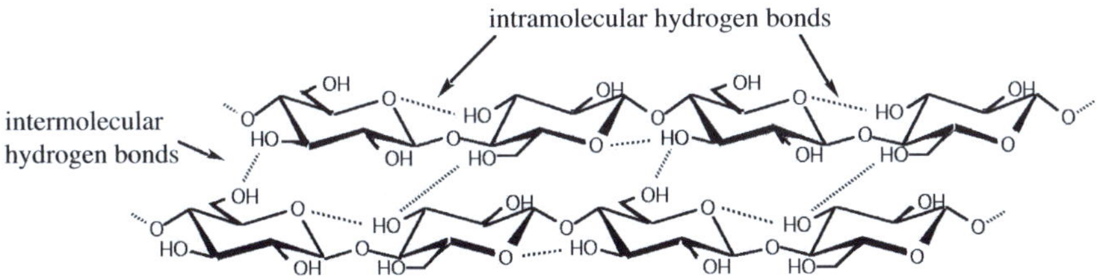

Fig. 3.40 Intra-molecular and inter-molecular hydrogen bonds between two adjacent chains of cellulose macromolecules.

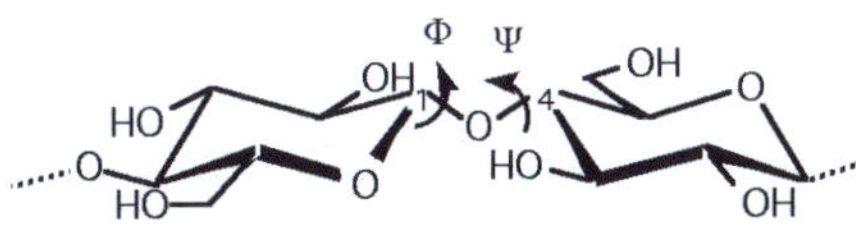

Fig. 3.41 Rotation angles Φ and Ψ around the glycosidic β- (1 $\rightarrow$ 4) bonds.

elementary fibrils according to Newman. In this diagram, each bar represents a chain of cellulose macromolecules. The bar with a dark grey color corresponds to the inner cellulose chain and the bar with a clear color stands for the surface of the elementary fibril. Newman (1998) has reported that Nuclear Magnetic Resonance (NMR) spectra of wood give signals assigned to cellulose chains with hydroxyl groups exposed on the surfaces of elementary fibrils, which can be distinguished from signals assigned to chains buried within the elementary fibrils. The existence of an important number of hydrogen bonds between the inner chains, associated with a strong lateral resistance, explains the formation of crystalline zones of fibrils approximately 60 to 150 nm in length. The remainder of the chains, which are less ordered, constitute the zones of transition between the crystalline parts of the fibrils, see Figure 3.43. The diameter of the elementary fibril can be estimated from the width of the peaks of an X-ray diffraction spectrum, whereas their length can be determined by selective hydrolysis of cellulose, since its crystalline part is resistant to dilute acid. During the selective

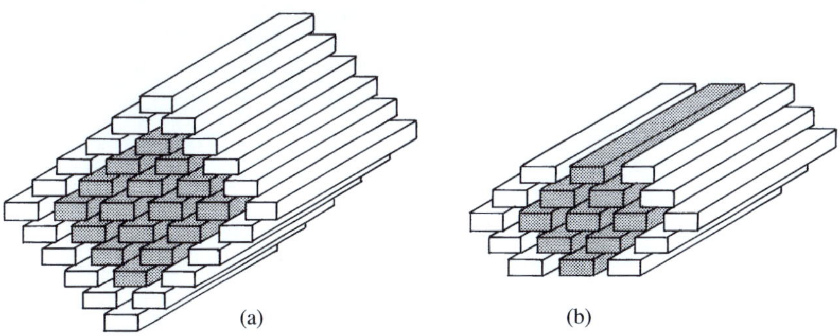

Fig. 3.42 Two models showing a portion of elementary fibrils with (a) 36 and (b) 19 chains. Each bar represents a cellulose chain. The clear grey indicates the cellulose chain on the surface of the fibril possessing free hydroxyl groups.

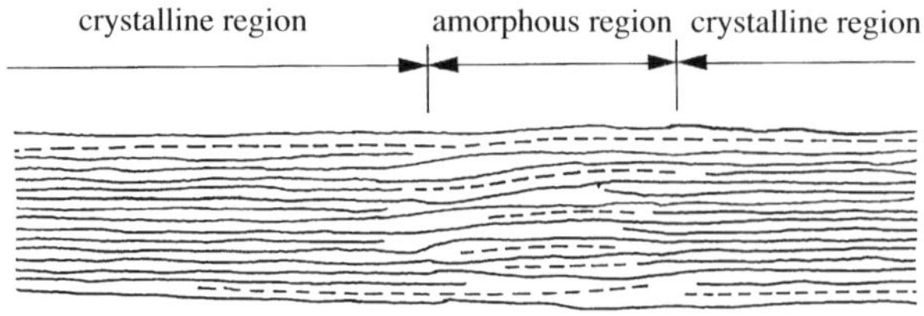

Fig. 3.43 Diagrammatic representation of a longitudinal section of a fibril consisting of crystalline and less ordered amorphous zones.

hydrolysis, the degree of polymerization (DP) decreases to a minimum value representing the length of a crystallite, which can be 150 to 300 nm in the case of wood.

In wood, the fibrils are associated with other fibrils to constitute entities called microfibrils, with a diameter of the order of 25 to 30 nm. The microfibril macromolecules are composed of crystalline as well as less ordered zones with a length of 2500 to 5000 nm. The microfibrils possess a high order of crystallinity varying in the range of 60 and 90%. The structure of microfibril macromolecules is not well known. Two suggested models for a microfibril in the longitudinal direction are given in Figure 3.44.

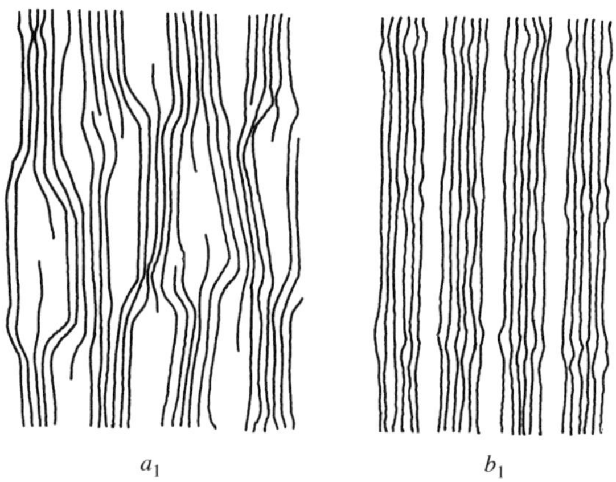

Fig. 3.44 A schematic representation of two suggested models for the arrangement of cellulose macromolecules inside the microfibrils (Fengel & Wegener, 1984).

3.7.2 Hemicelluloses

Hemicelluloses differ from cellulose in their chemical compositions, in their degree of polymerization (150-200) and in the ramification of the molecular chains. The sugar components of hemicelluloses can be divided into four groups: pentoses, hexoses, hexuronic acids and deoxy-hexoses, see Figure 3.45. The principal chain of hemicelluloses can be made up of a single type of monomer unit (homopolymer, e.g., xylans) or of several different units (heteropolymer, e.g., glucomannans). Some of the units are always or sometimes side groups of a main chain, e.g. 4-O-methylglucuronic acid or galactose.

Adopting the notation for the chemistry of proteins, abbreviations have also been introduced for the notation of sugars in order to simplify their chemical representation. Thus the first three letters of the name are generally taken: Glu for glucose, Xyl for xylose, Rha for rhamnose, etc. For the uronic acids, one adds either A or U after the abbreviation: GalA or GalU for the galacturonic acid. For the methyl acids, one adds the prefix Me: Me-GluA or Me-GluU for 4-O-methylglucuronic acid. A fourth letter is often added in the suffix to indicate the pyranose or furanose form of the sugar: Arap and Araf, respectively for arabinopyranose and arabinofuranose. The classification in Figure 3.45 shows the differences in chemical composition from cellulose, but it does not illustrate the specificity of hemicelluloses of the hardwood and softwood trees. For this reason, a classification based on the behavior of hemicelluloses during their separation from cellulose is used. Hemicelluloses that can be separated from holocellulose[2] are called "non-cellulosic glycosans" and the residue "cellulosic glycosans".

Aspinall (1973) has presented a general classification based on the various types of hemicelluloses of all plants. Its system includes

- hemicelluloses: xylans, glucomannans;
- pectic substances: galacturonans, arabinans, galactans and/or arabinogalactans I (primarily linear chains);

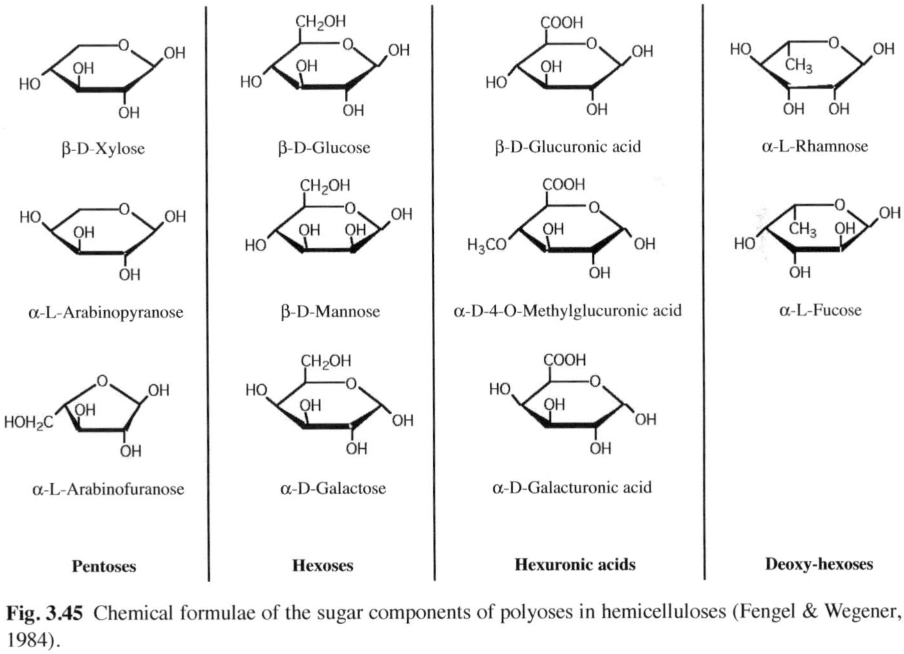

Fig. 3.45 Chemical formulae of the sugar components of polyoses in hemicelluloses (Fengel & Wegener, 1984).

2 Ritter and Kurth (1933) were the first to use the term "holocellulose" for the product obtained after the elimination of the lignin of wood. Ideal delignification would not attack polysaccharides chemically, but there exists no process of delignification that meets this requirement. Nevertheless, three important criteria define the holocellulose: a small residual percentage of lignin, a minimum polysaccharide loss and a minimum oxidizing and hydrolytic degradation of cellulose.

- other polysaccharides: arabinogalactans II (highly ramified), fuco- (or galacto-) xyloglucans; and
- glycoproteins.

Softwoods and hardwoods differ in their percentages of total and individual hemicelluloses. With regard to non-glucosic saccharides, softwoods contain a higher proportion of mannose and galactose units than hardwoods. The latter have a high portion of xylose units and contain more acetyl groups than softwoods. The more detailed compositions for two wood species, i.e., spruce (*Picea abies* L. Karst.) and beech (*Fagus sylvatica* L.) is given in Table 3.5.

Table 3.5 Non-glucosic units of spruce and beech hemicelluloses (Fengel & Wegener, 1984).

Species	Mannan (%)	Xylan (%)	Galactan (%)	Arbinan (%)	Uron-Acid (%)	Rhamnan (%)	Acetyl (%)
Spruce	13.6	5.6	2.8	1.2	1.8	0.3	1.2
Beech	0.9	19	1.4	0.7	4.8	0.5	3.8

To present the principal hemicelluloses of wood, the classification given in Fengel and Wegener (1984) will be used. Only xylans and the glucomannanes are presented here, since most of the hemicelluloses in wood are composed of these two.

Xylans

Xylans of the hardwood trees generally possess a principal homopolymeric chain of xylose units which are limked by β- (1 $\rightarrow$ 4)-glycosidic bonds. The principal chain comprises the groups 4-O-methylglucuronic acids with an α- (1$\rightarrow$ 2)-glycosidic linkage at the xylose unit, see Figure 3.46. The hardwoods contain 20 to 25% of xylans. Many hydroxyl groups at C2 and C3 of the xylose unit are substituted with O-acetyl groups. The majority of xylans of the hardwoods have a ratio (Xyl: Me-GluU) of 10:1 and a ratio (Xyl: acetyl) of 1:0.5-0.6. The average degree of polymerization of the

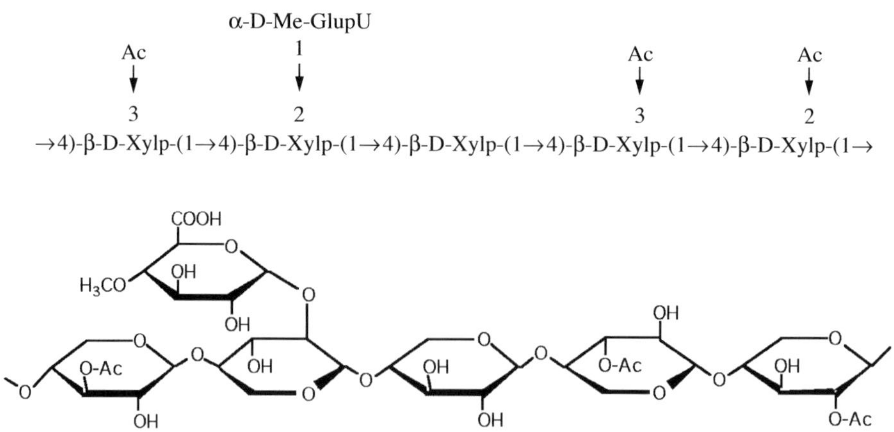

Fig. 3.46 Partial chemical structure of O-acetyl-4-O-methylglucuronoxylan of hardwood.

principal chain, which varies between 100 and 200, depends on the species and the mode of isolation.

Xylans of the coniferous trees differ from those of the hardwood trees in the absence of O-acetyl side groups and the presence of arabinofuranose units linked by α- (1 $\rightarrow$ 3)-glycosidic bonds to the principal chain. Thus the softwood xylans are arabino-4-O-methylglucuronic-xylans, see Figure 3.47. Their percentage in softwood varies between 7 and 14%. The average ratio of the three components (Xyl: Me-GluU: Ara) is 8:1.6:1. The chains of xylans in coniferous wood are appreciably shorter than those in hardwood trees with an average degree of polymerization between 70 and 130. On the other hand, the xylans of coniferous trees are branched with one or two side chains to the principal macromolecule chain.

α-D-Me-GlupU
1
↓
2

α-L-Araf
↓
3

→4)-β-D-Xylp-(1→4)-β-D-Xylp-(1→4)-β-D-Xylp-(1→4)-β-D-Xylp-(1→4)-β-D-Xylp-(1→

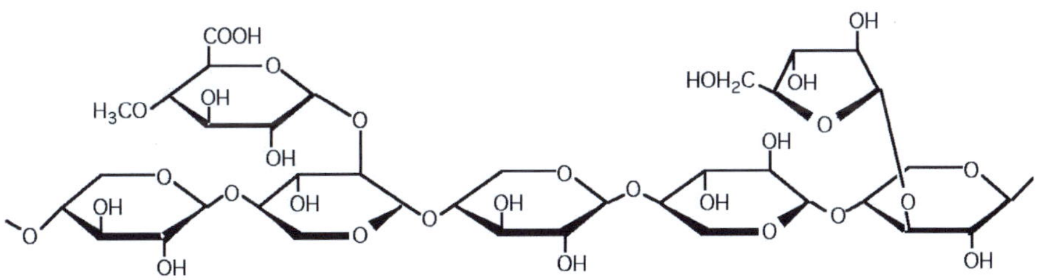

Fig. 3.47 Partial chemical structure of the arabino-4-O-methylglucuronoxylane of a softwood.

Glucomannans

The glucomannans of wood are characterized by a heteropolymeric principal chain of glucose and mannose units. The hardwood trees present the simplest chains of glucose and mannose linked by β- (1 $\rightarrow$ 4) glycosidic bonds connection without side groups, cf. Figure 3.48. The chains are however slightly branched. The percentage of glucomannans in hardwood is relatively low (3 to 5%). Generally, the ratio (Man:Glu) varies between 1.5-2.1. The avarage degree of polymerization of the glucomannans of the hardwood is approximately 60-70.

Softwood contains 20 to 25% of glucomannans with a slightly branched principal chain on which galactose side groups are joined by α-(1 $\rightarrow$ 6)-linkage and

→4)-β-D-Glup-(1→4)-β-D-Manp-(1→4)-β-D-Glup-(1→4)-β-D-Manp-(1→

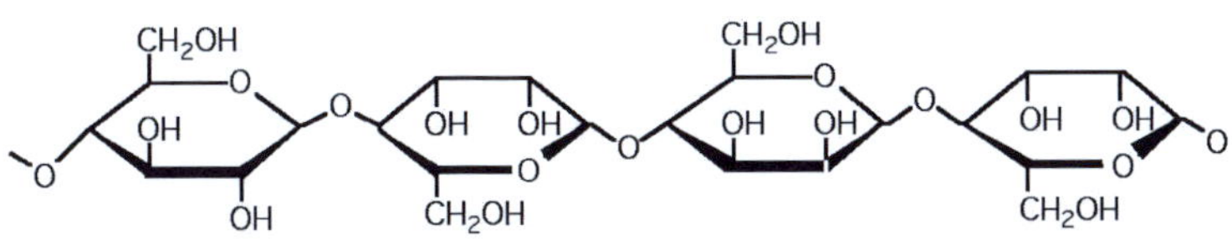

Fig. 3.48 Partial chemical structure of the glucomannan of hardwood.

α-D-Galp
Ac (→2)   1 (→6)   Ac (→3)

→4)-β-D-Manp-(1→4)-β-D-Glup-(1→4)-β-D-Manp-(1→4)-β-D-Manp-(1→4)-β-D-Glup-(1→

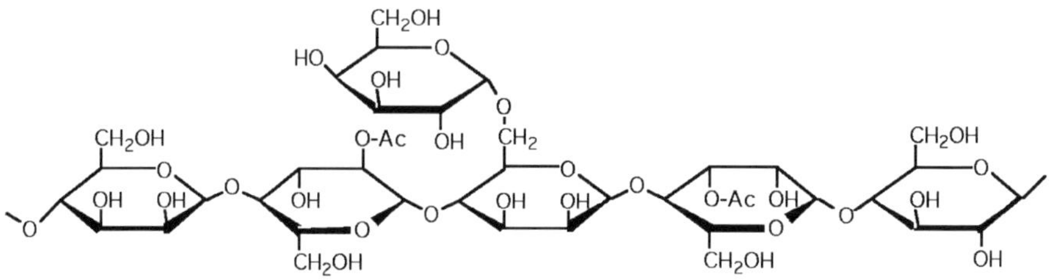

Fig. 3.49 Partial chemical structure of O-acetyl-galactoglucomannan of softwood.

O-acetyl side groups on the C2 and C3 carbons of mannose and glucose, see Figure 3.49. The mannose and glucose units are distributed at random along the chain and the (Man:Glu) ratio is approximately 3:1. The proportion of galactose differs according to the method of extraction. The water-soluble galactoglucomannan has a ratio (Man: Glu: Gal) of 3:1:1, whereas that in alkaline solution is 3:1:0.2. Some species like black spruce or red pine have a slightly branched principal chain, whereas that of Scots pine and of radiata pine is linear. Fengel and Wegener (1984) show that certain parts of the glucomannans are bonded to lignin. These connections are known as complex lignin-polysaccharides or lignin-carbohydrates.

3.7.3 Lignin

After cellulose, lignin is the most abundant organic substance in plants. The presence of lignin in the cell walls gives wood a good mechanical resistance and makes it possible for certain trees to reach a height of more than one hundred meters. The quantity of lignin present in the various wood species is variable. Hardwood varieties contain 18 to 25%, whereas softwood species contain 25 to 35%. The formation of the lignin macromolecules by the plants requires a complicated biochemical system. Although certain molecular models have been presented, see Figures 3.50 and 3.51, the molecular structure of lignin is not precisely known. The precursors of lignin, i.e.,

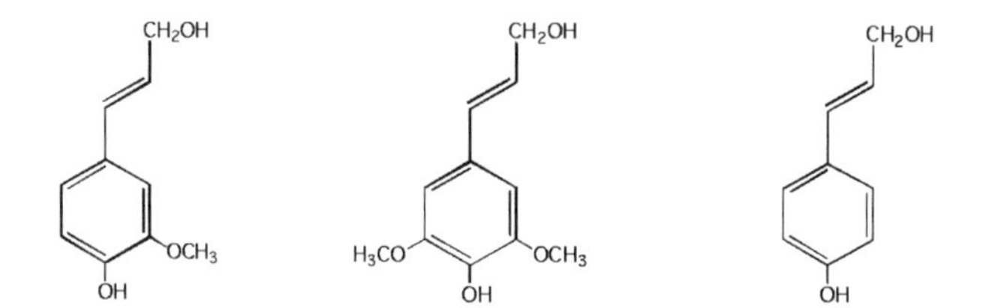

Coniferyl alcohol   Sinapyl alcohol   p-coumaryl alcohol

Fig. 3.50 Chemical formulae of the principal components of lignin.

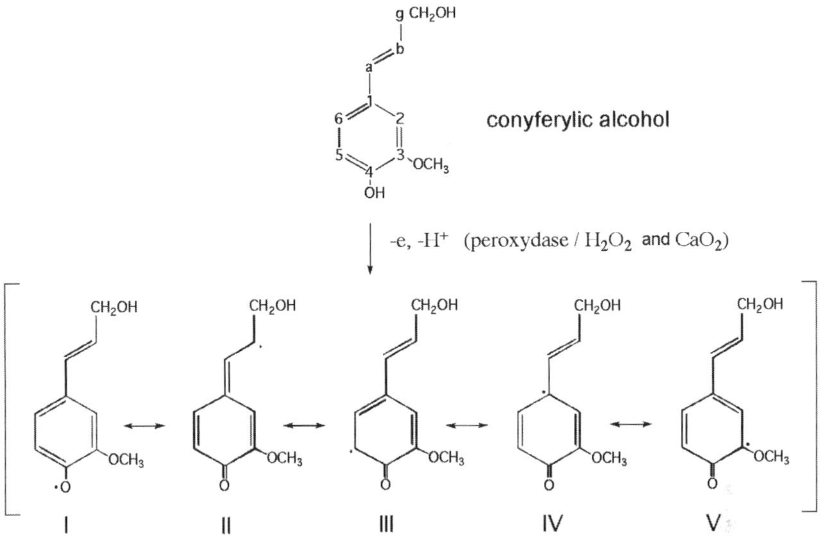

Fig. 3.51 Enzymatic dehydrogenation of coniferyl alcohol yielding phenoxy radicals.

the p-hydroxycinnamyl alcohols (coniferyl, sinapyl and p-coumaryl in Figure 3.50), are however well-known thanks to studies with ^{14}C.

The first stage in the polymerization of lignin consists of the enzymatic dehydrogenation of the p-hydroxycinnamyl alcohols, which leads to the formation of mesomeric phenoxy radicals stabilized by resonance, see Figure 3.51.

Only forms I to IV in Figure 3.51 are involved in the biosynthesis of lignin. This is due to the mesomeric form V being thermodynamically unfavorable. The principal modes of coupling of the radicals are presented in Table 3.6 and in Figure 3.52.

Table 3.6 Modes of coupling of the phenoxy radicals of coniferyl alcohols.

	I	II	III	IV
I	Unstable peroxide	β-O-4	4-O-5	1-O-4*
II	β-O-4	β-β	β-5	β-1*
III	4-O-5	β-5	5-5	1-5*
IV	1-O-4*	β-1*	1-5*	1-1*

* Other options possible.

From quantum mechanical calculations it was deduced that e.g. all phenoxy radicals have the highest π-electron densities at the phenoloc oxygen atom, thus favouring the formation of aryl ether linkages such as the β-O-4 linkage, the most frequent type of bond in softwood and hardwood lignins. Examples of polylignols and a model of the structure of spruce lignin are presented in Figures 3.53 and 3.54, respectively.

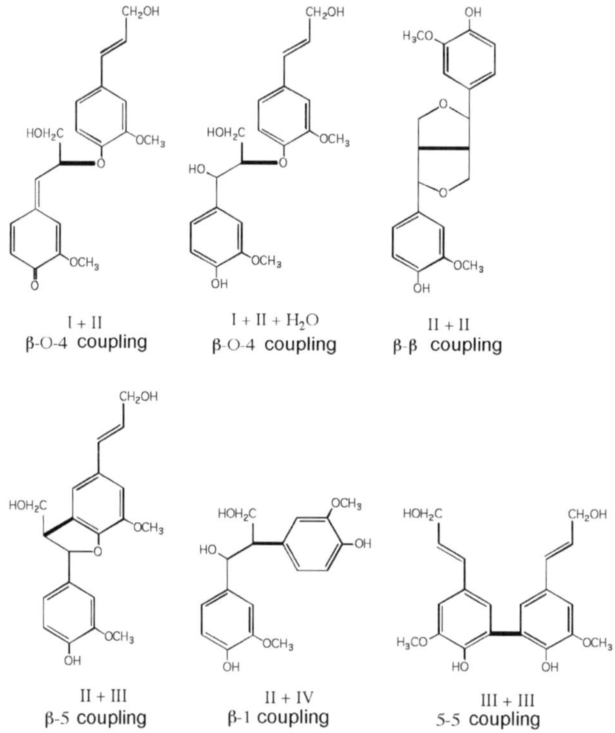

Fig. 3.52 Typical chemical structures of dilignols.

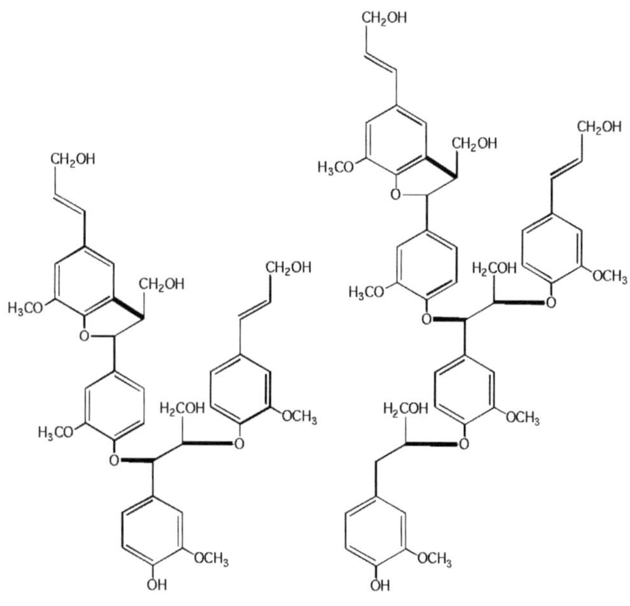

Fig. 3.53 Examples of tetra-alignols (left) and pentalignols (right).

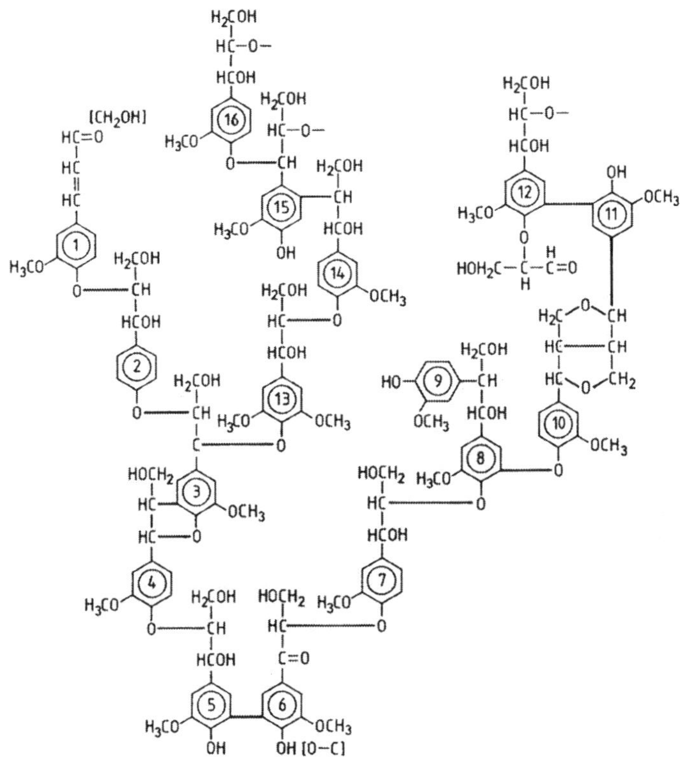

Fig. 3.54 Model of the structure of spruce lignin (Adler, 1977).

3.7.4 Extractives

Extractives are not an integral part of the cell walls and natural solvents easily eliminate them. They are composed of several different chemical substances such as greases, aromatic compounds, volatile oils, alcohols with high molecular weight and fatty-acids. The extractives do not contribute to the mechanical properties of wood. Rather, they pigment the heartwood and give wood an odor. They are toxic and increase the durability of heartwood against microbiological degradation. The term "extractives" covers a large range of compounds which can be extracted using various polar or non-polar solvents. Table 3.7 shows the type of extraction for different extractives.

Table 3.7 Classification of extractives according to the type of extraction (Fengel & Wegener, 1984).

Extraction process	Components
Steam distillation	Terpenes, phenols, hydrocarbons, lignans
Ether extraction	Fatty acids, fats, waxes, resins, resin acid, sterolds
Alcohol extraction	Colored matter, phlobaphenes, tannins, stilbenes
Water extraction	Carbohydrates, proteins alkaloids, inorganic materials (cations: Ca^+, K^+, Mg^+, Na^+, Fe^{2+}; anions: NO_3^-, SO_4^{2-}, PO_4^{3-})

3.7.5 Molecular model of association cellulose-hemicelluloses-lignin within the cell walls

The arrangement of the cellulose microfibrils, hemicelluloses and lignin within the cell walls is difficult to describe because of the multiplicity of the layers (P, S_1, S_2 and S_3). Moreover, the means of observation are still limited. Only the transmission electron microscope can provide clear images of the arrangement of the microfibrils. Figure 3.55 displays a model showing the association between the cellulose microfibrils, hemicelluloses and lignin. Moreover, Fengel & Wegener (1984) have presented a model that

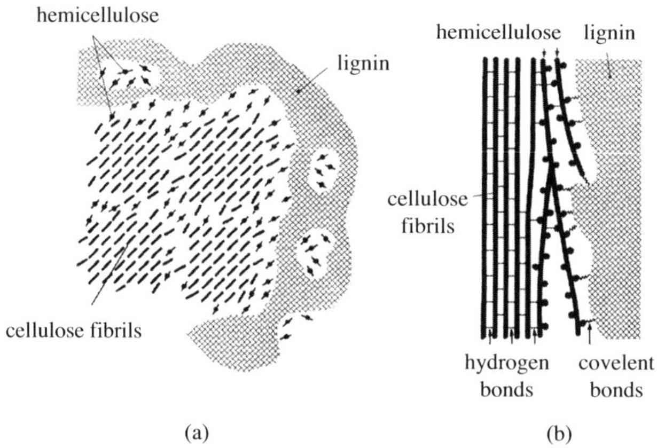

Fig. 3.55 Detailed model of the association of cellulose, hemicelluloses and lignin in the cell walls of wood; (a) a cross section, (b) a longitudinal section.

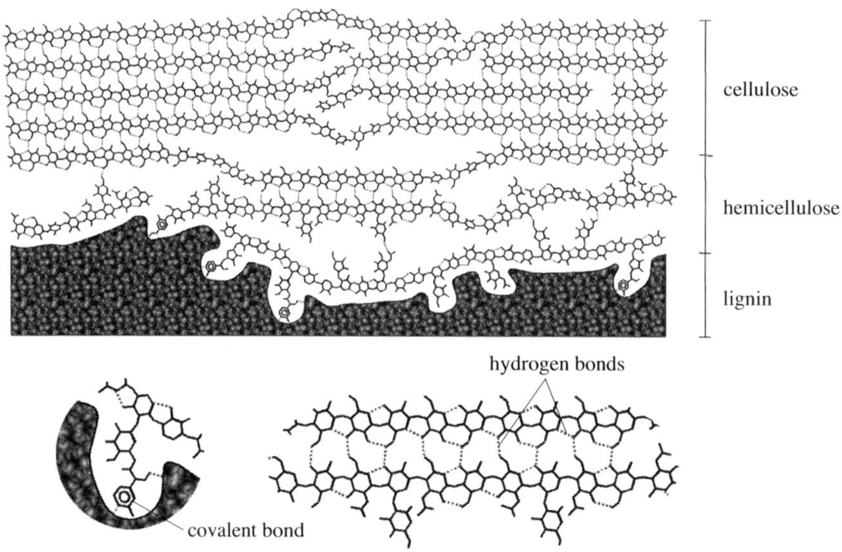

Fig. 3.56 Composite molecular model of a longitudinal section of (cellulose-hemicelluloses-lignin) of a cell wall in an anhydrous state.

considers the intimate bonds between hemicelluloses and cellulose, and between hemicelluloses and lignin. In this model, cellulose microfibrils, less ordered cellulose chains and hemicelluloses are bonded together through numerous hydrogen bridges. On the other hand, hemicelluloses are more strongly bonded to lignin by covalent bonds. Based on Fengel & Wegener (1984), a molecular model is proposed as in Figure 3.56.

The mechanical performance of wood and wood products is dependent on the structure as well as on the properties of the wood constituents within the fiber cell wall. Recently, more detailed knowledge has been obtained on the structure of wood at the micro- and nanostructural level. Some of these results has been reviewed by Salmén and Burget (2009).

3.8 REFERENCES

Adler, E. (1977). Lignin chemistry: past, present and future. *Wood Science and Technology, 11*(3):169-218.

Aspinall, G.O (1973). Carbohydratepolymers of plant cell walls. In: *Biogenesis of plant cell wall polysaccharides*. Loewus, F. (ed.). Academic Press, New York, London, pp. 95-115.

Barnett, J.R. & Jeronimidis, G. (2003). Reaction wood. In: *Wood quality and its biological basis*. Barnett, J.R & Jeronimidis, G. (eds.). Blackwell Publishing Ltd., Australia, ISBN 1-84127-319-8.

Bendtsen, B.A. (1978). Properties of wood from improved and intensively managed trees. *Forest Products Journal, 28*(10):61-72.

Bosshard, H.H. (1974). Holzkunde. Band 1. Mikroskopie und Makroskopie des Holzes.(Wood science. Volume 1. Microstructure and macrostructure of wood.) Birkhäuser Verlag, Basel, ISBN 3-7643-0702-1.

Bowyer, J.L., Shmulsky, R. & Haygreen, J.G. (2007). *Forest products & wood science*. (fifth edition), Blackwell Publishing, Ames, USA, ISBN 978-0-8138-2036-1.

Brändström, J. (2001). Micro- and ultrastructural aspects of Norway spruce tracheids: a review. *IAWA Journal, 22*(4):333-353.

Butterfield, B.G. & Meylan, B. (1980). *Three dimensional structure of wood*. (second edition), Chapman and Hall Ltd., London, ISBN 0-412 16320-9.

Carlqvist, S. (1988). *Comparative wood anatomy: systematic, ecological, and evolutionary aspects of Dicotyledon wood*. Springer-Verlag, Berlin, Heidelberg, ISBN3-540-18827-4.

Core, H.A., Côté, W.A. (ed.) & Day, A.C. (1976). *Wood, structure and identification*. Syracuse wood science series 6, Syracuse University Press.

Dickison, W.C. (2000). *Integrative plant anatomy*. Academic Press, San Diego, ISBN 0-12-215170-4.

Dinwoodie, J.M. (1981). *Timber, its nature and behaviour.* Van Nostrand Reinhold Co. Ltd., New York, ISBN 0-442-30445-5.

Dinwoodie, J.M. (1989). Wood: nature's cellular, polymeric, fibre-composite. Institute of Metals, London, ISBN 0-901462-35-7.

Fahlén, J. & Salmén, L. (2001). The lamella structure of the wood fibre wall-radial or concentric. In: *Proceedings of Cost action E20 workshop: Interaction between cell wall components, 26-28 April,* Swedish Agricultural University, Uppsala, Sweden, p. 25.

Fengel, D. & Wegener, G. (1984) *Wood: chemistry, ultrastructure, reactions*. Walter De Gruyter, Berlin, New York, ISBN 3-11-008481-3.

Fioravanti, M. (2001). The influence of age and growth factors on microfibril angle of wood. In: *Proceedings of the 1th International Conference of the European Society for Wood Mechanics*, Navi, P. (ed.). Lausanne, Switzerland, pp. 127-134.

Franklin, G.L. (1945). Preparation of thin sections of synthetic resins and wood resin composites, and a new method for macerating woods. *Nature 155*(3924):51.

Hardy B.J. & Sarko, A. (1996). Molecular dynamics simulations and diffraction-based analysis of the native cellulose fibre: structural modelling of the I-α and I-β phases and their interconversion. *Polymer, 37*(10):1833–1839.

HERMAN, M., DUTILLEUL, P. & AVELLA-SHAW, T. (1999). Growth rate effects on intra-ring and inter-ring trajectories of microfibril angle in Norway spruce (*Picea abies*). *IAWA Journal, 20*(1):3-21.

HILLIS, W.E. (1987). *Heartwood and tree exudutes*. Springer-Verlag, Berlin, Heidelberg, ISBN 3-540-17593-8.

HOADLEY, R.B. (1990). *Identifying wood: accurate results with simple tools*. The Taunton Press, Newtown, USA, ISBN 0-942391-04-7.

HON, D.N.-S. & SHIRAISHI, N. (eds.) (1991). *Wood and cellulosic chemistry*. Marcel Dekker, Inc., New York, Basel, ISBN 0-8247-8304-2.

HUANG, C.-L., KUTSCHA, N.P., LEAF, G.J. & MEGRAW, R.A. (1998). Comparsion of microfibril angle measurement techniques. In: *Proceedings of the IAWA/IUFRO International Workshop on the significance of microfibril angle too wood*. Butterfield, B.G. (ed.). University of Canterbury, Christchurch, New Zealand, pp. 177-205.

ILVESSALO-PFÄFFLI, M.-S. (1995). *Fiber atlas, identification of papermaking fibers*. Springer Verlag, Berlin, Heidelberg, ISBN 3-540-55392-4.

JAGELS, R. & DYER M.V. (1983). Morphometric analysis applied to wood structure. I. Cross-sectional cell shape area change in red spruce. *Wood and Fiber Science, 15*(4): 376-386.

JANE, F.W. (1956). *The structure of wood*. Adam & Charles Black, London.

JANG, H.F. (1998). Measurement of fibril angle in wood fibres with polarisation confocal microscopy. *Journal of Pulp and Paper Science, 24*(7):224-230.

KERR, A.J. & GORING, D.A. (1975). The ultrastructural arrangement of the wood cell wall. *Cellulose Chemistry and Technology, 9*(6):563-573.

KOLLMANN, F.F.P. & CÔTÉ, W.A. (1968). *Principles of wood science and technology: I. solid wood*. Springer-Verlag, Berlin, Heidelberg.

LARSEN, M.J., WINANDY, J.E. & GREEN, F. (1995). A proposal model of the tracheid cell wall of southern yellow pine having an inherent radial structure in the S2 layer. *Material und Organismen, 29*(3):197-210.

LEWIN, M. & GOLDSTEIN I. S. (1991). Wood structure and composition. Marcel Dekker, Inc., New York, Basel, ISBN 0-8247-8233-X.

Lichtenegger, H., Reiterer, A., Stanzl-Tschegg, S.E. & Fratzel, P. (1999). Variation of cellulose microfibril angles in softwoods and hardwoods: a possible strategy of mechanical optimisation. *Journal of Structural Biology, 128*(3):257-269.

LINNAEUS, C. (1735). *Systema naturae, sive Regna tria naturae. Systematice proposita per Classes, Ordines, Genera, & Species*. Lugduni Batavorum, Apud Theodorum Haak, Leiden. (See translation: Engel-Ledeboer, M.S.J. & Engel, H. (2003). Systema naturae 1735. Facsimile of the first edition with an introduction and a first English translation of the "observationes". Nieuwkoop Hes & De Graaf, ISBN 978-90-6004-104-8).

MORK, E. (1928). Die Qualität des Fichtenholzes unter besonderer Rücksichtnahme auf Schleif- und Papierholz. (The quality of spruce wood, with special consideration for pulpwood and paper.) *Der Papier-Fabrikant, 48*:741-747.

NAKASHIMA, J., MIZUNO, T., TAKABE, K., FUJITA, M. & SAIKI, H. (1997). Direct visualization of lignifying secondary wall thickening in *Zinnia elegans* cells in culture. *Plant and Cell Physiology, 38*(7):818-827.

NEWMAN, R. (1997). How stiff is individual cellulose microfibril? In: *Proceedings of the IAWA/IUFRO International Workshop on the significance of microfibril angle too wood*. Butterfield, B.G. (ed.). University of Canterbury, Christchurch, New Zealand, pp. 81-93.

NEWMAN, R.H. (1998). Evidence for assignment of 13C NMR signals to cellulose crystallite surfaces in wood, pulp and isolated celluloses. *Holzforschung, 52*(2):157-159.

PANSHIN, A.J., de ZEEUW, C. & BROWN, H.P. (1964). *Textbook of wood technology. Volume I: structure, identification, uses, and properties of the commercial woods of the United states*. (second edition) McGraw-Hill Book Company, New York, San Francisco.

PEREIRA, H. (2007). *Cork*. Elsevier, Amsterdam, ISBN 0-444-52967-5.

PHILLIPS, E.W.J. (1962). The beta ray method of determining the denisty of wood and the proportion of summer wood. *Journal of the Institute of Wood Science*, 5(1):16-28.

PHILLIPS, E.W.J., ADAMS, E.H. & HEARMON, R.F.S. (1962). The measurement of density variation within the growth rings in thin sections of wood using beta particles. *Journal of the Institute of Wood Science,* 10(11):11-28.

REITERER, A., JACOB, H.F., STANZL-TSCEGG, S.E. & FRATZEL, P. (1998). Spiral angle of elementary cellulose fibrils in cells walls of *Picea abies* determined by small-angle X-ray scattering. *Wood and Science Technology, 32*(5):335-345.

RITTER, G.J. & KURTH, E.F. (1933). Holocellulose, total carbohydrate fraction of extractive-free maple wood. *Industrial and Engineering Chemistry Research, 25*(11):1250-1253.

RUEL, K., BARNOUD, F. & GORING, D.A.I. (1978). Lamelllation in S2 layer of softwood tracheids as demonstrated by scanning transmission electron microscopy. *Wood Science and Technology, 12*(4):287-291.

SALMÉN, L. & BURGERT, I. (2009). Cell wall features with regard to mechanical performance. A review. *Holzforschung, 63*(2):121-129.

SCALLEN, A.M. (1974). The structure of cell wall of wood – a consequence of anisotropic inter-microfibrillar bonding? *Wood Science, 6*(3):266-271.

SCHWARZE, F. & ENGELS, J. (1998). Cavity formation and the exposure of peculiar structure in the secondary wall (S2) of tracheids and fibres by wood degrading Basidiomycetes. *Holzforschung, 52*(2):117-123.

SCHWEINGRUBER, F.H. (1990). *Anatomy of European woods. An atlas for the identification of European trees, shrubs and dwarf shrubs*. Paul Haupt, Berne, Stuttgart. ISBN: 3-258-04258-6.

SEDIGHI-GILANI, M., SUNDERLAND, H. & NAVI, P. (2005). Microfibril angle non-uniformities within normal and compression wood tracheids. *Wood Science and Technology, 39*(6):419-430.

SEDIGHI-GILANI, M., SUNDERLAND, H. & NAVI, P. (2006). Within-fiber nonuniformities of microfibril angle. *Wood and Fiber Science, 38*(1):132-138.

SELL, J. (1994). Mechanical aspects of new SEM observations on the fibrill-matrix structure of the S2 layer of softwood tracheids. In: *Proceddings of the International Congress on Plant Biomechanics*, Montpellier, France, pp. 163-164.

SELL, J. & ZIMMERMANN, T. (1993). Radial fibril agglomerations of the S2 on transverse-fracture surfaces of tracheids of tension-loaded spruce and white fir. *Holz als Roh- und Werkstoff, 51*(6):384.

SELL, J. & ZIMMERMANN, T. (1998). The fine structure of cell wall of hardwoods on transverse-fracture surfaces. *Holz als Roh- und Werkstoff, 56*(5):365-366.

SINGH, A.P. (1997). The ultrastructure of the attack of Pinus radiata mild compression wood by erosion and tunneling bacteria. *Canadian Journal of Botany, 75*(7):1095-1102.

SKAAR, C. (1988). *Wood-water relations*. Springer-Verlag, Berlin, Heidelberg, ISBN 3-540-19258-1.

TRENDELENBURG, R. & MAYER-WEGELIN, H. (1955). *Das Holz als Rohstoff*. (Wood as raw material.) (2 Auflage), Carl Hanser Verlag, München.

VINTILA, E. (1939). Untersuchungen über Raumgewicht und Schwindmass von Früh- und Spätholz bei Nadelhölzern. (Studies on density and shrinkage of earlywood and latewood in conifers.) *Holz als Roh- und Werkstoff, 2*(10):345-357.

WANGAARD, F.F. (ed.) (1981). *Wood: Its structure and properties*. Clark C. Heritages Memorial Series on Wood. The Pennsylvania State University.

ZIMMERMANN, T. & SELL, J. (2000). Comparison of the biomechanical properties of the fine structure of the cell wall of normal and reaction wood. In: *Proceedings of 3rd Plan Biomechanics Conference,* (eds.). Spatz, H.-C. & Speck, T. Freiburg-Badenweiler, Germany, pp. 186-192.

ZIMMERMANN, T. & SELL, J. (2001). Fungi as a tool to determine the arrangement of wood cell wall components. In: *Proceedings of Cost E20 workshop:* Methods to localise and characterise cell wall components, Grenoble, France.

ZOBEL, B.J. & BUIJTENEN, J.P. v. (1989). *Wood variation: Its causes and control*. Springer-Verlag, Berlin, Heidelberg, ISBN 3-540-50298-X.

ZOBEL, B.J. & SPRAUGE, J.R. (1998). *Juvenile wood in forest trees*. Springer-Verlag, Berlin, Heidelberg, ISBN 3-540-64032-0.

ZOBEL, B.J. & TALBERT, J. (1984). *Applied forest tree improvement*. John Wiley & Sons, New York, ISBN 0-471-09682-2.

CHAPTER 4

ELASTO-VISCOPLASTICITY OF WOOD

4.1 INTRODUCTION TO THE ELASTIC, PLASTIC AND VISCOELASTIC BEHAVIOR OF WOOD

When wood is subjected to an appropriate temperature and steam, its compliance properties increase, rendering it possible to form the wood through mechanical actions. During THM processing, wood can undergo a large transverse compressive deformation greater than 50%. For instance, through transverse hot pressing, a thickness reduction of more than 50% can be obtained. In wood surface densification, a thin region of the wood surface can be densified by much more than 50% through the compression of the lumens.

In wood bending, depending on the wood species and the treatment, one can achieve radii of curvature of approximately 4 cm for beech, 8 cm for birch and even 1.5 cm for elm in boards with a thickness of 2.5 cm (Stevens & Turner, 1970 and Table 10.1). This shows that large deformations, including elastic, plastic and visco-elastic behavior occur during THM treatment.

4.2 MECHANICAL BEHAVIOR OF WOOD UNDER LOADING

During its growth, the tree develops an organized porous structure to support its weight, growth stresses and applied external forces. The physical and mechanical properties of wood depend both on its structure and constituents. Wood has an

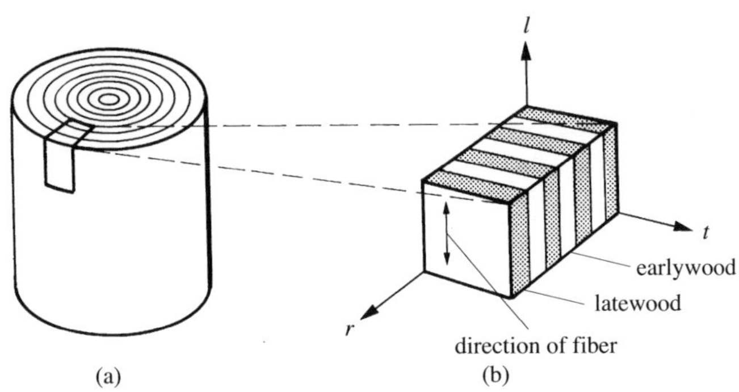

Fig. 4.1 (a) An idealized cylindrical trunk represented by a circular cylinder and circular annual rings. (b) A close-up view of a small element cut from the trunk. *l*, *r* and *t* represents the local principal axes in the longitudinal, radial and tangential directions, respectively.

anisotropic behavior, i.e., the strength and rigidity of wood is much greater in its longitudinal direction than in the transverse directions. The mechanical behavior of wood in the radial direction also differs from that in the tangential direction. In a tree trunk, which is more or less cylindrical, the circular shape of the annual rings and the organization of the longitudinal cells give the wood axis-symmetric mechanical properties. Apart from defects and natural growth imperfections, the two local symmetry planes *lr* and *rt* passing through each point signify that wood may be considered as a cylindrical orthotropic material. Figure 4.1 shows an idealized cylindrical trunk and a close-up element with principal directions of r, t, and l corresponding to a given point in the trunk.

4.2.1 Elastic and plastic behavior of wood

Under a longitudinal tensile force, the behavior of wood is typically elastic illustrating a quasi-linear elongation up to a breaking point, as given in Figure 4.2. Breaking occurs by a brittle fracture, under controlled force conditions when the ultimate strain is about 1-3%. Nevertheless, when the displacement is controlled, the response of a wood specimen to a simple tensile force as shown in Figure 4.3 is different. The force displacement curve shows a strain-softening behavior after the peak, as a consequence of strain localization (fracture with a damaged zone) in the specimen.

When wood is subjected to a tensile force in the transverse direction, it illustrates a force-displacement curve with features similar to those in the longitudinal direction, but with breaking and peak stresses that are much lower.

Many experimental results have shown that the behavior of wood under a compression force varies according to the direction of loading (longitudinal, radial,

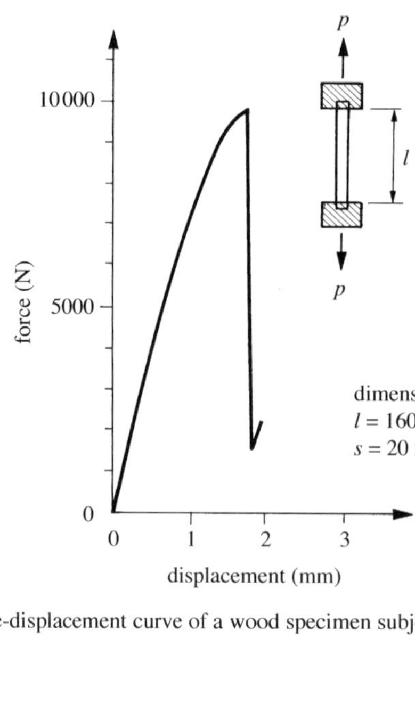

Fig. 4.2 Typical force-displacement curve of a wood specimen subjected to a controlled longitudinal tensile force.

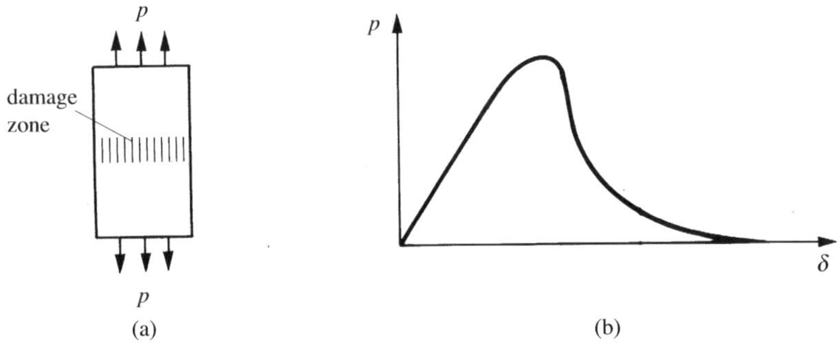

Fig. 4.3 Typical force-displacement curve of a specimen in tension in the longitudinal direction under controlled displacement; (a) specimen after damage, (b) curve illustrating the linear region as well as strain softening of the specimen after the peak force.

tangential). Roussel (1997) has examined the behavior of $25 \times 25 \times 5$ mm^3 specimens of poplar under simple compression in the three principal directions *l, r, t* under controlled deformation. Typical stress-displacement curves are shown in Figure 4.4.

Under axial compression, the specimen illustrates a 3 segmented stress-strain curve: a quasi-linear segment followed by the second segment non-linear curve with a negative slope. This segment can be interpreted as being due to localized longitudinal buckling of the cell walls and/or local fracture. The third segment shows an increase in the modulus of the specimen with increasing compression force.

Transverse compression stress-strain curves (corresponding to the radial or tangential direction) also consist of three segments. The first segment demonstrates a

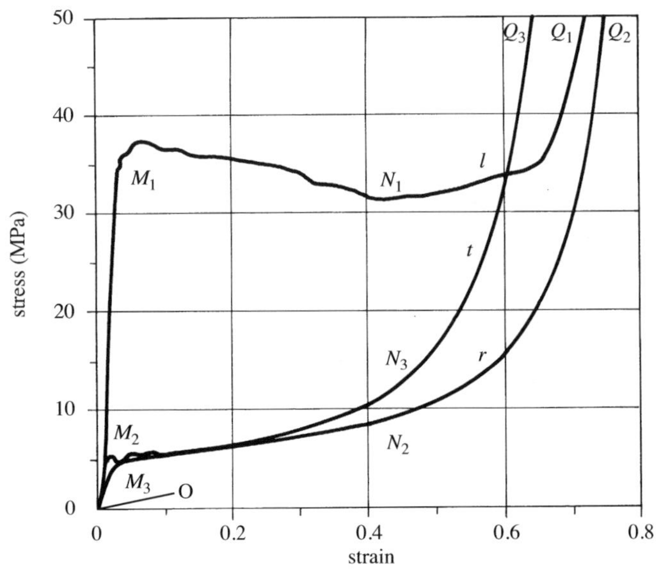

Fig. 4.4 Stress-strain curves of specimens of poplar (density 350 kg/m^3) with dimensions $25 \times 25 \times 5$ mm^3 subjected to compression in the *l, r, t* directions under controlled displacement at a rate of 1 mm/min. (Roussel, 1997).

quasi-linear behavior and is followed by a non-linear curve presenting a decreasing modulus corresponding to a transverse flexural buckling of the cell walls. The third segment is a quasi-linear curve indicating a progressive increase in the modulus, which can reach a value higher than in the first segment. During segments 2 and 3, the strain can attain values higher than 50%, indicating a large transversal deformation, which leads to a densification of the wood cells.

4.2.2 Physical mechanisms of large deformations under longitudinal and transverse compression at the cell level

Numerous researchers have studied the mechanisms of large deformations of softwood and hardwood specimens under compression at the cellular level, e.g., Bariska and Kučera (1985), Gibson and Ashby (1988), Hoffmeyer (1990), Boström (1992), François and Morlier (1993), Gril and Norimoto (1993), Roussel (1997) and Navi and Heger (2005). This section presents only the large deformation of coniferous wood. In softwood at the cellular level, the microstructure is more simple and uniform than that of hardwood. In softwood, longitudinal cells called tracheids, comprise approximately 90% of the volume. Figure 4.5 shows a typical diagrammatic representation of a small softwood specimen. The ray cells are directed in the radial direction in the *rl* plane.

As shown in Figure 4.4, the stress-strain response of a poplar wood specimen to a simple compression force consists of three distinct segments. The specimens behave almost linearly up to a certain limit that depends on the direction of the applied force. The slopes of the first segments of the curves; O-M_1, O-M_2 and O-M_3 give the effec-

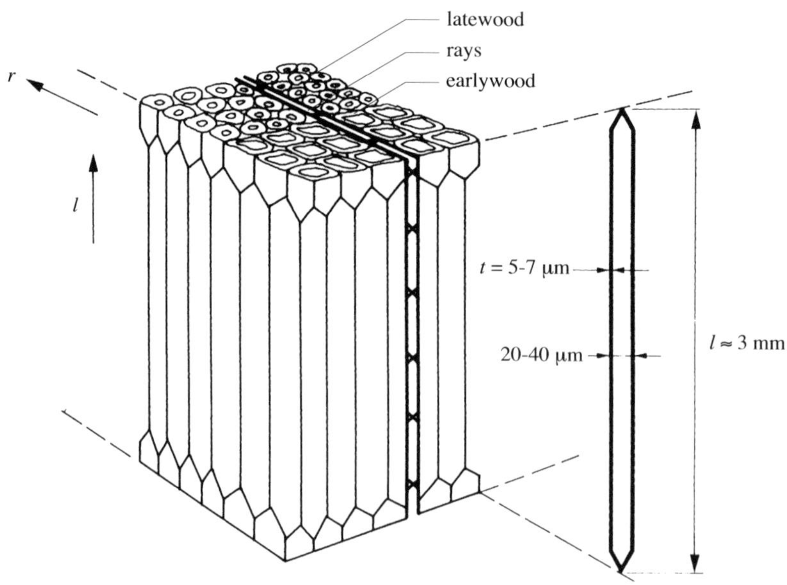

Fig. 4.5 Diagrammatic representation of a small softwood specimen as an assembly of tracheids and ray cells. The values given are for Norway spruce.

tive Young's modulus of the specimens in the l, r and t directions respectively, Figure 4.4. Above the linear limit, the wood behaves differently in the longitudinal direction as opposed to the other two directions. In the longitudinal direction, the specimen demonstrates a smoothening or strain softening behavior, segment M_1-N_1, followed by a rigidification or densification behavior, segment N_1-Q_1. In the r and t directions, on the other hand, the wood exhibits a standard plastic behavior with a positive work hardening, segments M_2-N_2 and M_3-N_3, followed by a rigidification or densification.

The softening behavior can be explained by various mechanisms of deformation at the cellular level. Many researchers have observed this behavior in various kinds of wood. For wood of low density (< 300 kg/m^3), Easterling *et al.* (1982) showed that the cause of the wood softening is the collapse of fibers by rupture at the ends of the cells. This mechanism is shown schematically in Figure 4.6(b). In wood of higher density, Kučera and Bariska (1982) have demonstrated that the softening of the wood can be due to a local Euler type buckling in the walls of the cells, illustrated in Figure 4.6(c), which generally leads to the formation of a shear band as shown in Figure 4.7 on the wood macro-level.

In the r and t directions, the first zone represents a linear elastic behavior, while the second zone (segments M_2-N_2 and M_3-N_3 in Figure 4.4) corresponds to a plastic behavior of the cell walls. In the third zone (segments N_2-Q_2 and N_3-Q_3), a densification of the cells of the specimen occurs. Various researchers have explained these phenomena. Under the application of a compressive force in the radial direction, deformation occurs as a flexural buckling (or crushing) of the fiber walls starting in the weakest layer of the material. The cells of earlywood are the first to buckle since their walls are thinner (~2 μm) than those of the cells of latewood (~10 μm). This buckling leads to a densification of the fibers, the establishment of points of contact between the cell walls and increasing the local rigidity so that this layer can support higher force sufficient to buckle the next layer which is the weakest one, Figure 4.8. These deformation mechanisms consisting of localization, buckling and densification have been observed by for instance Gibson and Ashly (1988). A diagrammatic representation

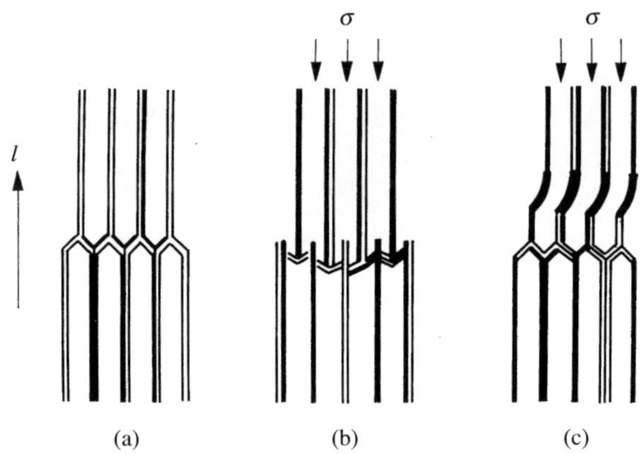

Fig. 4.6 Diagrammatic representation of the mechanisms of local large deformations of wood cells under a longitudinal compression force; (a) wood before the application of the force, (b) collapse of fiber by rupture, (c) collapse by the buckling of cell walls.

(a) (b)

Fig. 4.7 Collapse of the specimen by the buckling of fiber walls, which creates a shear band: (a) observed at the macro-level, (b) observed at the cell-wall level.

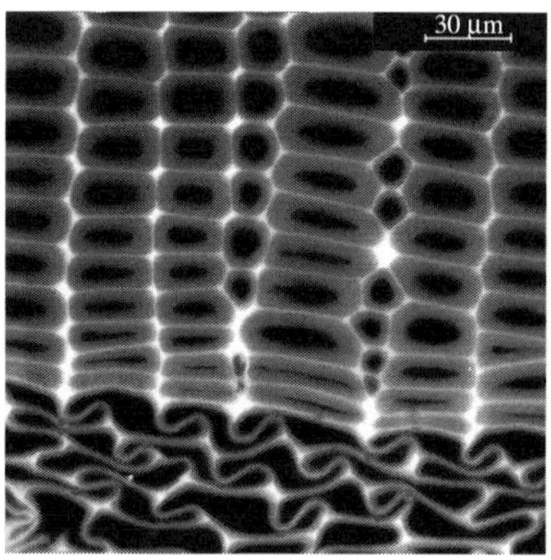

Fig. 4.8 Micrograph of a densified spruce sample in the radial direction at a ratio of 5% obtained by a confocal microscope. The contact points between cellular walls in the earlywood zone are shown by red dots.

of this phenomenon is given in Figure 4.9. After the yielding of the first cellular row, the phenomenon is propagated progressively to the other rows of earlywood cells and later to the latewood cells.

In softwoods in the radial direction, the earlywood rows yield first, followed by the latewood rows, whereas in the tangential direction, thanks to the micro-structure of wood, buckling of the cell walls occurs at the same time in both latewood and earlywood cells.

The slope of the curves in the second segment increases with an increasing compressive force, due to a closing of the lumens and multiplication of the points of contact between the walls of the cells of the initial wood. As the compression force is raised, the thickness of these bands of buckled cells increases and others start to yield. When all the lumens are closed, the stress necessary to continue the deformation increases exponentially. This phenomenon occurs in the third segment. Nevertheless, wood densification starts in segment two and continues in segment three. As a result of this linear and non-linear behavior, as shown in Figure 4.4, the wood can deform by more that 50%.

To model the behavior of wood under compression deformation, constitutive equations of wood undergoing linear and non-linear deformations therefore become important. These equations are presented in the following sections.

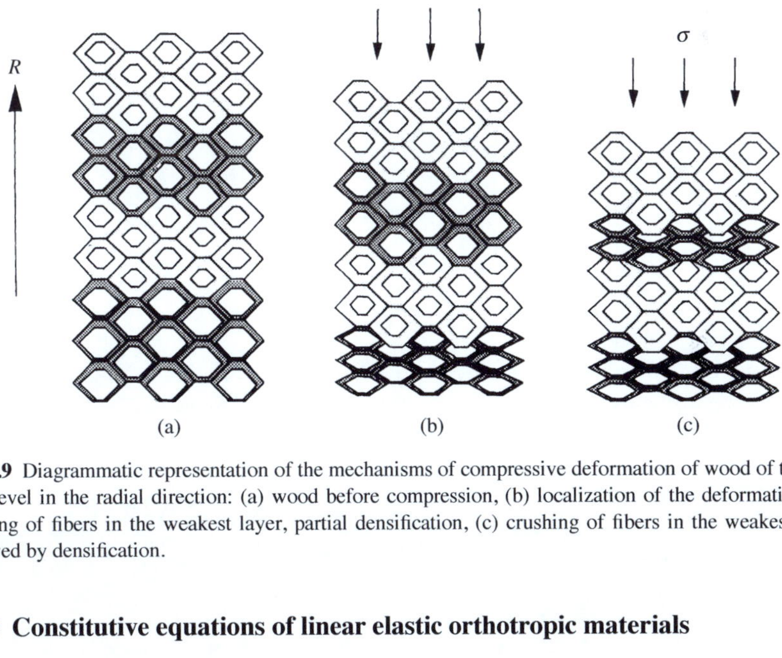

Fig. 4.9 Diagrammatic representation of the mechanisms of compressive deformation of wood of the cellular level in the radial direction: (a) wood before compression, (b) localization of the deformation and crushing of fibers in the weakest layer, partial densification, (c) crushing of fibers in the weakest layer followed by densification.

4.2.3 Constitutive equations of linear elastic orthotropic materials

When the strain is smaller than 3-5%, the elasticity of wood is considered as linear. Hence, the constitutive equation defining the relation between the strain tensor and the Cauchy stress tensor is given by

$$\{\varepsilon\}=[S]\{\sigma\} \tag{4.1}$$

where the relation between the strain and the stress is defined by a tensorial expression:

$$\varepsilon_{ij}=S_{ijkl}\,\sigma_{kl} \qquad \text{with } i,j,k,l\in(1,\ 2,\ 3) \tag{4.2a}$$

The relation (4.2a) includes the convention of summation (convention of Einstein) of a repeated index. For a numerical solution, it is often convenient to replace the tensorial notation (4.2a) with a matrix notation

$$\varepsilon_\alpha = S_{\alpha\beta}\,\sigma_\beta \qquad \text{with } \alpha, \beta \in (1,\ 2, \dots\ 6) \tag{4.2b}$$

Here, $\{\varepsilon\}$ represents the components of the strain vector

$$\{\varepsilon\} = \{\varepsilon_1, \varepsilon_2, \varepsilon_3, \gamma_4, \gamma_5, \gamma_6\}^T \tag{4.3}$$

defined as

$$\varepsilon_1 = \frac{\partial u}{\partial x} = \varepsilon_{11},\ \varepsilon_2 = \frac{\partial v}{\partial y} = \varepsilon_{22},\ \varepsilon_3 = \frac{\partial w}{\partial z} = \varepsilon_{33}$$

$$\gamma_4 = \frac{\partial w}{\partial y} + \frac{\partial v}{\partial z} = 2\varepsilon_{23},\ \ \gamma_5 = \frac{\partial u}{\partial z} + \frac{\partial w}{\partial x} = 2\varepsilon_{13} \text{ and } \ \gamma_6 = \frac{\partial v}{\partial x} + \frac{\partial u}{\partial y} = 2\varepsilon_{12} \tag{4.4}$$

where u, v and w are displacements in the x, y and z directions, respectively.

In the matrix notation (4.2b) $\{\sigma\}$ represents the components of the stress vector

$$\{\sigma\} = \{\sigma_1, \sigma_2, \sigma_3, \sigma_4, \sigma_5, \sigma_6\}^T \tag{4.5}$$

with $\sigma_4 = \sigma_{23}$, $\sigma_5 = \sigma_{13}$ and $\sigma_6 = \sigma_{12}$, and where $[\mathbf{S}]$ is symmetric and called the matrix of elastic compliance.

If wood is considered to be an orthotropic material, the matrix of elastic compliance with coordinates corresponding to the r, t and l axes is given by

$$[\mathbf{S}] = \begin{bmatrix} \frac{1}{E_r} & -\frac{\nu_{tr}}{E_t} & -\frac{\nu_{lr}}{E_l} & 0 & 0 & 0 \\ -\frac{\nu_{rt}}{E_r} & \frac{1}{E_t} & -\frac{\nu_{lt}}{E_l} & 0 & 0 & 0 \\ -\frac{\nu_{rl}}{E_r} & -\frac{\nu_{tl}}{E_t} & \frac{1}{E_l} & 0 & 0 & 0 \\ 0 & 0 & 0 & \frac{1}{G_{tl}} & 0 & 0 \\ 0 & 0 & 0 & 0 & \frac{1}{G_{rl}} & 0 \\ 0 & 0 & 0 & 0 & 0 & \frac{1}{G_{rt}} \end{bmatrix} \tag{4.6}$$

where, E_r, E_t, and E_l are the Young moduli corresponding respectively to the r, t and l axes, G_{rl}, G_{tl}, G_{rt} are the shear moduli corresponding respectively to the rl, tl, rt planes, and ν_{rl}, ν_{lr}, ν_{rt}, ν_{tr}, ν_{tl}, ν_{lt} are the Poisson ratios.

As the matrix [**S**] is symmetrical, i.e.,

$$\frac{\nu_{rt}}{E_r} = \frac{\nu_{tr}}{E_t}, \quad \frac{\nu_{rl}}{E_r} = \frac{\nu_{lr}}{E_l} \text{ and } \frac{\nu_{tl}}{E_t} = \frac{\nu_{lt}}{E_l} \tag{4.7}$$

The elastic properties of each species of wood are therefore completely defined by nine parameters, E_r, E_t, E_l, G_{rl}, G_{tl}, G_{rt} and ν_{rl}, ν_{tr}, ν_{tl}. It should be noted that these characteristics are functions of the moisture content and density of the wood, and differ for each species. Elastic moduli of some wood species are given by Kollmann and Côté (1968), Bodig and Goodman (1973), Allaili, *et al.* (1986), Huet *et al.* (1986). A detailed list of the characteristics of various species is also given in Guitard (1987). Table 4.1 gives indicative values of the elastic characteristics of certain species.

Table 4.1 Elastic characteristics of some softwoods and hardwoods.

Elastic constants	Spruce (Picea sitchensis)	Douglas (Pseudotsuga menziesii)	Fir (Abies alba)	Poplar (Populus alba)	Oak (Quercus rubra)	Beech (Fagus silvatica)
Moisture content (%)	12	12	13	14	12	11
Density (kg/m^3)	360	590	310	400	600	740
E_l (N/mm^2)	10 700	16 550	8020	6830	14 800	14 000
E_r (N/mm^2)	649	1300	816	1190	1500	2280
E_t (N/mm^2)	348	900	304	493	828	1160
G_{rl} (N/mm^2)	533	1200	558	1000	967	1640
G_{rt} (N/mm^2)	41	80	48	200	398	470
G_{tl} (N/mm^2)	438	929	461	900	695	1080
ν_{rl}	0.02	0.03	0.03	0.04	0.06	0.07
ν_{tr}	0.3	0.41	0.25	0.37	0.33	0.36
ν_{lt}	0.4	0.38	0.33	0.45	0.69	0.52

Experimental results on many wood species show the following relationship between the magnitudes of the Young moduli of wood in the three principal directions

$$E_l > E_r > E_t \tag{4.8}$$

In general, ν_{rl} and ν_{tl} have very small values and the shear moduli follow the order

$$G_{rl} > G_{tl} > G_{rt} \tag{4.9}$$

4.2.4 Constitutive equations of large nonlinear deformation materials

Many experimental results, e.g., Figure 4.4, show that wood subjected to compression in the transverse direction can deform by more that 50%. When the deformation is greater than 3-5%, the behavior is nonlinear, and to account for such large deformations, appropriate stress and strain tensors and constitutive relations must be defined. To model the large deformation of wood, depending on the nature of the problem, various models can be implemented on the micro- or macroscopic levels.

Nairn (2006) and Rangsri *et al.* (2004) have for example, studied the transverse compressive deformation of wood at the cellular level under ambient conditions with different wood under transverse compression by applying an analytical model derived by Gibson and Ashby (1988) for foam made of a regular array of hexagonal cells. In another study, Nairn (2006) implemented the Material Point Method (MPM) to numerically simulate the transverse compression of wood at the cellular level. Figure 4.10 shows the large deformation of a piece of loblolly pine (0.832×0.541 mm^2 in cross section) at different strain levels subjected to a compression force in the radial direction.

Nairn showed that the MPM technique could take into account material heterogeneity, material nonlinearity, large deformations and the contacts between cell walls during the wood densification. Sulsky *et al.* (1994) and Sulsky and Schreyer (1996) developed this numerical method for solving problems in dynamic solid mechanics. In this method, the solid material is represented by an array of points, where each point is assigned specific material properties, such as mass, stiffness and toughness. To evaluate the bulk material properties, a mechanical test simulation is carried out through a series of time steps, where in each step the properties of each point such as position, velocity, and stress are calculated by imposing the current point properties onto a background grid for computation.

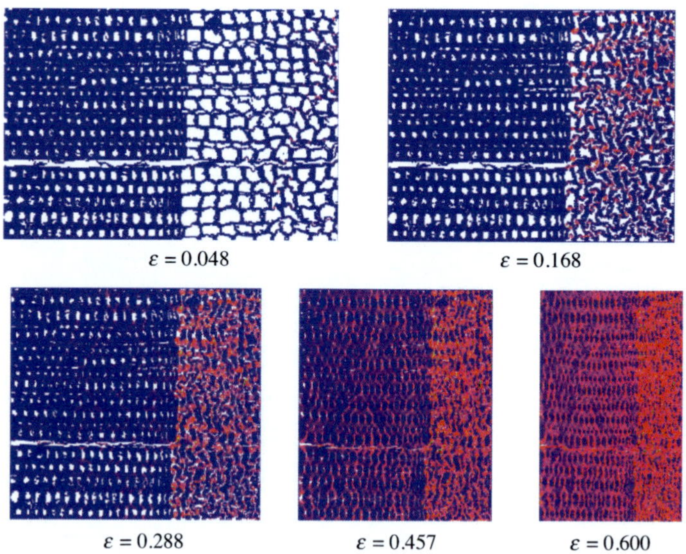

Fig. 4.10 Wood structure at a cellular level during strain-controlled radial compression (Nairn, 2006).

Rangsri (2004) has interpreted the indentation test on a wood element by applying the finite element technique and have studied the large deformation of the cells subjected to transverse compressive forces. He has verified experimentally the potential of the finite element method for modeling the large deformation of the wood. Figure 4.11 compares the results obtained experimentally and by the finite element simulation.

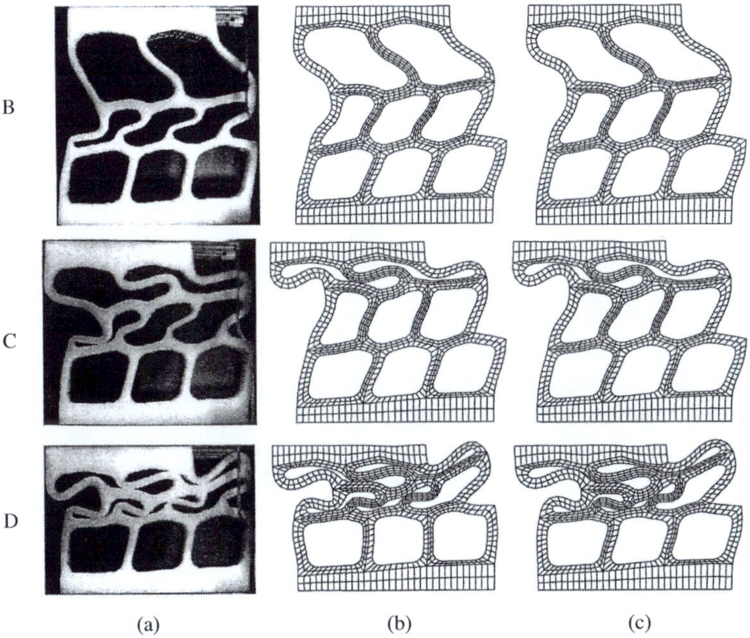

Fig. 4.11 Comparison of the experimental results with those obtained using numerical simulation by two different finite element codes; (a) experiments on Teflon, (b) result obtained by CASTEM2000 code, (c) results obtained by MSC. Marc code (Rangsri *et al.*, 2003, 2004).

None of these investigations concerned the buckling of the specimens during a large transversal deformation. Recently, De Magistris and Salmén (2008) investigated the large transverse deformation of small, thin wood tissues subjected to compression and to combined shear and compressive forces using finite element methods. Although the simulated tissues were small, consisting of only 4 to 12 fibers, they successfully implemented a semi 3-D finite element method to study the transversal deformation and analyzed the possibility of the occurrence of buckling modes of the tissues under the compressive loading. They also confirmed their modeling results with experimental data.

The application of these techniques is usually limited to small-scale 2-D problems, whereas wood elements used for instance in wood shaping are large. In addition, the overall "macroscopic" characterization of wood is much easier than "microscopic" characterization. We shall therefore try to explain the large deformation of wood on the macroscopic level. It is worth noting that there exists no macroscopic modeling developed particularly for the study of large deformations of wood. On the other hand, phenomenological models have been developed for studying the behaviour of non-linear elastic materials such as: those commonly known as hyperelastic. Materials such as elastomers, porous solids like foams, and biological tissues are referred to as quasi-elastic materials, and may be considered to be hyperelastic. Generally, the modeling of these materials by a finite element method is based on a strain energy density function. To present this approach, we shall first present the derivation of the principal relations for modeling large deformations, namely the Green-Lagrange strain tensor, the first Piola-Kirchoff stress tensor, the second Piola-Kirchoff stress tensor, constitutive equations of large nonlinear materials and a strain energy function for neo Hookean isotropic materials. All this information can be found in Rivlin, (1948), Ogden (1972), Treloar (1975), Valanis and Landel (1978), Marsden and Hughes (1994), Curnier (2005, 2011), BME/ME 506 (A course on Computational Modeling of Biological Tissues in Internet).

4.2.5 Finite strain continuum mechanics

Definition of the deformation gradient tensor

To define a large deformation strain in a Lagrange description, we first define the relationship between the deformed and initially undeformed configuration of the solid. It is important to note that the solid may undergo rigid body motion in addition to finite strain when it is subjected to a load. Figure 4.12 illustrates the initial undeformed configuration of a solid as well as its deformed configuration. In this 3D space, a vector x represents a particle position in the reference or undeformed configuration and a vector y corresponds to the position of the particle in the deformed configuration. The relationship between the two position vectors in the space is the displacement vector u.

The position of y on the deformed configuration is considered to be unknown. It is a function of the position x on the undeformed configuration and time t according to

$$y = \mathrm{y}(x, t) \tag{4.10}$$

$$y = \mathrm{y}_i(x_1, x_2, x_3, t) \qquad m, i = 1, 2, 3 \tag{4.11}$$

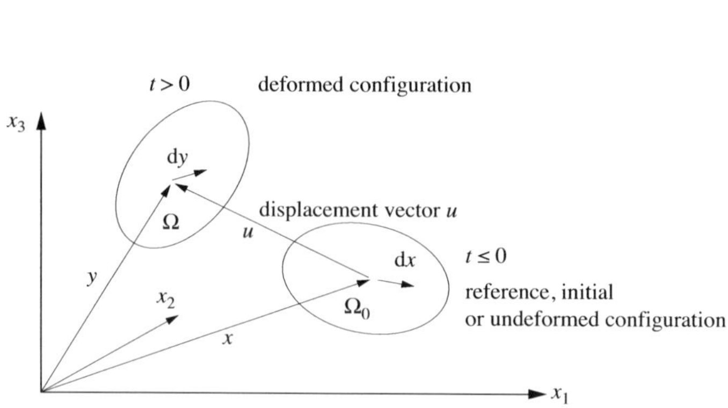

Fig. 4.12 The initial undeformed configuration, the deformed solid and the displacement vector u.

The displacement equation can be obtained by relating the position vectors in the initial and the deformed configurations:

$$y_i = x_i + u_i \tag{4.12}$$

where u is the displacement vector.

Consider two vectors dx and dy describing an infinitesimal element of material in respectively the initial configuration and the deformed configuration, Figure 4.12. The linear mapping between vectors dx and dy can be written as:

$$\mathrm{d}y_i = \frac{\partial y_i}{\partial x_j}\,\mathrm{d}x_j \tag{4.13}$$

$$\mathrm{d}y_i = F_{im}\,\mathrm{d}x_m$$

where

$$F = F(x, t) = \frac{\partial y}{\partial x}(x, t)$$

or $F_{im} = \left(\partial y_i / \partial x_m\right) = y_{i,m}$, is called the deformation gradient tensor.

$$F = \begin{bmatrix} F_{11} & F_{12} & F_{13} \\ F_{21} & F_{22} & F_{23} \\ F_{31} & F_{32} & F_{33} \end{bmatrix} = \begin{bmatrix} \dfrac{\partial y_1}{\partial x_1} & \dfrac{\partial y_1}{\partial x_2} & \dfrac{\partial y_1}{\partial x_3} \\ \dfrac{\partial y_2}{\partial x_1} & \dfrac{\partial y_2}{\partial x_2} & \dfrac{\partial y_2}{\partial x_3} \\ \dfrac{\partial y_3}{\partial x_1} & \dfrac{\partial y_3}{\partial x_2} & \dfrac{\partial y_3}{\partial x_3} \end{bmatrix} \tag{4.14}$$

It is important to note that F is a second order tensor and that it is not symmetrical i.e. $F_{ij} \neq F_{ji}$. Finally, derivatives are taken from the displacement equation given in (4.12), to define the deformation gradient tensor:

$$\frac{\partial y_i}{\partial x_m} = \frac{\partial x_i}{\partial x_m} + \frac{\partial u_i}{\partial x_m} \tag{4.15}$$

In (4.15), the first term on the right-hand side is a second-order identity tensor, represented by the "Kronecker delta", δ_{im}. The deformation gradient tensor (F_{im}) can thus be defined as the displacement gradient plus the identity:

$$F_{im} = \frac{\partial u_i}{\partial x_m} + \delta_{im} \tag{4.16}$$

There are other properties of F that are important for deriving information on the strains, stresses and constitutive equations of the mechanics of large deformation. For example, as with all second-order tensors, three invariants of the deformation gradient tensor can be defined according to the following:

$$\text{First invariant or trace of } F\text{: } F_{11} + F_{22} + F_{33} = F_{ij}\delta_{ij} \quad \text{(a)}$$

$$\text{Second invariant: } 1/2\ (F_{ij}F_{ij} - F_{ii}F_{jj}) \quad \text{(b)} \qquad (4.17)$$

$$\text{Third invariant: det } F = J > 0 \quad \text{(c)}$$

Using the third invariant or the determinant of the deformation gradient tensor facilitates the mapping of a volume from an undeformed into a deformed configuration as

$$\mathrm{d}V = \det\ F\ \mathrm{d}V' = J\ \mathrm{d}V' \qquad (4.18)$$

where $\mathrm{d}V'$ is an infinitesimal volume in the reference configuration and $\mathrm{d}V$ is its corresponding volume in the deformed configuration. The value of J should remain positive and bounded, i.e., ($0 < J < \infty$) to avoid any fracturing and auto-penetration during the deformation.

When an area from the initial configuration with normal vectors N is mapped into the deformed configuration with the normal vector n, the relation between two areas can be written as

$$n_i\ \mathrm{d}A = J\ F_{im}^{-1}\ N_m\ \mathrm{d}A' \qquad (4.19)$$

where $\mathrm{d}A$ and $\mathrm{d}A'$ are the areas of the surfaces in the deformed and undeformed configurations, respectively, whereas n_i and N_m are the components of the vectors normal to these surfaces in respectively the deformed and undeformed configurations.

4.2.6 Concepts of strain measures for large deformation mechanics

Since the deformation gradient tensor may contain a rotation, it cannot be used directly for strain definition. To define a strain measure, it is then necessary to determine the change in the square of a length of a material vector going from an un-deformed to a deformed configuration. Note that this measure should be independent of rigid body rotation. The square of a vector is a dot product with itself. In the initial configuration it is written as

$$\|\mathrm{d}x\|^2 = \mathrm{d}x_m \mathrm{d}x_m \qquad (4.20)$$

Similarly, in the deformed configuration it is expressed as

$$\|\mathrm{d}y\|^2 = \mathrm{d}y_n \mathrm{d}y_n,\ n = 1, 2, 3 \qquad (4.21)$$

A strain measure should indicate how much the length of a material has changed when going from the initial to the deformed configuration. Therefore, the following relation can be written as the definition of the strain tensor E_{mn}:

$$\|\mathrm{d}y\|^2 - \|\mathrm{d}x\|^2 = \mathrm{d}y_n \mathrm{d}y_n - \mathrm{d}x_m \mathrm{d}x_m = 2\mathrm{d}x_m E_{mn} \mathrm{d}x_n \qquad (4.22)$$

Note, however, that the material vector dy can be replaced by dx using the relation defined in (4.13) so that this equation becomes

$$\mathrm{d}x_m F_{im} F_{in} \mathrm{d}x_n - \mathrm{d}x_m \mathrm{d}x_m = 2\mathrm{d}x_m E_{mn} \mathrm{d}x_n \tag{4.23}$$

Consider now the second term in (4.22), which can be rewritten as

$$\mathrm{d}x_m \mathrm{d}x_m = \mathrm{d}x_m \delta_{mn} \mathrm{d}x_n \tag{4.24}$$

Using this equation, the strain equation (4.23) can be expressed as

$$\mathrm{d}x_m F_{im} F_{in} \mathrm{d}x_n - \mathrm{d}x_m \delta_{mn} \mathrm{d}x_n = \mathrm{d}x_m (F_{im} F_{in} - \delta_{mn}) \mathrm{d}x_n = 2\mathrm{d}x_m E_{mn} \mathrm{d}x_n \tag{4.25}$$

from which the strain tensor can be directly written in terms of the deformation gradient tensor by matching the strain E_{mn} to the quantities in parentheses:

$$E_{mn} = \frac{1}{2}(F_{im} F_{in} - \delta_{mn}) \tag{4.26}$$

In (4.26), the strain tensor gives the deformation, independent of rigid body rotation. Noting this, as nothing has been assumed regarding the magnitude of the deformation during derivation of relation (4.26), the strain tensor is thus exact for any deformation. This strain tensor is known as the Green-Lagrange strain tensor. It is also possible to derive the small deformation strain from equation (4.26). Substitution of the displacement equation given in (4.16) for the deformation gradient in (4.25) gives

$$E_{mn} = \frac{1}{2}\left[\left(\delta_{km} + \frac{\partial u_k}{\partial x_m}\right)\left(\delta_{kn} + \frac{\partial u_k}{\partial x_n}\right) - \delta_{mn})\right] \tag{4.27}$$

If in this equation the terms in parentheses are expanded, we obtain

$$E_{mn} = \frac{1}{2}\left[\delta_{mn} + \frac{\partial u_m}{\partial x_n} + \frac{\partial u_n}{\partial x_m} + \frac{\partial u_k}{\partial x_m}\frac{\partial u_k}{\partial x_n} - \delta_{mn})\right] =$$
$$\frac{1}{2}\left[\frac{\partial u_m}{\partial x_n} + \frac{\partial u_n}{\partial x_m} + \frac{\partial u_k}{\partial x_m}\frac{\partial u_k}{\partial x_n}\right] \tag{4.28}$$

This equation demonstrates that the large deformation strain tensor contains a quadratic term. Consequently, all large deformation analyses are nonlinear, whereas the small deformation strain tensor is defined as

$$\varepsilon_{mn} = \frac{1}{2}\left(\frac{\partial u_m}{\partial x_n} + \frac{\partial u_n}{\partial x_m}\right) \tag{4.29}$$

Thus, if it is assumed that the deformation is small, the quadratic terms from the Green-Lagrange strain tensor can be dropped. It is important to note that the gradients in the Green-Lagrange strain tensor are defined with respect to the initial or un-deformed configuration, and thus that all the strain measures are calculated with respect to the initial configuration.

To define a constitutive equation for large deformations, a tensor called the right Cauchy deformation tensor is often used. This tensor is defined as

$$C_{mn} = F_{im} F_{in} \tag{4.30}$$

Using the right Cauchy deformation tensor, the Green-Lagrange strain tensor is rewritten as

$$E_{mn} = \frac{1}{2}\left(C_{mn} - \delta_{mn}\right) \tag{4.31}$$

The right Cauchy deformation tensor is symmetrical and positively definite and this tensor thus has three real eigenvalues. The square roots of these eigenvalues are the principal strains of the material.

4.2.7 Concept of stress measures for large deformation mechanics

By analogy with the strain measures for large deformation, it is possible to define similar stress measures. These stress measures must be appropriate for use with the strain defined in the reference configuration. The Cauchy stress tensor for a small elastic deformation is given as

$$t_i = T_{ij} n_j, \; T_{ij} \equiv \sigma_{ij} \tag{4.32}$$

where T_{ij} is the Cauchy stress tensor, t is the Cauchy traction vector and n is the unit normal vector according to Figure 4.13.

The strain tensor that is appropriately used with the Cauchy stress tensor is that for small deformation. The problem when using the Cauchy stress tensor for a solid undergoing a large deformation is that the area in the deformed configuration is not generally known, and it is thus necessary to define a stress measure that can be used in the un-deformed configuration.

To derive a stress tensor with respect to the reference configuration from the one in the deformed configuration, we utilize the principle that the Cauchy stress tensor in

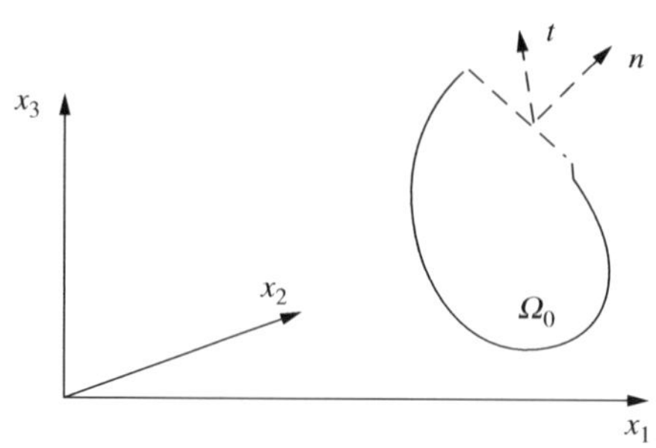

Fig. 4.13 Cauchy traction vector t and the unit normal vector n.

the reference configuration should give the same force as the stress tensor defined in the deformed configuration. Since the traction is a force on the surface and the product of the stress and the normal vector to the surface, the force dq_i (the Cauchy traction vector) in the deformed configuration is determined by

$$dq_i = \sigma_{ij} n_j dA \tag{4.33}$$

where σ_{ij} is the Cauchy stress tensor and n_j is the component of the unit normal vector to the surface of dA in the deformed configuration. Moreover, the same force should be developed by the stress tensor, T_{ij}, in the reference configuration with normal N_i and surface dA', and we can thus write

$$dq_i = T_{ij} N_j dA' \tag{4.34}$$

where the new stress tensor T_{ij} is called the first Piola-Kirchoff stress tensor. The mapping between reference and deformed configuration areas (defined in 4.19) thereby leads to

$$n_j dA \;=\; JN_k F_{jk}^{-1} dA' \qquad \sigma_{ij} n_j dA = \sigma_{ij} JN_k F_{jk}^{-1} dA' \tag{4.35}$$

Since the expressions for the total force are consistent, the two expressions can be subtracted and the result set equal to zero, viz.:

$$\left(T_{im} - \sigma_{ik} JF_{mk}^{-1}\right) N_m dA' = 0 \tag{4.36}$$

Since the term in parentheses must be zero, the relationship between the first Piola-Kirchoff and the Cauchy stress tensor is given by

$$T_{im} = \sigma_{ik} JF_{mk}^{-1} \tag{4.37}$$

and the Cauchy stress tensor is expressed in terms of the first Piola-Kirchoff stress tensor as

$$\sigma_{ij} = J^{-1} T_{im} F_{mj} \tag{4.38}$$

The first Piola-Kirchoff stress tensor thus gives the actual forces in the un-deformed surface area.

Two problems arise when the first Piola-Kirchoff stress tensor is used. Firstly, it is not energetically appropriate to use it together with the Green-Lagrange strain tensor, i.e., the scalar product of the first Piola-Kirchoff stress tensor and the Green-Lagrange strain tensor does not give the same strain energy density as the product of the Cauchy stress tensor and the small deformation strain tensor. Secondly, the first Piola-Kirchoff stress tensor is not symmetrical which makes it difficult to use with numerical analyses such as the finite elements method. As a result, it is necessary to look further for an appropriate stress tensor.

To derive a new stress tensor in the un-deformed configuration, we try the "push forward/pull back" operation as shown diagrammatically in Figure 4.14.

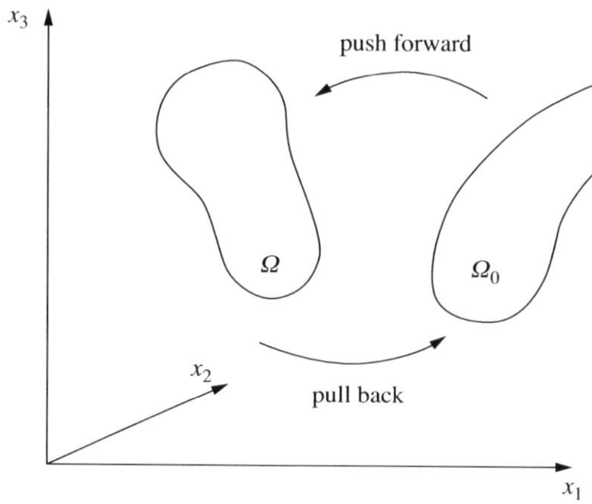

Fig. 4.14 Diagrammatically representation of the push forward/pull back operation (Marsden & Hughes, 1994).

The force $\mathrm{d}q_i$ can then be mapped back into the undeformed configuration using the inverse of the deformation gradient tensor, i.e.,

$$\mathrm{d}q'_m = F_{im}^{-1}\mathrm{d}q_i \tag{4.39}$$

We define another stress tensor, S_{mn}, that gives the total force in the undeformed configuration on the area in the un-deformed configuration, and this leads to

$$\mathrm{d}q'_m = S_{mn}N_n\mathrm{d}A' \tag{4.40}$$

The Cauchy stress tensor σ_{ij} that gives the total force $\mathrm{d}q'_m$ in the undeformed configuration can then be transformed

$$\mathrm{d}q'_m = F_{im}^{-1}\mathrm{d}q_i = F_{im}^{-1}\sigma_{ik}n_k\mathrm{d}A \tag{4.41}$$

As before, substitution for the normal vector in the deformed surface area gives

$$\mathrm{d}q'_m = F_{im}^{-1}\sigma_{ik}JF_{rk}^{-1}N_r\mathrm{d}A' \tag{4.42}$$

Insertion of equation (4.42) into equation (4.40) leads to

$$\left(S_{mn} - F_{im}^{-1}\sigma_{ik}JF_{nk}^{-1}\right)N_n\mathrm{d}A' = 0 \tag{4.43}$$

and the second Piola-Kirchoff stress tensor S_{mn} can be expressed as a function of the Cauchy stress tensor

$$S_{mn} = F_{im}^{-1}\sigma_{ik}JF_{nk}^{-1} \tag{4.44}$$

This can then be written

$$\sigma_{ik} = F_{mi} S_{mn} J^{-1} F_{kn} \tag{4.45}$$

This 2nd Piola-Kirchoff stress tensor S_{mn} is symmetrical and energetically consistent with the Green-Lagrange strain, and the product is the same as that calculated from the Cauchy stress tensor and the small deformation strain tensor:

$$\sigma_{ij}\varepsilon_{ij} = S_{mn} E_{mn} \tag{4.46}$$

The advantage of this relationship is that the expression on the right-hand side is computed in the reference configuration while that on the left-hand side is calculated in the deformed configuration.

It should be noted that the 2nd Piola-Kirchoff stress tensor has no physical meaning. It is merely a means of solving the large deformation problem, after which the Cauchy stress tensor is computed from the 2nd Piola-Kirchoff stress tensor. Curnier (2011) has defined the Cauchy stress σ, 1st PK stress, T and 2nd PK stress S, like:

$$\sigma = \frac{\text{act. force}}{\text{act. area}}, \; T = \frac{\text{act. force}}{\text{orig. area}}, \; S = \frac{\text{orig. force}}{\text{orig. area}}$$

where act. = actual = spatial = deformed configuration, orig. = original = material = undeformed configuration.

4.2.8 Constitutive formulations for nonlinear elastic materials under large deformations

This section presents the constitutive equations for nonlinear elastic materials under large deformations where it is assumed that the expression for the strain energy function exists. These constitutive equations are written in terms of the second Piola-Kirchoff stress tensor S_{mn} and the Green-Lagrange strain tensor E_{mn}.

Consider first a linear elastic material under a small deformation. The strain energy function is defined as

$$W = \tilde{W}(\varepsilon) = \frac{1}{2} D_{ijkl} \varepsilon_{ij} \varepsilon_{kl} \tag{4.47}$$

where D_{ijkl} is a 4th order stiffness tensor and its components are constant, $\tilde{W}$ is the elastic strain energy and ε_{ij} is the strain tensor.

Taking the first derivative of the strain energy function (4.47) with respect to the strain, we obtain Hooke's law:

$$\sigma_{ij} = \frac{\partial W}{\partial \varepsilon_{ij}} = D_{ijkl} \varepsilon_{kl} \tag{4.48}$$

where σ_{ij} is the stress tensor.

Taking the second derivative of (4.48) with respect to the strain, we find the material elastic coefficients.

$$D_{ijkl} = \frac{\partial^2 W}{\partial \varepsilon_{ij} \partial \varepsilon_{kl}} \tag{4.49}$$

The principles illustrated in (4.48) and (4.49) also hold for non-linear materials.

The first derivative of the strain energy function with respect to the Green-Lagrange strain tensor gives the second Piola-Kirchoff stress tensor with respect to the Green-Lagrange strain tensor:

$$W = \tilde{W}(E) \tag{4.50}$$

$$S_{mn} = \frac{\partial W}{\partial E_{mn}} \tag{4.51}$$

The derivative of (4.51) with respect to the strain gives the tangent stiffness tensor:

$$D^t_{mnrs} = \frac{\partial \tilde{W}(E)}{\partial E_{mn} \partial E_{rs}}, \; D^t_{mnrs} \neq \text{Constant (generally)}, \; r,\, s = 1,\, 2,\, 3 \tag{4.52}$$

Equation (4.51) can in terms of the right Cauchy deformation tensor be expressed as

$$S_{mn} = \frac{\partial W}{\partial C_{rs}} \frac{\partial C_{rs}}{\partial E_{mn}} \tag{4.53}$$

where the relationship between E_{mn} and C_{mn} is given by equation (4.31).

To illustrate a non-quadratic form of strain-energy function $\tilde{W}(E)$ for a nonlinear material under large strain, we first prensent a simple example of the form of quadratic energy of isotropic hyperelastic linear material. This material corresponds to classical elastic isotropic material of Saint Venant-Kirchhoff.

$$\tilde{W}(E) = \frac{1}{2}\lambda (tr\ E)^2 + \mu\ tr\ E^2 \tag{4.54}$$

where λ and μ are the constants of Lamé and E is the strain tensor. The stress tensor S corresponds to:

$$S_{mn} = \frac{\partial \tilde{W}(E)}{\partial E_{mn}} = \lambda\ tr\ E\ \delta_{mn} + 2\mu\ E \tag{4.55}$$

For an isotropic solid the strain energy function W must take a form which does not depend on the choice of direction of the coordinate axes. This means that W must be a function of three invariants $E, (E_I, E_{II}, E_{III})$ or the three principal invariants $\lambda_I, \lambda_{II},$

λ_{III}, which are three principal dilatation, (Ward 1985). Consider now an isotropic elastic nonlinear material under large deformation. Under the condition of incompressibility, the strain energy function W is given in terms of λ_I , $\lambda_{II,}$ λ_{III} (Treloar, 1975) by

$$W = \frac{1}{2}\, G(\lambda_I^2 + \lambda_{II}^2 + \lambda_{III}^2 - 3) \tag{4.56}$$

where λ_I , $\lambda_{II,}$ λ_{III} can be measured or calculated from a given deformation configuration (Ogden, 1972) and G is the shear modulus. Among different types of strain energy functions, proposed by Rivlin (1948), or Valanis and Landel (1978) for instance, there is a function consisting of three functions of three principal dilations λ_I , $\lambda_{II,}$ λ_{III}:

$$W = w(\lambda_I) + w(\lambda_{II}) + w(\lambda_{III}) \tag{4.57}$$

where $w(\lambda_i)$ has the same form for all $i = I, II, III$. The form usually used for $w(\lambda_i)$ is the one proposed by (Ogden, 1972):

$$W = \sum_{n=1}^{N} \frac{\mu_n}{\alpha_n} (\bar{\lambda}_I^{\alpha_n} + \bar{\lambda}_{II}^{\alpha_n} + \bar{\lambda}_{III}^{\alpha_n} - 3) \tag{4.58}$$

where $\bar{\lambda}_i^{\alpha_n} = J^{-\alpha_n/3} \lambda_i^{\alpha_n}$, $J = \lambda_I \lambda_{II} \lambda_{III}$ (volume dilation), μ_n and α_n are material constants which are determined experimentally in axial, bi-axial and shear tests. However, since this equation applies for an elastic non-compressible material, it is necessary to adapt this strain energy function when it is implemented for a compressible material.

For example, the form of the strain energy function for a solid elastomer that demonstrates a small compressibility is given in numerical code “MSC. Marc”:

$$W = \sum_{n=1}^{N} \frac{\mu_n}{\alpha_n} \left[J^{-\frac{\alpha_n}{3}} (\lambda_I^{\alpha_n} + \lambda_{II}^{\alpha_n} + \lambda_{III}^{\alpha_n} - 3) \right] + 4{,}5K(J^{\frac{1}{3}} - 1)^2 \tag{4.59}$$

where K is the compressibility, and N is the number of terms ($N = 2$ or 3). The model based on the formulation of Ogden (1972) used in the numerical code takes the following form:

$$W = \sum_{n=1}^{N} \frac{\mu_n}{\alpha_n} \left[J^{-\frac{\alpha_n}{3}} (\lambda_I^{\alpha_n} + \lambda_{II}^{\alpha_n} + \lambda_{III}^{\alpha_n} - 3) \right] + \sum \frac{\mu_n}{\alpha_n} (1 - J^{\alpha_n}) \tag{4.60}$$

where μ_n, α_n, β_n are material constants. When the strain function is evaluated, the second Piola-Kirchoff stress can be calculated from the relation given in (4.53). In the absence of any strain energy function developed for wood under large deformation, the Ogden type of function may be used. The application of this function for wood under compression has been employed and verified for a 2-D problem by Rangsri (2004) and Rangsri *et al.* (2003, 2004).

4.3 INTRODUCTION TO THE VISCOELASTIC BEHAVIOR OF WOOD

Like all polymeric materials, wood exhibits a viscoelastic behavior (Ferry, 1980; Morlier, 1994). The time-dependent behavior of wood depends on the loading history, the temperature, and the moisture content as well as on moisture variations. These parameters may interact together and produce coupling effects. Under conditions of moderate loading, constant temperature and constant moisture content, the time-dependent characteristics of wood can be measured by static (creep, stress relaxation) and dynamic tests (Salmén & Hagen, 2001). For the characterization of the linear viscoelastic behavior of wood, creep and stress relaxation tests are generally preferred to dynamic measurements. In static tests, the creep and relaxation functions are obtained explicitly on a time scale, whereas dynamic tests give a complex function in terms of frequency. Conversion of dynamic complex functions to static creep or relaxation functions is not always possible. Nevertheless, within the limits of linear viscoelasticity, the results obtained from these three methods are mathematically related.

The viscoelasticity of wood derived by any of these three methods is strongly dependent on the moisture content, on the temperature and on their variations. In addition, the viscoelastic behavior of wood is anisotropic. In order to characterize and model it, we shall present creep, relaxation and cyclic testing under both constant and variable climatic conditions.

4.3.1 The creep and recovery test

The linear viscoelastic behavior of a material can be characterized by a creep test, shown diagrammatically in Figure 4.15.

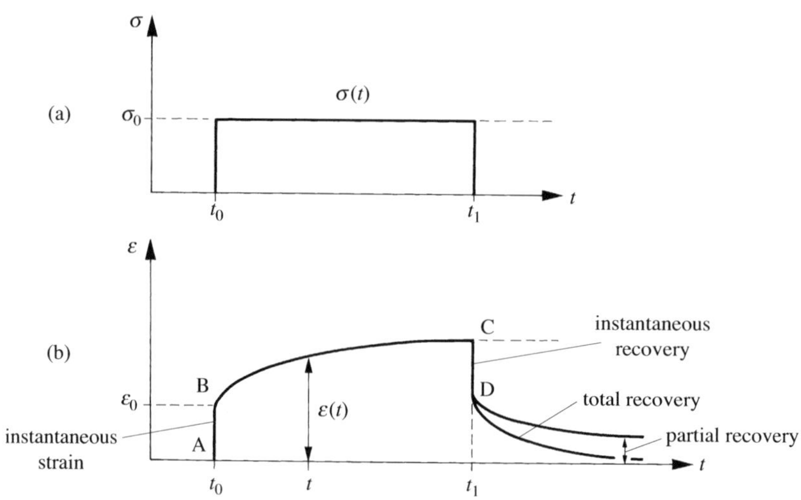

Fig. 4.15 An ideal representation of a creep-recovery test: (a) stress loading, (b) response of a viscoelastic material.

In a creep-recovery test, the shape of the loading program in stress consists of three parts:

- The application of an instantaneous stress σ_o at time $t = t_o$ produces an instantaneous strain ε_o.
- The maintenance of a constant stress σ_o during $t > t_o$, leads to a strain $\varepsilon(t)$ that is a function of time, where $\varepsilon(t) - \varepsilon_o > 0$.

Withdrawal of the stress σ_o at time $t = t_1$ leads to an instantaneous recovery. In a linear and non-ageing material where the loading gives rise to no damage in the material, the instantaneous creep recovery may be equal to the instantaneous strain ε_o, whereas the continued creep recovery is a time dependent strain, $\varepsilon_r(t - t_1)$. The creep recovery can be total or partial, depending on the type of the material and on the level of the applied stress.

Creep function – unidimensional, constant climate

In a material exhibiting viscoelastic behavior under constant stress σ_o, the strain $\varepsilon(t)$ is a function of time and of σ_o:

$$\varepsilon(t) = \tilde{\varepsilon}(t, \sigma_0) \qquad (4.61)$$

When the viscelasticity is linear, the function $\tilde{\varepsilon}$ in (4.61) is proportional to σ_0:

$$\tilde{\varepsilon}(t, \sigma_0) = S(t)\sigma_0 \qquad (4.62)$$

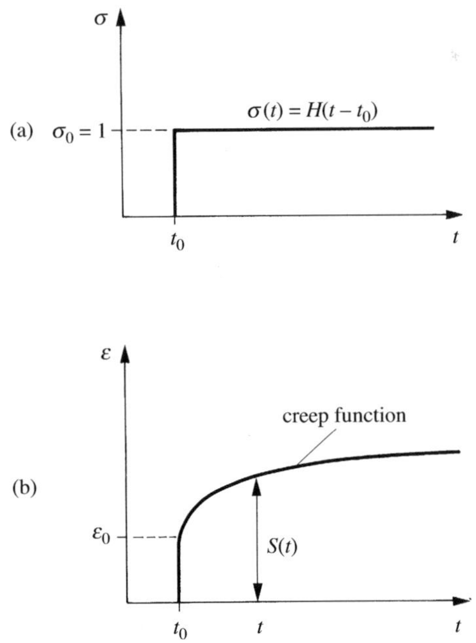

Fig. 4.16 Schematic presentation of a creep function; (a) application of a stress step function $\sigma(t) = H(t - t_0)$, $H(t)$ is the Heaviside function, where $H(t) = 0$ for $t < 0$ and $H(t) = 1$ for $t \geq 0$, (b) the deformation response (creep function).

where $S(t)$ is the creep function or creep compliance defined as

$$S(t) = \frac{\varepsilon(t)}{\sigma_0} \tag{4.63}$$

$S(t)$ represents as the growth of the strain in time t divided by the applied constant stress σ_o. The typical shape of a creep function is given in Figure 4.16.

For a non-ageing viscoelastic material like wood, the creep function is defined by:

$$S(t - t_0) = \frac{\varepsilon(t - t_0)}{\sigma_0} \tag{4.64}$$

where t_0 is the time of application of the stress σ_0.

Boltzmann's superposition principle

In a general case, the response of a viscoelastic material at time t depends on the history of the mechanical loading. In a particular case when the viscoelasticity is linear, based on Boltzmann's superposition principle, the response in strain to the stresses $\Delta\sigma_0, \Delta\sigma_1, \ldots \Delta\sigma_n$ applied respectively at times $t_0, t_1, \ldots t_n$ is expressed by the expression

$$\varepsilon(t) = \Delta\sigma_0 S(t - t_0) + \Delta\sigma_1 S(t - t_1) + \ldots\ldots \Delta\sigma_n S(t - t_n), \quad t > t_n \tag{4.65}$$

or more generally

$$\varepsilon(t) = \sum_{i=0}^{i=n} \Delta\sigma_i S(t - t_i) \tag{4.66}$$

where $\Delta\sigma_0 = \sigma_0$.

Expression (4.66) can be generalized in an integral form as

$$\varepsilon(t) = \int_{t_o}^{t} S(t - u) \frac{d\sigma(u)}{du} du \tag{4.67}$$

where u is a time variable ranging from t_o to t, and the derivative includes the finite steps, and

$$\varepsilon_0 = S_0 \sigma_0 \tag{4.68}$$

Equation (4.67) represents the Boltzmann superposition integral of the unidimensional iso-hygro-thermal stress-strain equation for a linear viscoelastic material, and S_o is the instantaneous elastic compliance of the material. Equation (4.67) expresses the strain ε at time t as a function of the history of the stress $\sigma(u)$. To calculate $\varepsilon(t)$, it is necessary to know the analytical expression of the creep function $S(t)$ as well as the value of the instantaneous elastic compliance S_0. Partial integration of equation (4.67) leads to

$$\varepsilon(t) = \sigma(t)\, S(0) + \int_{t_o}^{t} \sigma(u) \frac{\mathrm{d}S(t-u)}{\mathrm{d}u}\, \mathrm{d}u \tag{4.69}$$

For a stress history where $\sigma(t) = \sigma_o H(t)$, $H(t)$ is the Heaviside step function the value of $\varepsilon(t)$ obtained from equation (4.67), corresponds to $S(t)\sigma_o$, which is given in (4.63) for a linear viscoelastic material.

4.3.2 The stress relaxation-recovery test

Another test, which makes it possible to characterize the viscoelastic behavior of a material, is the stress relaxation-recovery test. It is similar to the creep-recovery test and is illustrated in Figure 4.17.

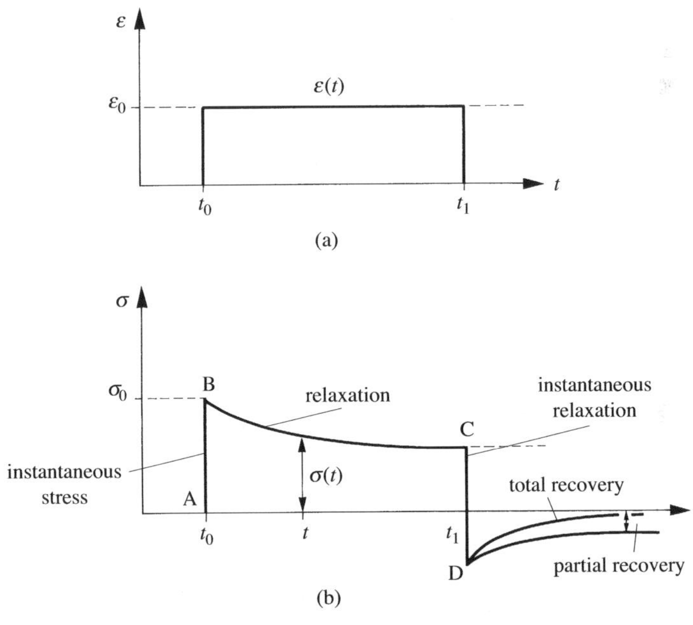

Fig. 4.17 An ideal representation of a stress relaxation-recovery test (a) typical strain loading, (b) response of a viscoelastic material to stress.

This test, which studies the response in stress of a specimen subjected to a constant strain loading ε_o between times t_0 and t_1 includes three parts:

- The application of an instantaneous strain ε_o at t_0 produces an instantaneous stress σ_o.
- The maintenance of a constant strain ε_o between times t_0 and t_1 leads to a time-dependent stress relaxation $\sigma_r(t)$.
- A rapid withdrawal of the deformation ε_o at time $t = t_1$ results in an instantaneous recovery followed by a stress recovery $\sigma_r(t - t_1)$ as a function of time. This stress recovery can be partial or total. When the recovery is total, the behavior is

purely viscoelastic, whereas if the recovery is only partial, leading to a permanent stress σ_p after a long period of time, the behavior is considered to be elasto-viscoplastic. In practice, it is easier to maintain a constant stress than a constant strain. Therefore, the creep test is usually preferred to the stress relaxation test.

Relaxation function – unidimensional, constant climate

Like the creep function given by equation (4.64), the relaxation function for a non-ageing and linear material is

$$R(t-t_o)=\frac{\sigma(t-t_o)}{\varepsilon_o} \tag{4.70}$$

where R is the relaxation function or modulus, ε_o is an applied constant strain applied at $t = t_0$ and $\sigma(t)$ is the response in stress. A typical relaxation function is shown in Figure 4.18.

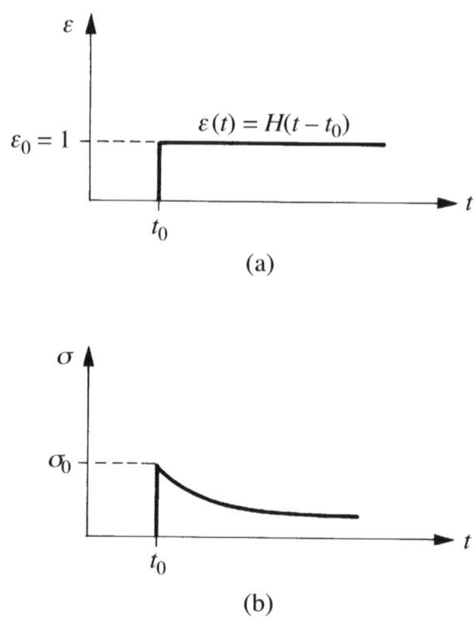

Fig. 4.18 A typical relaxation function: (a) applied strain in the form of a step function, where $\varepsilon(t)$ is defined by the Heaviside step function $H(t-t_0)$, (b) response in stress or relaxation function.

The relaxation function, $R(t)$ is thus defined as the stress at time t divided by the applied strain ε_o. In a linear and non-ageing material, the response in stress to a strain varying with time is expressed by the Boltzmann superposition principle:

$$\sigma(t)=\int_{t_o}^{t} R(t-u)\frac{\mathrm{d}\varepsilon(u)}{\mathrm{d}u}\,\mathrm{d}u \tag{4.71}$$

where the derivative includes the finite steps, and

$$\sigma_o = R_o\varepsilon_o \tag{4.72}$$

R_o is instantaneous elastic modulus of the material. It is known per definition that there is an inverse relationship between the instantaneous elastic modulus and the instantaneous elastic compliance:

$$R_o = 1/S_o \tag{4.73}$$

In a linear viscoelastic material, there is also an inverse relationship between $S(t)$ and $R(t)$.

Based on the definition of a relaxation function, it follows that when $\varepsilon(t)$ is a step unit function $H(t - t_o)$, the response in stress will be a relaxation function, i.e.,

$$\varepsilon(t) = H(t - t_o) \qquad \Rightarrow \qquad \sigma(t) = R(t - t_o) \tag{4.74}$$

If then $\sigma(t)$ in (4.67) is replaced by $R(t - t_o)$ as an applied stress, $\varepsilon(t)$ should be the response as $H(t - t_o)$, and the integral will be

$$H(t - t_0) = \int_{t_o}^{t} S(t-u) \frac{\mathrm{d}R(u - t_0)}{\mathrm{d}u} \mathrm{d}u \tag{4.75}$$

where t_0 is a constant and u is the time variable ranging from t_o to t. Equation (4.75) gives an integral relationship between $S(t)$ and $R(t)$ in the unidimensional case.

4.3.3 Dynamic behavior of viscoelastic materials

The mechanical dynamic behavior of a viscoelastic material can be measured by applying an oscillatory force or deformation. It is experimentally verified that, for materials with linear viscoelastic behaviors, for which the applied stress is sinusoidal, the response in strain is also sinusoidal where the strain lags the stress. Mathematically speaking, when the applied stress is

$$\sigma(t) = \sigma_0 \sin \omega t \tag{4.76}$$

the response in strain is

$$\varepsilon(t) = \varepsilon_0 \sin(\omega t - \delta) \tag{4.77}$$

Here, ω is the angular frequency, δ is the phase angle between stress and strain, and σ_o and ε_o are respectively the maximum stress and strain. For mathematical convenience, the strain and stress are usually represented by the real part of a complex value as follows:

$$\varepsilon = \mathrm{Re}\left[\varepsilon^c\right] \text{ and } \sigma = \mathrm{Re}\left[\sigma^c\right],$$

where

$$\begin{aligned} \sigma^c &= \sigma_0(\sin(\omega t) + i\cos\omega t) \\ \varepsilon^c &= \varepsilon_0(\sin(\omega t + \delta) + i\cos(\omega t + \delta)) \end{aligned} \tag{4.78}$$

where $i = \sqrt{-1}$. The complex modulus E^c of the viscoelastic material is by definition

$$E^c = \frac{\sigma^c}{\varepsilon^c} = E' + iE'' = \frac{\sigma_0}{\varepsilon_0}(\cos\delta + i\sin\delta) \tag{4.79}$$

where E' is the storage modulus, E'' the loss modulus and tan δ the loss factor. The most important quantities determined by the dynamic mechanical analysis are then the values of E', E'', and tan δ as a function of ω or temperature T. The loss factor tan δ is the tangent of the phase angle between the applied stress and strain response (Figure 4.19) or the ratio of the loss modulus to the storage modulus:

$$\tan\delta = \frac{E''}{E'} \tag{4.80}$$

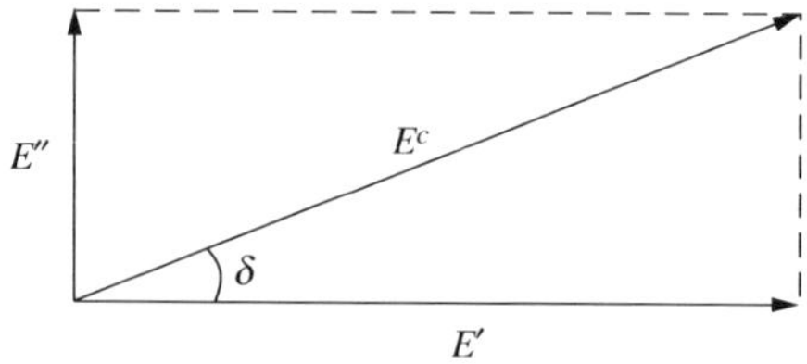

Fig. 4.19 The complex modulus E^c and loss factor tan δ.

In a viscoelatic material, E'' is usually much smaller than E', and E is thus approximately equal to E'. The variation in the components of the complex modulus E^c and the loss factor tan δ for a typical viscoelastic solid has been measured as a function of ω and demonstrated by several authors.

Figure 4.20 shows E', E'' and tan δ as a function of frequency ω for a polymer. At low frequency, the polymer behaves as a rubber with a low modulus, at high frequency the polymer is a solid glassy with a high modulus and at intermediate frequencies it behaves as a viscoelastic solid. The loss factor tan δ, is a measure of the relative dissipated energy and is associated with the molecular relaxation processes.

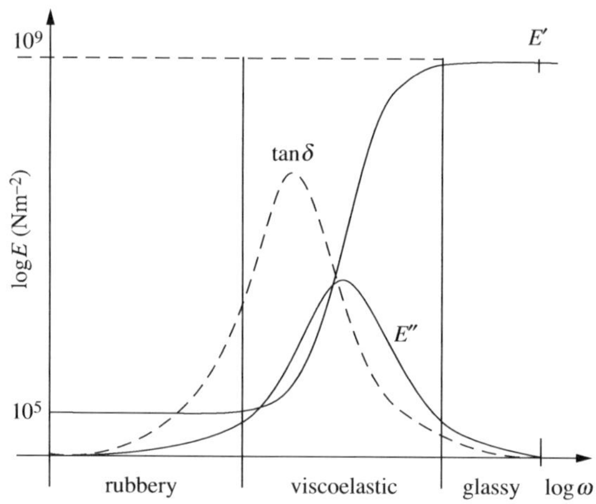

Fig. 4.20 Variation of E', E'' and tan δ with frequency ω (Ward, 1985).

The complex compliance S^c of a viscoelastic material is defined in a manner analogous to the relation given in equation (4.79):

$$S^c = \frac{\varepsilon^c}{\sigma^c} = \frac{1}{E^c} = S' - iS'' \tag{4.81}$$

From the above relation, the values of S' and S'' are obtained in terms of E' and E'':

$$S' = \frac{E'}{E'^2 - E''^2} \quad \text{and} \quad S'' = \frac{E''}{E'^2 - E''^2} \tag{4.82}$$

It is for instance shown by (Ferry, 1980) that the components of the complex dynamic modulus can be determined from the relaxation function $R(t)$ of the material by using the Fourier transforms:

$$E'(\omega) = R(0) + \omega \int_0^\infty \left[R(t) - R(0)\right] \sin \omega t \ \mathrm{d}t \tag{4.83}$$

$$E''(\omega) = \omega \int_0^\infty \left[R(t) - R(0)\right] \cos \omega t \ \mathrm{d}t \tag{4.84}$$

where $R(0)$ is the value of the relaxation function at $t = 0$. The value of $R(t)$ can be obtained in terms of E' or E'' as

$$R(t) = R(0) + \frac{2}{\pi} \int_0^\infty \left[\frac{E' - R(0)}{\omega}\right] \sin \omega t \ \mathrm{d}\omega \tag{4.85}$$

and

$$R(t) = \frac{2}{\pi} \int_0^\infty \frac{E''}{\omega} \cos \omega t \ \mathrm{d}\omega \tag{4.86}$$

In practice, these integrals can be performed numerically if E' and E'' are known over a wide range of frequencies.

4.3.4 Influence of moderate temperature and humidity on the viscoelastic behavior of wood: experimental data

The viscoelasticity of wood and its modeling under climatic conditions have been the subject of numerous experimental and theoretical studies. A large amount of information is available on the behavior of wood under constant and variable ambient air. Schniewind (1968) has presented a review of earlier work. A vast number of experimental studies on the time-dependent behavior of wood, as well as materials derived from wood, are also described by Dinwoodie *et al.* (1990), showing that the creep of wood depends on the level of loading and on the climatic conditions. This dependence varies from one species of wood to another. A summary of information on the creep of structural timber and the coupling effect between variations in the moisture content of wood and the mechanical loading (the mechanosorptive effect) is given by Huet *et al.* (1988) and Morlier (1994). This section presents the influence of the stress level, moisture content and temperature, as well as their variations, on the behavior of wood. The following experimental work and investigations were carried out mainly on Norway spruce (*Picea abies* (L.) Karst.) by Hoffmeyer (1990) for three creep modes in compression, tension and bending mode, and the study involved the short and long-term mechanical behavior of 2050 wood specimens. The dimensions of the specimens were $49 \times 95 \times 1800$ mm^3. The configuration of the bending test is given in Figure 4.21a.

(a)

(b)

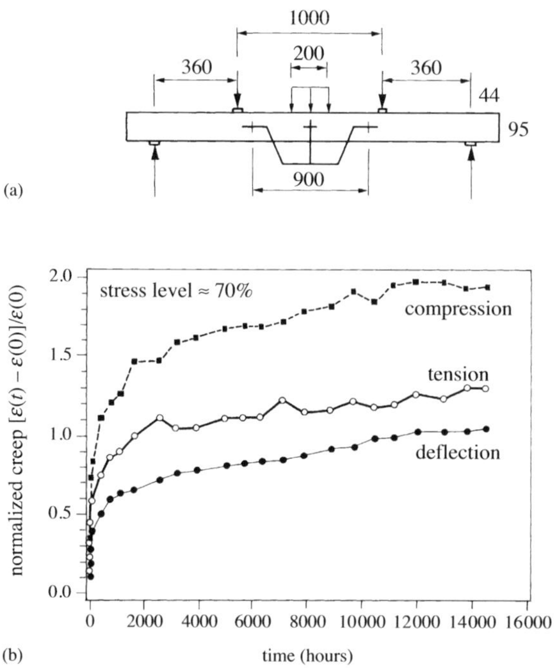

Fig. 4.21 Typical normalized creep for three modes. (a) Configuration of the bending test, (b) testing in compression, tension and bending. The moisture content of the test pieces during creep testing was kept constant at about 20% (Hoffmeyer, 1990).

Long-term creep tests were carried out at wood moisture contents of 20 and 11% and the results are given in respectively Figures 4.21b, and 4.22. The variation of the normalized creep was defined as $C(t) = (\varepsilon(t) - \varepsilon(0)) / \varepsilon(0)$, for three creep modes. In this relation $\varepsilon(0)$ is the instantaneous strain.

The results in these figures show that:

a) Creep in bending is smaller than creep in both compression and tension.

b) Creep is much greater in a moist as opposed to in a dry beam.

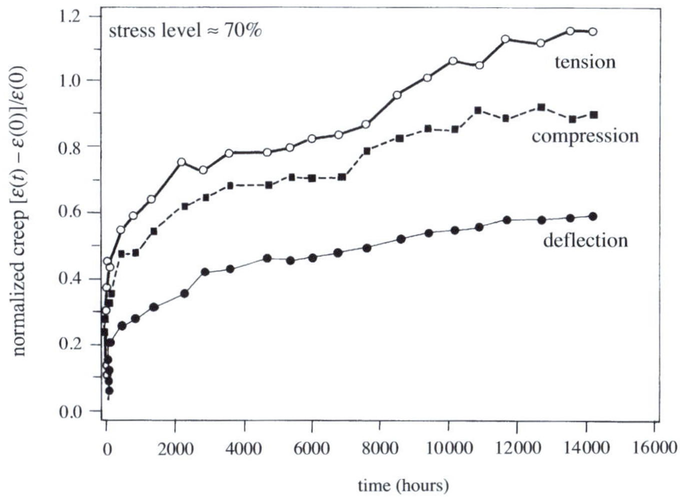

Fig. 4.22 Typical example of creep test in compression, tension and bending, at a moisture content of about 11% (Hoffmeyer, 1990).

Figure 4.23 presents creep results for four beams under varying relative humidity. In this test, the ambient relative humidity was varied between 90 and 54%, at a constant temperature of 20 °C. Each cycle lasted for 8 weeks and the total duration of the experiment was 92 weeks. The four diagrams show the mechanosorptive creep behavior during 1, 2, 3 and 7 drying-humidification cycles. The variation in moisture content of the test pieces is given in Figure 4.24.

It is evident from Figure 4.23 that the effect of mechanosorption (viscoelasticity under varying moisture content) is significant in all three creep modes, compression, tension and bending. It is important to note that the creep of the beams during drying- humidification cycles (between 90 and 54%) is much greater than the creep at a constant relative humidity of 90%.

To be able to quantify and predict the viscoelastic behavior of composite materials based on wood, Dinwoodie *et al.* (1991) have carried out series of tests on various types of particle-boards under constant and variable climate conditions. In their study, the creep in beams under 4-point bending was measured when the ratio of span to beam height was 16:1. The tests included various types of panels at two stress levels, viz. 30 and 75% of the short-term breaking stress, at relative humidies of 30 and 90% and at temperatures of 10 and 30 °C. One set of results at temperature 20 °C is presented in Figure 4.25.

One may observe that;

a) Creep in bending under a constant climate increases with increasing temperature, increasing relative humidity and an increasing level of applied stress.
b) The influence of these parameters depends strongly on the wood species.
c) Creep in sawn timber and panels is highly sensitive to variations in relative humidity between 65 and 90%, but the degree of sensitivity is dependent on the material. The panels show a linear increase in normalized creep with an increasing applied stress up to 75% of the short-term breaking stress, whereas

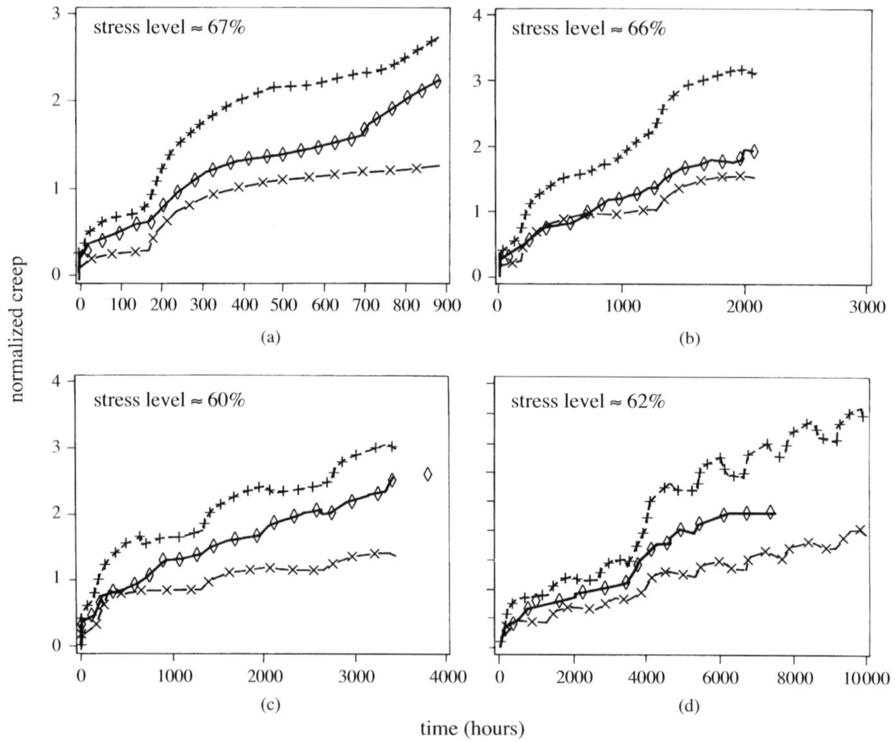

Fig. 4.23 Normalized creep of various beams, during relative humidity cycling between 90 and 54%. The period of each cycle is 8 weeks. The level of loading corresponds to approximately 65% of the instantaneous rupture loading and the symbols are –+——+—+– for compression, –◊–◊–◊ for tension and -x-x-x for bending; (a) one drying-humidification cycle, (b) two drying–humidification cycles (c) three drying-humidification cycles, (d) seven drying-humidification cycles (Hoffmeyer, 1990).

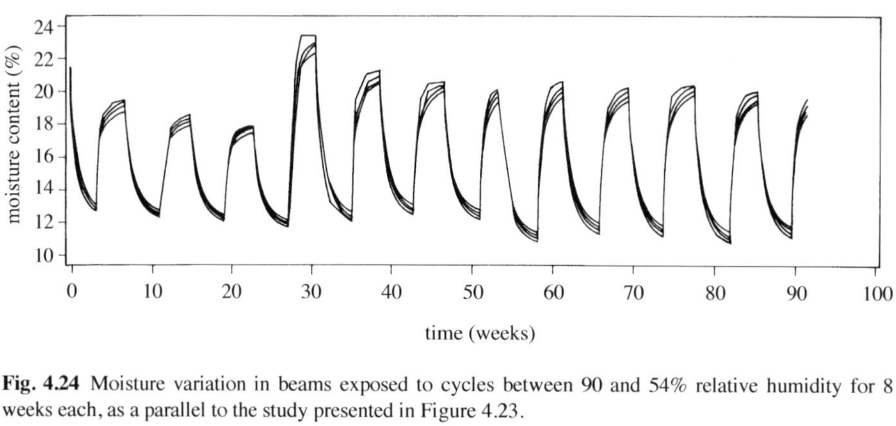

Fig. 4.24 Moisture variation in beams exposed to cycles between 90 and 54% relative humidity for 8 weeks each, as a parallel to the study presented in Figure 4.23.

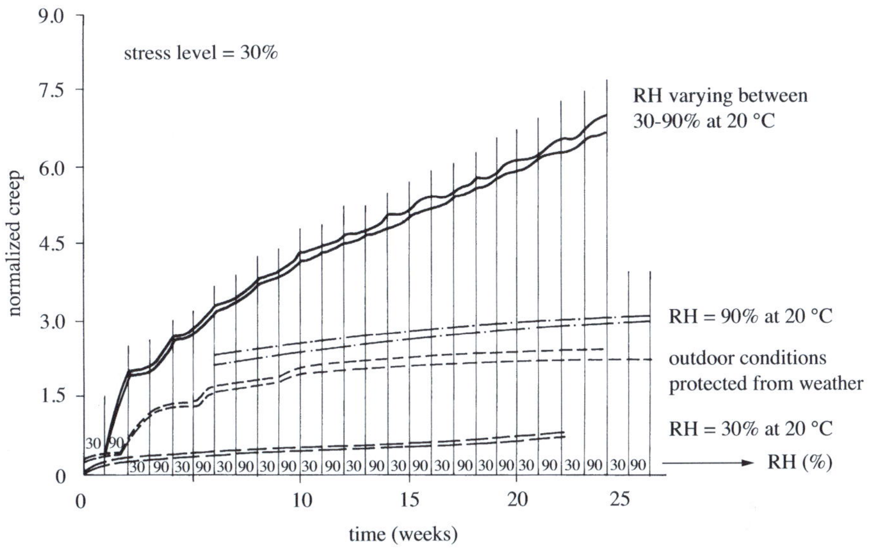

Fig. 4.25 The normalized creep of beams of melamine-urea-formaldehyde resin (MUF) particle boards, under constant and cyclic relative humidity (Dinwoodie *et al.*, 1991).

the behavior of sawn timber is non-linear above a stress of approximately 45 to 50%. However, the strong dependence of these results on the wood species signifies that it is impossible to perform generalizations. It is nevertheless clear that cycling the relative humidity considerably increases the creep of wood and of materials derived from wood. The creep is much smaller than the creep when varying the moisture content of wood as well as wood derivatives, which can occur at a constant high relative humidity.

4.3.5 Influence of temperature on the creep of wood

The influence of the temperature on the creep of wood is complex. Davidson (1962), Kingston and Budgen (1972), and Bach and McNatt (1990) showed that an increase in temperature reduces the modulus of elasticity of wood. Figure 4.26 presents results obtained by Kingston & Clarke (1961) on pine at temperatures of 21.5 and 41.5 °C, under an applied stress of 55 to 70% of the breaking stress. The figure shows that the normalized creep increases with increasing temperature. Other experimental results have shown that the effect of temperature on creep is more complex, and it is speculated that this may be due to two interactions: one between the stress and the temperature, the other between the stress and the history of the applied temperature.

In general, the creep of a wooden beam under a constant temperature increases with increasing temperature, especially when this is higher than 55 °C, Davidson (1962), Arima (1967) and Huet & Zaoui (1981). The results obtained by Huet & Zaoui are given in Figure 4.27.

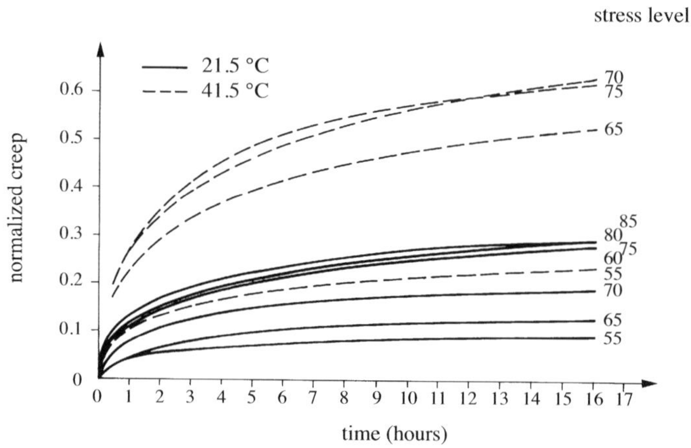

Fig. 4.26 Normalized creep of pine at two temperatures under various stress levels (Kingston & Clarke, 1961).

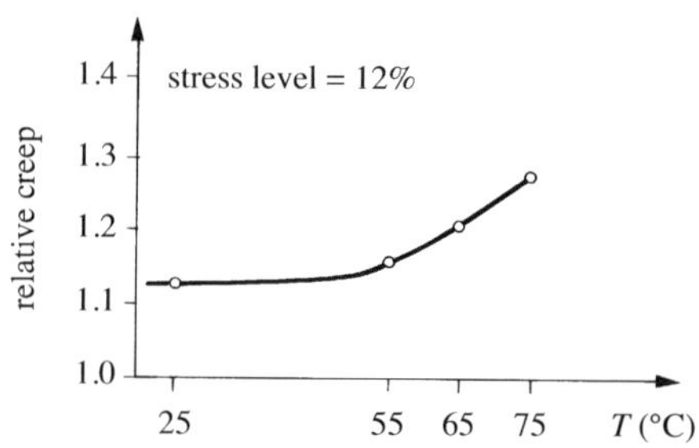

Fig. 4.27 The relative creep of spruce in bending during one week ($\varepsilon(t)/\varepsilon(t_0)$), t = one week) at various temperatures and at a moisture content of 12% (Huet & Zaoui, 1981).

In general, variations in temperature lead to a more complex deformation, which is difficult to predict from creep tests under constant temperature, Jouve and Sales (1986). Figure 4.28 shows the creep of beams of a tropical species (Mélénthiéra) tested under three-point bending at a moisture content higher than the fiber saturation point.

In this species the creep was much greater when the temperature was varied between 20 and 90 °C as opposed to when the temperature was kept constant at 90 °C. These results also show that the time-dependent behavior of wood under variable moisture content and variable temperature is complex. To express the constitutive equation of wood, it is usually suggested that its behavior be studied under constant and variable climatic conditions separately.

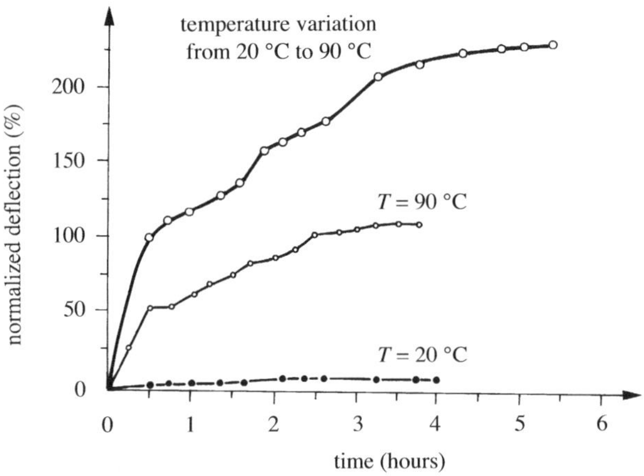

Fig. 4.28 Normalized creep of Mélénthiéra (tropical species), at a moisture content higher than the FSP at temperatures of 20 and 90 °C, and during temperature variations between 20 and 90 °C (Jouve & Sales, 1986).

4.4 CONSTITUTIVE EQUATIONS OF A VISCOELASTIC ORTHOTROPIC MATERIAL

It should be noted that wood is often considered to be an orthotropic material and that it is often used in three-dimensional applications. The constitutive equations (equations relating the stresses and the strain at time must therefore be expressed in three-dimensional space. In order to present these equations, the local principal axes defined in Figure 4.1 for linear elasticity will be used for the time-dependent behavior of wood.

The extension of the linear uni-dimensional viscoelastic behavior given in (4.67) to a three-dimensional case in a tensorial system leads to

$$\varepsilon_{ij}(t)=\int_{t_o}^{t} S_{ijkl}(t-\tau)\frac{\mathrm{d}\sigma_{kl}(\tau)}{\mathrm{d}\tau}\mathrm{d}\tau \quad \text{with } i,j,k,l\in(1,\ 2,\ 3) \tag{4.87}$$

where $\varepsilon_{ij}(t)$, $\sigma_{ij}(t)$ and $S_{ijkl}(t)$ are respectively the components of the strain tensors, stress tensors and creep compliance tensors. For an orthotropic material, the creep compliance tensor $S_{ijkl}(t)$ has 9 independent components. If the axes of the coordinates coincide with the axes of material symmetry, and the double subscript is transformed into a single subscript, equation (4.87) can be written in the form

$$\varepsilon_{\alpha}(t)=\int_{t_o}^{t} S_{\alpha\beta}(t-\tau)\frac{\mathrm{d}\sigma_{\beta}(\tau)}{\mathrm{d}\tau}\mathrm{d}\tau \quad \text{with } \alpha,\ \beta\ \in\ (1,\ 2,\ 3,\ 4,\ 5,\ 6) \tag{4.88}$$

The constitutive equation for an orthotropic material given in (4.88) takes the following form in a matrix notation:

$$\begin{bmatrix} \varepsilon_1(t) \\ \varepsilon_2(t) \\ \varepsilon_3(t) \\ \gamma_4(t) \\ \gamma_5(t) \\ \gamma_6(t) \end{bmatrix} = \int_{t_o}^{t} \begin{bmatrix} S_{11}(t-\tau) & S_{12}(t-\tau) & S_{13}(t-\tau) & 0 & 0 & 0 \\ S_{21}(t-\tau) & S_{22}(t-\tau) & S_{23}(t-\tau) & 0 & 0 & 0 \\ S_{31}(t-\tau) & S_{32}(t-\tau) & S_{33}(t-\tau) & 0 & 0 & 0 \\ 0 & 0 & 0 & S_{44}(t-\tau) & 0 & 0 \\ 0 & 0 & 0 & 0 & S_{55}(t-\tau) & 0 \\ 0 & 0 & 0 & 0 & 0 & S_{66}(t-\tau) \end{bmatrix} \frac{\partial}{\partial \tau} \begin{bmatrix} \sigma_1(\tau) \\ \sigma_2(\tau) \\ \sigma_3(\tau) \\ \sigma_4(\tau) \\ \sigma_5(\tau) \\ \sigma_6(\tau) \end{bmatrix} d\tau \tag{4.89}$$

where $\gamma_4 = 2\varepsilon_{23}$, $\gamma_5 = 2\varepsilon_{13}$, $\gamma_6 = 2\varepsilon_{12}$.

Similar equations can be written relating stresses to strains by relaxation functions. It can be shown that the matrix for the creep compliance given in (4.89) is symmetrical, i.e., $S_{12} = S_{21}, S_{13} = S_{31}, S_{23} = S_{32}$. Consequently, for a full characterization of wood as an orthotropic linear material, nine independent creep or relaxation functions must be determined. To find these functions for an orthotropic material exhibiting a linear viscoelastic behavior, extensive experimental work is needed (Schniewind & Barrett, 1972; Cariou, 1987; Hayashi *et al.*, 1993).

4.4.1 Linearity limit of wood viscoelasticity and the creep function under constant climatic conditions

Experimental results have shown that when wood under moderate and constant climatic conditions is subjected to stresses below a certain limit, it behaves like a linear viscoelastic material. In this case, the response of wood to a stress $\sigma(t)$ is determined by the Boltzmann superposition principle defined in expression (4.67).

This stress limit has been studied by various authors under constant climatic conditions and has been expressed as a percentage of the short-term breaking stress. Table 4.2 illustrates the levels of stress below which various wood species present a linear viscoelastic behavior.

These results indicate that the limit of linearity in bending in the longitudinal direction decreases with an increasing moisture content and is quite low, about 30 to 40% of the breaking stress. In tension, however, it is fairly high (about 50%). In compression, it is difficult to determine the stress limit. Under a moderate constant ambient climate and moderate loading, wood viscoelasticity may therefore be considered to be linear (Whale, 1988).

Table 4.2 Stress level, defined as a percentage of the short-term breaking stress, below which wood illustrates linear viscoelastic behavior in a moderate constant climate.

Direction	Authors	Species	Testing	Creep time	% breaking stress level	Climatic conditions
L	Le Govic, (1988)	Spruce	Bending	1 week	30%	18%, 25 °C
9 components	Cariou, (1987)	Maritime pine	Tension and shear	24 hours	Shear 30% Tension 25%	20%, 25 °C
L	Hoyle *et al.*, (1986)	Douglas fir	Bending	400 hours	40%	12%, 21 °C
L	Foudjet, (1986)	Tropical, Hardwood	Bending	2 weeks	35%	18%
L	Miller & George, (1974)	Eastern spruce	Bending	30 days	40%	12%
L	Bhatnagar, (1964)	Teak	Tension	5 hours	50%	12%, 30 °C
L	Hayashi *et al.*, (1993)	Spruce	Tension	4 days	60%	12, 20 °C
L	Mukudai, (1983)	Hinoki	Bending	10 hours	35-45%	11%, 25 °C
L	King, (1961)	Basswood	Tension	30 min.	30-40%	12%, 25 °C

4.4.2 Creep functions

Considering the viscoelasticity of wood, with the same axes of symmetry as for the elastic case, the number of independent components (creep functions) is reduced to 9, where the suffix 1 indicates the radial direction, the suffix 2 the tangential direction and the suffix 3 the longitudinal direction. Moreover, $S_{44}(t), S_{55}(t)$ and $S_{66}(t)$ are creep functions in the *rl*, *tl* and *rt* planes, respectively. All these components are functions of time, temperature and moisture content and need to be determined for each wood species.

There are few experimental results that give all 9 creep functions. Cariou (1987) has studied the evolution of 9 creep compliances $S_{ij}(t)$ in Maritime pine in terms of moisture content at a constant temperature of 25 °C. Based on this experimental work, Cariou has proposed a mathematical expression for all 9 components of the creep function.

In the literature, a large number of mathematical expressions have been proposed to predict uni-dimensional creep or stress relaxation functions. For a linear viscoelastic behavior, these models can be divided into two principal groups. In the first "empirical method", after selecting a mathematical expression, the unknown coefficients of creep function are determined through experimental creep curves using curve-fitting techniques. Examples of such creep functions are

$$S(t) = S_o + A_1 \log(t+1) \tag{4.90}$$

$$S(t) = S_o + A_1 \log(t+1) + A_2 \log^2(t+1) \tag{4.91}$$

$$S(t) = S_o + mt^n \tag{4.92}$$

where, A_1, A_2, S_o, m and n depend on the species, on creep direction, and on the temperature and moisture content of the wood. These parameters are calculated from experimental data.

One of the most commonly used expressions for the creep of wood is the power law or parabolic model proposed by Nielsen (1984), which for one and 3-dimensional cases leads to

$$S(t)=S_0\left(1+\left(\frac{t}{\tau}\right)^k\right) \tag{4.93}$$

$$S_{\alpha\beta}(t)=S_{0,\alpha\beta}\left(1+\left(\frac{t}{\tau}\right)^k\right);\ \alpha = 1, 2, \ldots 6 \text{ and } \beta=1, 2, \ldots 6 \tag{4.94}$$

Equation (4.93) defines a creep function with three unknown coefficients (S_0, k, τ) for a unidimensional medium, and equation (4.94) presents 36 functions containing only 9 independent creep functions for a three-dimensional orthotropic material. Each function has three unknowns that must be determined through experimental creep tests. In these equations, τ appears as a constant called the doubling time $(S(\tau)=2S_0)$, S_0, $S_{0,\alpha\beta}$ are the elastic compliances, and k is a creep kinetics factor varying between $0<k<1$.

Table 4.3 Parameters of the "power law" model reported by various authors (Le Govic, 1991).

Author	Temp. MC	Testing	Compliance	k	τ minutes	S_o, MPa^{-1}
Cariou, 1987[1]	25 °C, 12.4%	Tension *rt*	$S_{11}(t)$	0.28	1 E5	0.939 E–3
Cariou, 1987[1]	25 °C, 12.7%	Tension *tl*	$S_{22}(t)$	0.16	2.6 E6	1.362 E–3
Cariou, 1987[1]	25 °C, 13.7%	Tension *tr*	$S_{23}(t)$	0.22	0.13 E6	1.427 E–3
Cariou, 1987[1]	25 °C, 12.5%	Tension *lt, r*	$S_{33}(t)$	0.16	(4.8-6.9) E10	(9.69-8.07) E–5
Cariou, 1987[1]	25 °C, 13.3%	Shear	$S_{44}(t)$	0.15	5.6 E8	0.5778 E–3
Cariou, 1987[1]	25 °C, 12.7%	Shear	$S_{55}(t)$	0.25	3.5 E6	0.7381 E–3
Cariou, 1987[1]	25 °C, 11.8%	Shear	$S_{66}(t)$	0.2	8.9 E5	4.734 E–3
Cariou, 1987[1]	25 °C, 14.0%	Tension	$S_{31}(t)$	0.17	1.0 E12	2.640 E–5
Cariou, 1987[1]	25 °C, 12.5%	Tension *lt*	$S_{23}(t)$	0.15	1.9 E10	5.075 E–5
Cariou, 1987[1]	25 °C, 13.7%	Tension *tr*	$S_{21}(t)$	0.18	1.4 E6	0.603 E–3
Cariou, 1987[1]	25 °C, 12.4%	Tension *rt*	$S_{21}(t)$	0.25	0.1 E6	0.539 E–3
Hayashi *et al.*, 1993[2]	25 °C, 12%	Tension	$S_{ll}(t)$	0.21	2.2 E5	0.552 E–3
Hayashi *et al.*, 1993[2]	25 °C, 12%	Tension	$S_{13}(t)$	0.13	7.3 E7	2.470 E–5
Hayashi *et al.*, 1993[2]	25 °C, 12%	Tension	$S_{55}(t)$	0.23	9.4 E5	0.212 E–3
Hayashi *et al.*, 1993[2]	25 °C, 12%	Tension *lr*	$S_{55}(l)$	0.27	5.9 E5	0.743 E–3
Hayashi *et al.*, 1993[2]	20 °C, 12%	Tension *l*	$S_{33}(t)$	0.89	5.9 E6	6.540 E–5
Le Govic, 1991[2]	25 °C, 12%	Bending[3] *l*	$S_{33}(t)$	0.11	6.3 E10	5.870 E–5
Nielsen, 1984	20 °C, 15%	Bending[3] *l*	$S_{33}(t)$	0.2-0.25	1.5 (E7-E8)	

[1] Maritime Pine.

[2] Spruce.

[3] The test includes various temperatures.

Equation (4.93), which is the same as equation (4.94), has several advantages compared to other empirical models. According to Nielsen (1984), the power k in this model when applied to wood is relatively independent of the climatic conditions and varies between 0.2 and 0.25 for the longitudinal direction l and between 0.23 and 0.33 for the directions r and t. In contrast to k, the doubling time of the deformation τ is very sensitive to the climatic conditions and to the direction r, t, or l. The value of τ under constant climatic conditions (moisture content 15% and temperature 20 °C) lies between 10^5 and 10^4 days for the longitudinal direction but is only 50 days for the tangential or radial direction. Under equilibrium conditions, at high moisture and high temperature (constant climate), τ decreases, but for a thin sample under variable climatic condition, τ can decrease to 10^{-4} and 10^{-2} days in the radial and longitudinal directions, respectively. Table 4.3 presents values of k and τ for several creep functions under constant climatic conditions as obtained by Cariou (1987), Hayashi *et al.* (1993) and Nielsen (1984).

Other types of uni-dimensional creep functions based on rheological models have been developed. These models consist of series and parallel combinations of elastic springs and viscous dashpots (Bodig & Jane, 1982; Bach & McNatt, 1990), and can properly represent the viscoelastic behavior of wood under constant climatic conditions. The coefficients of the rheological models are also evaluated from the creep curve.

One of the rheological approaches commonly used to describe the long-term viscoelastic behavior of wood is the generalized Kelvin-Voigt model presented in Figure 4.29. The corresponding time-dependent creep function is given by

$$S(t) = S_0 + \sum_{i=1}^{n} S_i \left(1 - e^{-\frac{t}{\tau_i}}\right) + \frac{t}{\eta_\infty} \tag{4.95}$$

where, S_0, η_∞, $S_i (i = 1, ... n)$ and $\tau_i (i = 1, ... n)$ are unknown coefficients.

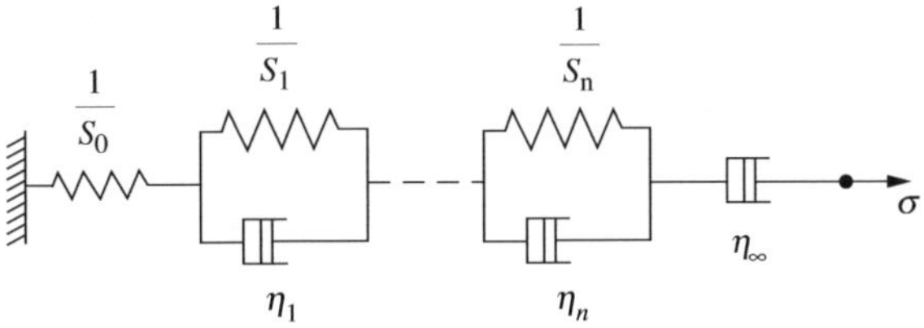

Fig. 4.29 Elements of the generalized Kelvin-Voigt model.

When the stress level is high, wood may exhibit a non-linear viscoelastic behavior. In this case, the creep and stress relaxation functions become dependent on respectively the stress and strain loadings. Consequently, the Boltzmann superposition principle, which is valid for linear behavior, loses its meaning. In this case, the relationship between the time-dependent strain and stress components in a uni-dimensional medium may be described by the Green-Rivlin expression summarized in Morlier (1994):

$$
\begin{aligned}
\varepsilon(t) = & \int_{0^-}^{t} S_1(t-\tau_1)\frac{\partial\sigma(\tau_1)}{\partial\tau_1}\mathrm{d}\tau_1 \\
& + \int_{0^-}^{t}\int_{0^-}^{t} S_2(t-\tau_1,\, t-\tau_2)\frac{\partial\sigma(\tau_1)}{\partial\tau_1}\frac{\partial\sigma(\tau_2)}{\partial\tau_2}\mathrm{d}\tau_1\mathrm{d}\tau_2 \\
& + \int_{0^-}^{t}\int_{0^-}^{t}\int_{0^-}^{t} S_3(t-\tau_1,\, t-\tau_2,\, t-\tau_3)\frac{\partial\sigma(\tau_1)}{\partial\tau_1}\frac{\partial\sigma(\tau_2)}{\partial\tau_2}\frac{\partial\sigma(\tau_3)}{\partial\tau_3}\mathrm{d}\tau_1\mathrm{d}\tau_2\mathrm{d}\tau_3 \\
& + \ldots\ldots
\end{aligned}
\tag{4.96}
$$

In a test under a constant stress σ, the Green-Rivlin formulation becomes a nonlinear polynomial function of stress and its third-order reduced function takes the form

$$
\varepsilon(t) = \sigma S_1(t) + \sigma^2 S_2(t,\, t) + \sigma^3 S_3(t,\, t,\, t) \tag{4.97}
$$

In practice, it is usually difficult to calibrate the Green-Rivlin formulation by performing experiments even for the third-order equation given in relation (4.97). Other simplified alternative mathematical expressions for nonlinear formulations have been proposed in the literature by Nakada (1960), Schapery (1966) and Findely and Lai (1967).

4.4.3 Modeling the thermo-viscoelasticity of wood

Wood is a natural composite of biological origin consisting essentially of three polymers; cellulose, hemicelluloses and lignin. The resemblance between polymeric materials and wood means that certain concepts validated experimentally on polymeric material may also be valid for wood. Among these concepts, we will briefly present that known as the principle of time-temperature equivalency (time-temperature superposition) in polymeric materials (Ferry, 1980).

Salmén (1984) has reported on the applicability of this concept on the in situ lignin under water saturated conditions. More recently, Bardet and Gril (2002) also investigated the applicability of this concept on the transverse mechanical properties of tropical species in the green condition in the complex plane. Dlouha *et al.* (2009) have validated the time-temperature equivalency principle on green tropical species in the longitudinal direction by predicting their long-term viscoelastic behavior from short-term creep tests at temperatures ranging from 30 to 70 °C. The results obtained from this investigation have demonstrated the validity of this concept on wood albeit with some discrepancy. The origin of this discrepancy was found to depend mainly on the fact that the increase in temperature not only accelerates the viscoelastic processes of wood but also slightly increases its compliance. Placet *et al.* (2007) have shown that this concept could not be applied to the whole viscoelastic range but might be valid within each transition state.

Time-temperature equivalency

In certain particular ranges of time and temperature, which correspond to a transition, an explicit equivalency between the effects of time and temperature can be shown experimentally and quantified by the functions of thermo-activation. This is called the principle of time-temperature. Polymers obeying this principle are considered to be thermo-rheologically simple materials.

According to this principle, the viscoelastic behavior of thermo-rheologically simple materials at high temperature for short times is equivalent to that observed at low temperatures for long times. The principle can be presented by the idealized diagram given in Figure 4.30. In this figure, which illustrates the creep compliance $S(t, T)$ under temperatures T_0, T_1, and T_2, the time scale is logarithmic.

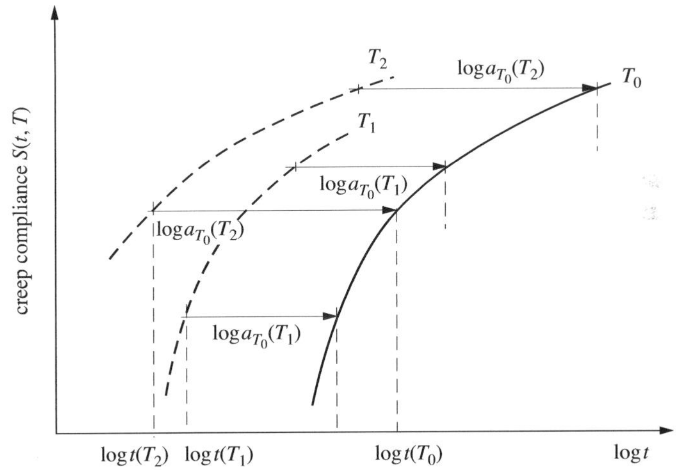

Fig. 4.30 Idealized diagram presenting time-temperature equivalency using isothermal creep compliance curves. T stands for temperature, $\log a_T$ for shifts along the log time axis and three temperatures $T_0 < T_1 < T_2$.

Figure 4.30 shows that the time-temperature equivalency can be written as

$$\log t(T_0) = \log t(T_1) + \log a_{T_0}(T_1) \tag{4.98}$$

where, $a_{T_0(T_1)}$ is a translation factor (or shift factor) between the creep compliance of temperatures T_0 and T_1, depending on the nature of the transition. In general, this relation can be expressed as

$$\log t(T_0) = \log t(T_m) + \log a_{T_0}(T_m) \tag{4.99}$$

where $m = 1, 2, \ldots N$.

The assumption of time-temperature equivalency is useful in numerous applications. Two applications are given in sections 4.4.4 and 4.4.5.

4.4.4 Prediction the long-term viscoelasticity behavior based on short-term tests

Equation (4.99) shows that on a logarithmic time scale, the isothermal creep compliances for two different temperatures are parallel and that one can be obtained from the other by shifting along the time axis with an amplitude corresponding to $\log a_{T_0}(T_m)$.

The application of the time-temperature equivalency assumption is also illustrated by Tschoegl (1997) for the relaxation modulus $R(t, T)$ by shifting the short-term relaxation functions of isotherms T_1, T_2, T_3 and T_{-1}, T_{-2}, T_{-3} to obtain the long-term reference curve of isotherm T_0 on a logarithmic scale. The curve corresponding to the isotherm T_0 is called the *reference or master curve*, and is presented in Figure 4.31.

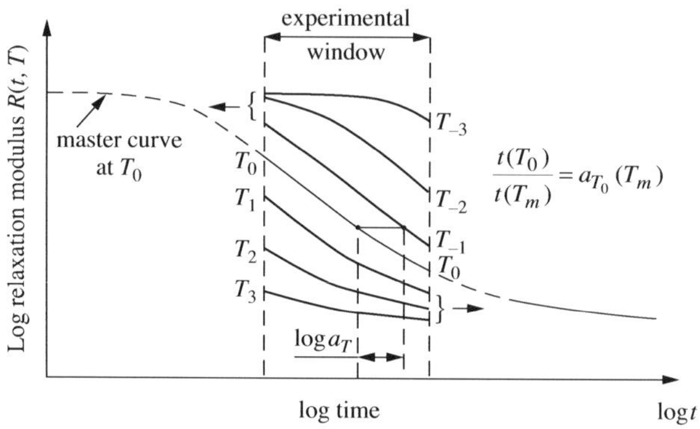

Fig. 4.31 Ideal diagram representing the application of the principle of time-temperature equivalency to determine the long-term reference curve of the relaxation function on a logarithmic scale from short-term experimental curves as presented by Tschoegl (1997).

4.4.5 Reduced variables of time – temperature

The principle of time-temperature equivalency makes it possible to introduce the concept of reduced variables in the formulation of creep function. This means that the creep compliance $S(t, T)$ which depends on two variables t and T, can be written as a one variable dependent on temperature T alone. Based on the power law expression, equation (4.93), the linear time-temperature viscoelastic behavior of materials is modeled as

$$S(t,T) = S_0\left(1 + \left(\frac{t}{\tau(T)}\right)^k\right) \tag{4.100}$$

Applying the time-temperature equivalency principle to equation (4.100), one can express S for two instances t_1 and t_2 and two corresponding temperatures T_1 and T_2, gives the following relations:

$$\frac{t_1}{\tau(T_1)} = \frac{t_2}{\tau(T_2)} = \frac{t}{\tau(T)} \tag{4.101}$$

Based on the relations (4.99) and (4.101), the value of the translation factor is given by:

$$\frac{t_1}{t} = \frac{\tau(T_1)}{\tau(T)} = a_{T_1}(T) \tag{4.102}$$

For amorphous polymers, the relation between temperature and time is given by the shift factor a_T, which obeys the thermo-activation law. This law has been defined in the form of an Arrhenius equation and also by the WLF law (Williams *et al.*, 1955). The Arrhenius equation may be applied at temperatures below the glass transition temperature (Tg), whereas the WFL law is more suitable at temperatures between Tg and Tg +50 °C (Salmén & Hagen, 2001).

Based on the Arrhenius equation, a_T is given by:

$$\ln(a_T) = \frac{W}{R}\left(\frac{1}{T} - \frac{1}{T_0}\right) \tag{4.103}$$

where W is an activation energy, R is the ideal gas constant and T_0 is the reference temperature. In this case, the variation of $\tau(T)$ with temperature T is given by:

$$\tau(T) = \tau_0 e^{\frac{W}{RT}} \tag{4.104}$$

where τ_0 is the value of τ when the temperature is very high, and T is the temperature in Kelvin.

At higher temperatures, above the glass transition temperature, the mathematical expression of the shift factor based on WLF law is written

$$\ln(a_T) = \frac{C_1^o(T - T_0)}{C_2^o + (T - T_0)} \tag{4.105}$$

where T_0 is the reference temperature and C_1^o, C_2^o are two constants depending on T_0. To experimentally determine the values of these two constants, the glass transition temperature is usually taken as the reference temperature:

$$\ln(a_T) = \frac{C_1^g(T - T_g)}{C_2^g + (T - T_g)} \tag{4.106}$$

The WLF law is written as

$$\tau(T) = \tau_0 e^{\frac{C_1^g(T - T_g)}{C_2^g + T - T_g}} \tag{4.107}$$

Equations (4.104) and (4.107) give the dependence of τ on temperature in the form of respectively the Arrhenius equation and the WFL equation. Insertion of one of these expressions in the creep compliance equation (4.100) gives a model presenting the uni-dimensional thermo-creep behavior of wood. This thermo-activation power law model possesses 4 parameters, S_0, k, τ_0, W, to be determined experimentally and two variables t and T.

Experimental studies have shown that the thermo-viscoelasticity of wood may involve not one but several second-order transitions (Poliszko, 1986; Le Govic *et al.*,

1988). Based on these results, (Huet, 1987) proposed a multi-transition model for the thermo-viscoelasticity of wood using a complex creep compliance. This model consists of a serial link of parabolic elements (Huet, 1987; Huet & Navi, 1990), and such modelling has shown a strong potentiality for identifying all the parameters in the complex rheological model through experimental results.

The investigation and characterization of the thermoactivated-viscoelastic behavior of green wood at high temperature have become essential in order to be able to better control the process of transformation of wood. This type of research is important for the thermo-hygro-mechanical processing of wood. The influence of temperature on the viscoelastic behavior of saturated green wood in the transverse and longitudinal directions has recently been studied both experimentally and theoretically (Bardet, 2001; Vincent, 2006; Dlouha *et al.*, 2009).

4.5 TIME-DEPENDENT BEHAVIOR OF WOOD UNDER VARIABLE CLIMATIC CONDITIONS – THE MECHANO-SORPTIVE EFFECT

The term "mechano-sorption" is often used to express the effect of the coupling between mechanical stress and variations in moisture content in wood. Experimental results have shown that a variation in the moisture content of a wood specimen subjected to a mechanical load can lead to important deformations. Under high load and extreme variations in moisture content, a high deformation can lead to fracturing of the wood. Figure 4.32 illustrates the effect of various cyclic variation in the relative humidity combined with different stress levels on the delayed deformation in bending of small specimens.

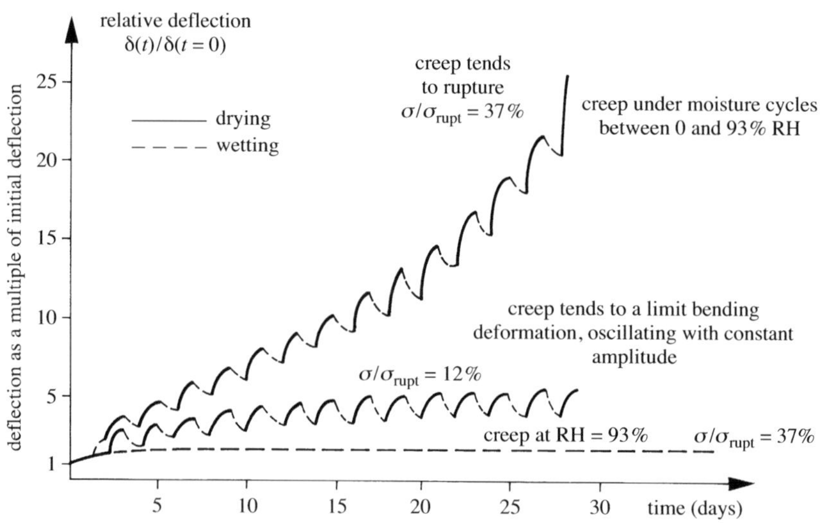

Fig. 4.32 Creep in bending of small specimens of beech with dimensions ($2 \times 2 \times 60$ mm^3) under cyclic relative humidity variations (Hearmon & Paton, 1964).

Phenomena observed under varying moisture conditions present several characteristics that have been recapitulated by Grossman (1976). These phenomena are presented in Figure 4.33 which shows a typical mechano-sorptive creep curve after deduction of the free swelling or addition of the free shrinkage observed under zero loading.

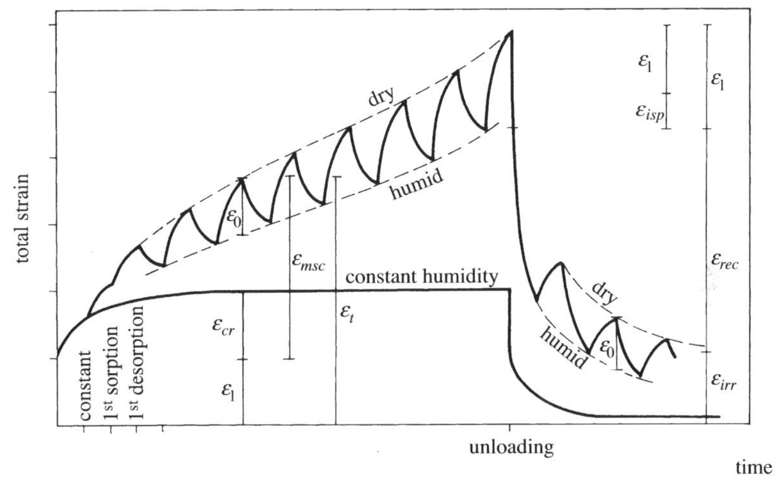

Fig. 4.33 Typical curve showing the mechano-sorptive creep of wood (Grossman, 1976).

The main features are given below.

- The deformation increases during drying. This has been shown by Armstrong and Kingston (1962), Hearmon and Paton (1964), Pittet (1996) and others.
- The first re-humidification causes an increase in the deformation. Subsequent re-humidification leads to very little reduction in deformation when the applied compressive load is low and to an increase in deformation under higher compressive stresses (Armstrong & Kingston, 1962; Hearmon & Paton, 1964; Navi *et al.*, 2002).
- The deformation due to mechano-sorption is independent of time; it is determined by the degree of variation in moisture content while it is below the fiber saturation point (FSP) (Armstrong & Kingston, 1962; Leicester, 1971).
- When the stress level is less than 15-20% of the short-term ultimate stress, the deformation due to mechano-sorption seems to be linear
- A constant flow of moisture through the wood without producing any local change in the moisture content does not lead to any mechano-sorptive effect (Armstrong, 1972).
- After unloading and recovery of the instantaneous elasticity, most of the total deformation is irrecoverable. New humidification-drying cycles decrease the residual deformation. The recovery is greater during drying than during humidification (Armstrong & Kingston, 1962; Pittet, 1996).

It is important to note that the dimensions of the specimens have a considerable influence on the kinetics of the development of the mechano-sorptive effect. This is partly related to the time needed to reach moisture equilibrium in the specimen. For example, for two specimens, one $1 \times 1 \times 60$ mm^3 and the other $20 \times 20 \times 900$ mm^3 in size, the mechano-sorptive effect is similar, but the time necessary to reach moisture equilibrium in the thin specimen is only 2 to 3 hours whereas about 50 hours are required for the thick specimen.

4.5.1 Modeling of the mechano-sorptive effect in wood

Modeling of the viscoelastic behavior of wood under varying moisture content, i.e., the mechano-sorptive effect, began in 1970. These models are based either on purely phenomenological (macroscopic) considerations or on physical interpretations of the deformation at the wood molecular level.

Randriambolona (2003) has studied the existing phenomenological models which rely on two assumptions. Certain authors, Leicester (1971), Ranta-Maunus (1975), Toratti (1992), Mårtensson (1992), deem that the origins of the time-dependent viscoelastic creep and of the mechano-sorptive effect are not the same, and that they are not strongly coupled together. The total deformation is thus regarded as the sum of the various deformations: instantaneous strain, viscoelastic deformation, deformation due to mechano-sorption, deformation due to free shrinkage-swelling and deformation due to the influence of the stress on free shrinkage-swelling.

They have also assumed that the value of each deformation can be measured independently. The models based on this assumption are called ***models with independent activation***. On the other hand, several authors, such as Bazant (1985), Gril (1988), Hanhijärvi (1995), Hunt (1989, 1999), Randriambolona (2003), consider that there is a strong interaction between the viscoelastic mechanism and that of mechano-sorption, and that there is a strong coupling between the deformations due to the effect of time (viscoelastic creep) and those due to the variations in moisture (mechano-sorptive creep). These models are called ***models with combined activation***. Table 4.4 shows the partition of the deformations according to these two assumptions.

Table 4.4 Partition of the deformations

Model with independent activation	$\varepsilon(t) = \varepsilon_e(t) + \varepsilon_{fl}(t) + \varepsilon_{ms}(t) + \varepsilon_{uo}(t) + \varepsilon_{u\sigma}(t)$
Model with combined activation	$\varepsilon(t) = \varepsilon_e(t) + \varepsilon_{fl+ms}(t) + \varepsilon_{uo}(t) + \varepsilon_{u\sigma}(t)$

In this table, $\varepsilon(t)$ = total strain, $\varepsilon_e(t)$ = instantaneous elastic strain, $\varepsilon_{fl}(t)$ = viscoelastic deformation, $\varepsilon_{uo}(t)$ = free deformation of shrinkage-swelling, $\varepsilon_{u\sigma}(t)$ = strain induced by the applied stress σ on the free deformation, $\varepsilon_{ms}(t)$ = deformation due to mechano-sorption, $\varepsilon_{fl+ms}(t)$ = viscoelastic deformation due to the effect of time and the non-separable moisture variation, according to the assumption of the model with combined activation.

Ranta-Maunus and Kortesmaa (1988) proposed a rheological model based on the assumptions considered in the model of independent activation, where the total strain (in the uni-dimensional case) is the sum of the elastic strain, the viscoelastic deformation, the deformation due to free shrinkage-swelling and the mechano-sorptive deformation.

$$\varepsilon(t) = \varepsilon_e(t) + \varepsilon_{fl}(t) + \varepsilon_{ms}(t) + \varepsilon_{uo}(t) \tag{4.108}$$

In this model, the modulus of elasticity of wood, $E(u(t))$, is considered to be a function of moisture content u at constant temperature. The kinetics of elastic strain are given by

$$\varepsilon_e^\circ(t) = \frac{\sigma^\circ(t)}{E(u(t))} - \frac{\sigma(t)E(u^\circ(t))}{E^2(u(t))} \tag{4.109}$$

The kinetics of free shrinkage-swelling are expressed as

$$\varepsilon_{uo}^\circ(t) = \alpha u^\circ(t) \tag{4.110}$$

where α is the coefficient of moisture expansion of wood. The kinetics of mechano-sorptive deformation are written

$$\varepsilon_{ms}^\circ(t) = m\sigma(t)u^\circ(t) \tag{4.111}$$

Here, the coefficient m takes the values (m^{ms+}, m^{ms++}) during humidification and (m^{ms-}) during drying. Indeed, in agreement with experimental results, by taking m^{ms++} different from m^{ms+}, this model distinguishes the effect of the first humidification from that of other humidifications. The kinetics of viscoelastic deformation are defined by

$$\varepsilon_{ft}^\circ(t) = S(t)\sigma^\circ \tag{4.112}$$

According to the model proposed by Ranta-Maunus and Kortesmaa (1988), the following relation gives the sum of the kinetics of the deformations:

$$\varepsilon^\circ(t) = \varepsilon_e^\circ(t) + \varepsilon_{ft}^\circ(t) + \varepsilon_{ms}^\circ(t) + \varepsilon_{u0}^\circ(t) \tag{4.113}$$

The extention of the constitutive equations defined in (4.108-4.113) for a linear uni-dimensional mechano-sorptive viscoelastic material to a three-dimensional orthotropic material lead to the sum of vectors

$$\left\{\varepsilon^\circ(t)\right\} = \left\{\varepsilon_e^\circ(t)\right\} + \left\{\varepsilon_{fl}^\circ(t)\right\} + \left\{\varepsilon_{ms}^\circ(t)\right\} + \left\{\varepsilon_{u0}^\circ(t)\right\} \tag{4.114}$$

where each vector has six components defined for the three orthogonal directions r, t, l:

$$\left\{\varepsilon^\circ(t)\right\} = \left\{\varepsilon_r^\circ(t),\ \varepsilon_t^\circ(t),\ \varepsilon_l^\circ(t),\ \gamma_{rt}^\circ(t),\ \gamma_{rl}^\circ(t),\ \gamma_{tl}^\circ(t)\right\}^T \tag{4.115}$$

The mechano-sorptive deformation given in (4.114) can be written as the constitutive equation for a three-dimensional case:

$$\left\{\varepsilon_{ms}^{\circ}\right\}=\left[S_{ms}\right]\{\sigma\}u^{\circ} \tag{4.116}$$

where

$$\left\{\varepsilon_{ms}^{\circ}\right\}=\left\{\varepsilon_{ms(r)}^{\circ},\ \varepsilon_{ms(t)}^{\circ},\ \varepsilon_{ms(l)}^{\circ},\ \gamma_{ms(rt)}^{\circ},\ \gamma_{ms(rl)}^{\circ},\ \gamma_{ms(tl)}^{\circ}\right\}^{T} \tag{4.117}$$

$$\{\sigma\}=\left\{\sigma_{r},\ \sigma_{t},\ \sigma_{l},\ \tau_{rt},\ \tau_{rl},\ \tau_{tl}\right\}^{T} \tag{4.118}$$

and $[S_{ms}]$ is defined as the matrix of mechano-sorptive compliance. The value of the components of the matrix depend on the sign of the variation in moisture content $u^{\circ}=u^{\circ}(t)$, as follows

$$\left[S_{ms}\right]=\left\{\begin{array}{l}\left[S_{ms^{+}}\right]\text{for } u^{\circ}>0\\ \left[S_{ms^{-}}\right]\text{for } u^{\circ}<0\end{array}\right\} \tag{4.119}$$

where, for example,

$$\left[S_{ms^{+}}\right]=\begin{bmatrix} m_{rr}^{+} & v_{tr}m_{tt}^{+} & -v_{lr}m_{ll}^{+} & 0 & 0 & 0\\ -v_{rt}m_{rr}^{+} & m_{tt}^{+} & -v_{lt}m_{ll}^{+} & 0 & 0 & 0\\ -v_{rl}m_{rr}^{+} & -v_{tl}m_{tt}^{+} & m_{ll}^{+} & 0 & 0 & 0\\ 0 & 0 & 0 & m_{rt}^{+} & 0 & 0\\ 0 & 0 & 0 & 0 & m_{rl}^{+} & 0\\ 0 & 0 & 0 & 0 & 0 & m_{tl}^{+} \end{bmatrix} \tag{4.120}$$

The components of the matrix $\left[S_{ms^{-}}\right]$ are similar to the components of the matrix, $\left[S_{ms^{+}}\right]$, with the parameters v_{ms+} and m^{+} replaced by respectively v_{ms-} and m^{-}. It is important to note that, in this model, as for the matrix of mechano-sorptive compliance, it is necessary to determine 12 parameters in addition to the components due to the first humidification.

The effect of mechano-sorption $\{\varepsilon_{ms}\}$ can be evaluated by the following integral:

$$\left\{\varepsilon_{ms^{+}}\right\}=\int_{o}^{t}\left[S_{ms^{+}}\right]\sigma u^{\circ}\mathrm{d}t \tag{4.121}$$

It should also be noted that in the model of Ranta-Maunus, the delayed deformation due to the effect of mechano-sorption is irreversible.

4.6 REFERENCES

ALLAILI A., JARDIN, P. & PLUVINAGE, G. (1986). Propagation dynamique d'une fissure dans un matériau orthotrope: le bois. (Dynamical propagation of cracks in an orthotropic material: wood.) In: *Rhéologie des matériaux anisotropes*. (Rheology of anisotropic materials.) Huet, C,. Bourgoin, D. & Richemond, S. (eds.). Groupe Français de Rhéologie. Comptes rendus du 19e Colloque National Annuel, Editions Cepadues, Toulouse, France.

ARIMA, T. (1967). The influence of high temperature on compressive creep of wood. *Journal of the Japan Wood Research Society, 13*(2):36-40.

ARMSTRONG, L.D. (1972). Deformation of wood in compression during moisture movement. *Wood Science, 5*(2):81-86.

ARMSTRONG, L.D. & KINGSTON, R.S.T. (1962). Effect of moisture changes on the deformation of wood under stress. *Australian Journal of Applied Science, 13*(4):257-276.

BACH, L. & MCNATT, J.D. (1990). Creep of OSB with various strand alignments. In: *proceedings of IUFRO S5.02 Meeting in Saint John.*

BARDET, S. (2001). *Comportement Thermoviscoélastique Transverse du Bois Humide*. (Visco-plastic behaviour in the transverse direction of wet wood.) PhD. Thesis, Université Montpellier II, France.

BARDET, S. & GRIL, J. (2002). Modelling the transverse viscoelasticity of green wood using a combination of two parabolic elements. *Comtes Rendus Mécanique, 330*(8):549-556.

BARISKA, M. & KUČERA, L.J. (1985). On the fracture morphology in wood. Part II. Macroscopical deformations upon ultimate axial compression in wood. *Wood Science and Technology, 19(1)*:19-34.

BAZANT, Z.P. (1985). Constitutive equation of wood at variable humidity and temperature. *Wood Science and Technology, 19*(2):159-177.

BHATNAGAR, N.S. (1964). Kriechen von Holz bei Zugbanspruchung in Faserrichtung. (Creep of wood in tension parallel to grain.) *Holz als Roh- und Werkstoff, 22*(8):296-299.

BME/ME 506: Large deformation nonlinear mechanics definition http://www.engin.umich.edu/class/bme5062000/.

BODIG, J. & GOODMAN, J.B. (1973). Prediction of elastic parameters for wood. *Wood Science, 5*(4):249–264.

BODIG, J. & JAYNE, B.A. (1982). *Mechanics of wood and wood composites*. Van Nostrand Reinhold Company, New York, Cincinnati, ISBN 0-442-00822-8.

BOSTRÖM, L. (1992). *Method for determination of the softening behaviour of wood and the applicability of a nonlinear fracture mechanics model*. PhD. Thesis No. TVBM-1012, Lund University of Technology.

CARIOU, J.L. (1987). *Caractérisation d'un Matériau viscoélastique anisotrope*. (Characterization of anisotropic properties of visco-elastic materials.) PhD. Thesis, University of Bordeaux, France.

CURNIER, A. (2005). *Méthodes numériques en mécanique des solides*. (Numerical methods in solid mechanics.) Presses Polytechniques et Universitaires Romandes.

CURNIER, A. (2011). *Mécanique des solides déformables*. (Mechanics of deformable solids.) In press by Presses polytechniques et universitaires romandes.

DAVIDSON, R.W. (1962). The influence of temperature on creep in wood. *Forest Products Journal, 12*(August):377-381.

DE MAGISTRIS, F. & SALMÉN, L. (2008). Finite element modelling of wood cell deformation transverse to the fibre axis. *Nordic Pulp & Paper Research Journal, 23*(2):240-246.

DINWOODIE, J.M., HIGGINS, J.A., PAXTON, B.H. & ROBSON, D.J. (1990). Creep research on particleboard. *Holz als Roh- und Werkstoff, 48*(1):5-10.

DINWOODIE, J.M., HIGGINS, J.A., PAXTON, B.H. & ROBSON, D.J. (1991). Quantifying, predicting and understanding the mechanism of creep in board materials. In: *Proceedings of Cost 508 action: Fundamental Aspects on Creep in Wood,* Mårtensson, A., Ranta-Maunus, A. & Seoane, I. (eds.). Lund, Sweden, pp. 99-119.

DLOUHA, J., CLAIR, B., ARNOULD, O., HORACEK, P. & GRIL, J. (2009). On the time-temperature equivalency in green wood: Characterisation of viscoelastic properties in longitudinal direction. *Holzforschung 63*(3):327-333.

EASTERLING, K.E., HARRYSSON, R., GIBSON, L.J & ASHBY, M.F. (1982). *On the mechanics of balsa and other woods. Proceedings of the Royal Society, A383*(1784):31-41.

FERRY, J.D. (1980). *Viscoelastic properties of polymers*. John Willy & Son, Inc.

FINDELY, W.N. & LAI, J.S.Y. (1967). A modified superposition principle applied to creep of nonlinear viscoelastic material under abrupt changes in state of combined stress. *Transaction of the Society of Rheology, 11*(3):361-380.

FOUDJET, A. (1986). *Contribution à l'étude rhéologique du matériau bois.* (Contribution to the study of rheological wood material.) PhD. Thesis, University of Lyon, France.

FRANÇOIS, P. & MORLIER, P. (1993). Plasticité du bois en compression simple. (Plasticity of wood in simple compression.) *Matériaux et Techniques, 81*(12):5-14.

GIBSON, L.J. & ASHBY, M.F. (1988). *Cellular solids: Structure and properties.* Pergamon Press, Oxford, ISBN 0-08-036607-4.

GRIL, J. (1988). *Une modélisation du comportement hygro-rhéologique du bois à partir de sa microstructure.* (A model of hygro-rheological behaviour of wood from its microstructure.) PhD. Thesis, Université de Paris, France.

GRIL, J. & NORIMOTO, M. (1993). Compression of wood at high temperature. In: *Proceedings of workshop on wood plasticity and damage. In: COST 508 Wood Mechanics, Workshop on Wood: plasticity and damage,* Birkinshaw, C. (ed.). University of Limerick, Ireland, pp. 135-144.

GROSSMAN, P.U.A. (1976). Requirements for a model that exhibits mechano-sorptive behaviour. *Wood Science and Technology, 10*(3):163-168.

GUITARD, D. (1987). *Mécanique matériau bois et composites.* (Mechanics of wood and composite materials.) Cepadues Edition, FRANCE.

HANHIJÄRVI, A. (1995). *Modelling of creep deformation mechanism in wood.* PhD. Thesis Espoo University, Finland, Technical Research Centre of Finland (VTT) Publications 231.

HAYASHI, K., FELIX, B. & LE GOVIC, C. (1993). Wood viscoelastic compliances determination with special attention to measurement problems. *Materials and Structures, 26*(160):370-376.

HEARMON, R. & PATON, J. (1964). Moisture content changes and creep of wood. *Forest Products Journal, 14*(8):357-359.

HOFFMEYER, P. (1990). *Failure of wood as influenced by moisture and duration of load.* PhD. Thesis, State University of New York, College of Environmental Science and Forestry, Syracuse, New York.

HOYLE, R.J., ITANI, R.Y. & ECKARD, J.J. (1986). Creep of Douglas-fir beams due to cyclic humidity fluctuations. *Wood and Fiber Science, 18*(3):468-477.

HUET, C. (1987). Nouvelle méthode de représentation graphique des résultats de fluage, relaxation et vitesse de propagation du son pour l'identification des modèles mathématiques multitransition de la thermo-viscoélasticité du bois: un space de phase modifié. (New method of graphical representation of the results of creep, and relaxation for the identification of the multi-transition mathematical models of the thermo-viscoelasticity of wood: a space of phase modification.) Note technique CH3 destinée au CTBA, France.

HUET, C & ZAOUI, A. (eds.) (1981). *Reological behaviour and structures of materials.* 15th Annual Colloquium of the French Group of Rheology, Presses ENPC, Paris.

HUET C., BOURGOIN, D. & RICHEMOND, S. (eds.) (1986). Rhéologie des matériaux anisotropes. (Rheology of anisotropic materials.) Groupe Français de Rhéologie. Comptes rendus du 19e Colloque National Annuel, Editions Cepadues, Toulouse, France.

HUET, C., GITTARD, D. & MORLIER, P. (1988). *Le bois en structure son comportement différé.* (Time dependent behaviour of wood.) Annales de l'Institut Technique du Batiment et des Travaux Publics.

HUET, C. & NAVI, P. (1990). Multiparabolic multitransition model for thermoviscoelastic behaviour of wood. In: *Mechanics of Wood and Paper Materials*, Dallas, Texas; ASME, pp. 17-24.

HUNT, D. (1989). Linearity and non-linearity in mechano-sorptive creep of softwood in compression and bending. *Wood Science and Technology, 23*(4):323-333.

HUNT, D. (1999). A unified approach to creep of wood. *The Royal Society of London Academy, Proceedings: Mathematical, Physical and Engineering Science, 455*(1991):4077-4095.

JOUVE, J.H. & SALES, C. (1986). *Influence de traitements physico-chimiques sur le fluage du matériau bois et la relaxation des contraintes* (Influence physicochemical treatments on the creep of the wood and the stress relaxation.) In: *Séminaire interne 22-24 sept. (Nogent/Vernisson) GS Rhéologie du bois CNRS.*

KING, E.R. Jr. (1961). Time-dependent strain behaviour of wood in tension parallel to grain. *Forest Products Journal, 11*(3):156-165.

KINGSTON, R.S.T & CLARKE, L.N. (1961). Some aspects of the rheological behaviour of wood. I. The effect of stress with particular reference to creep. II. Analysis of creep data by reaction-rate and thermodynamic methods. *Australian Journal of Applied Science*, 12(2):211-240.

KINGSTON, R.S.T. & BUDGEN, B. (1972). Some aspects of the rheological behaviour of wood. Part IV: Non-linear behaviour at high stresses in bending and compression. *Wood Science and Technology, 6*(3):230-238.

KOLLMANN, F.F.R. & CÔTÉ, W.R. (1968). *Principles of wood science and technology. Vol. 1: Solid wood.* Springer-Verlag, New York.

KUČERA, L.J. & BARISKA, M. (1982). On the fracture morphology in wood. Part 1: A SEM study of deformations in wood of spruce and aspen upon ultimate axial compression load. *Wood Science and Technology, 16*(4):241-259.

LE GOVIC, C. (1988). *Le comportement viscoélastique du bois en liaison avec sa constitution plymérique.* (The viscoelastic behavior of wood in connection with its polymeric constitution.) Rapport bibliographique réalisé avec les concours financiers du Ministères de la Recherche et de l'Enseignement Supérieur, CTBA – Centre Technique du Bois et de L'Ameublement.

LE GOVIC, C. (1991). Modelling wood linear viscoelastic behaviour in constant climate. In: *Proceedings of Cost 508 action: Fundamental Aspects on Creep in Wood,* Mårtensson, A., Ranta-Maunus, A. & Seoane, I. (eds.). Lund, Sweden, pp. 44-55.

LE GOVIC, C., HADJ HAMOU, A., ROUGER, F.L. & FELIX, B. (1988). Modélisation du fluage sur la base d'une équivalence temps-température, Acts du 2e Colloque Sciences et Industries du Bois. (Modelling of creep based on a temperature-time equivalence) In: *Proceedings of the 2nd scientific Conference and Industry of Wood,* Nancy 1987, A.R.B.O.L.O.R.

LEICESTER R. (1971). A reological model for mechano-sorptive deflection of beams. *Wood Science and Technology, 5*(3):211-220.

MARSDEN, J.E. & HUGHES, T.J.R. (1994). *Mathematical foundation of elasticity.* Dover Publications, Inc., New York, ISBN 0-486-67865-2.

MILLER, D.G. & GEORGE, P. (1974). Effect of stress level on the creep of eastern spruce in bending. *Wood Science, 7*(1):21-24.

MORLIER, P. (ed.) (1994). *Creep in timber structures.* E & FN Spon, London, ISBN 0-419-18830-4.

MUKUDAI, J. (1983). Evaluation of linear and non-linear viscoelastic bending loads of wood as a function of prescribed deflections. *Wood Science and Technology, 17*(3):203-216.

MÅRTENSSON, A. (1992). *Mechanical behaviour of wood exposed to humidity variation.* PhD. Thesis No. TVSM-1006, Lund University of Technology, Sweden.

NAKADA, O. (1960). Theory of non-linear responses. *Journal of the Physical Society of Japan, 15*(12):2280-2288.

NAIRN, J.A. (2006). Numerical simulations of transverse compression and densification in wood. *Wood and Fiber Science, 38*(4):576-591.

NAVI, P., PITTET, V. & PLUMMER, C.J.G. (2002). Transient moisture effect on wood creep. *Wood Science and Technology, 36*(6):447-462.

NAVI, P. & HEGER, F. (2005). *Comportement thermo-hydromécanique du bois.* (Behaviour of thermo-hydromechanical processed wood.) Presses polytechniques et universitaires romandes, Lausanne, ISBN 2-88074-620-5.

NIELSEN, L.F. (1984). Power law creep as related to relaxation, elasticity, damping, rheological spectra and creep recovery, with special reference to wood. In: *IUFRO, Timber Engineering Group meeting,* Xalapa, Mexico.

OGDEN, R.W. (1972). Large deformation isotropic elasticity – On the correlation of theory and experiment for incompressible rubber like solids. In: *Proceedings of the Royal Society of London. Series A, Mathematical and Physical Sciences, 326*(1567):565-584.

PITTET, V. (1996). *Etude expérimental des couplages mécanosorptifs dans le bois soumis à variations hygrométriques contrôlées sous chargements de longue durée.* (Experimental study of mechanosorptive effect in wood subjected to moisture variations under controlled loads of long duration.) PhD. Thesis No. 1526, Ecole Polytechnique Fédérale de Lausanne (EPFL), Switzerland.

PLACET, V., PASSARD, J. & PERRÉ, P. (2007). Viscoelatic properties of green wood across the grain measured by harmonic tests in the range 9-95 °C: Hardwood vs. softwood and normal wood vs. reaction wood. *Holzforschung, 61*(5):548-557.

POLISZKO, S. (1986). Anisotropy of dynamic wood viscoelasticity. In: *Rheology of anisotropic materials,* Huet, C., Bourgouin,. & Richemond, S. (eds.). Cepadues Editions, Toulouse, pp. 453-460.

RANDRIAMBOLONA, H. (2003). *Modelisation du comportement différé du bois en environnement variable*. (Modeling the behavior of wood in different variable environment.) PhD. Thesis, University of Limoges, Faculty of Sciences, France.

RANGSRI, W. (2004). *Etude expérimentale et numérique de l'essai d'indentation du bois*. (Experimental and numerical study of the indentation test of wood.), PhD. thesis, University of Montpellier 2, France

RANGSRI, W., GRIL, J. & JERONIMIDIS, G. (2003). Finite element simulation of the transverse compressive deformation of some tropical hardwoods. In: *2nd International Conference of the European Society for Wood Mechanics*, Salmen, L. (ed.). Stockholm, Sweden.

RANGSRI, W., GRIL, J. & JERONIMIDIS, G. (2004). Multiscale modelling of the transverse compression of a tropical hardwood. In: *Euromech colloquium 459 "Mechanical behaviour of cellular solids"*, Ganghoffer, J.F. & Onck, P. (eds.). Nancy, France.

RANTA-MAUNUS, A. (1975). The viscoelasticity of wood at varying moisture content. *Wood Science and Technology, 9*(3):189-205.

RANTA-MAUNUS, A. & KORTESMAA, M. (1988). An analysis of the state of stress in timber caused by moisture gradient. In: *Proceeding of the IUFRO Timber Engineering Meeting*, Turku, pp. 113–116.

RIVLIN, R.S. (1948). Large elastic deformations of elastic materials: IV. Further development of the general theory. *Philosophical Transactions of the Royal Society, A241*(835):379–397.

ROUSSEL, M.O. (1997). *Conception et caractérisation d'un bois reconstitué pour absorbeur d'énergie*. (Design and characterization of a reconstituted wood for absorbing energy). PhD. Thesis, University of Bordeaux I, France.

SALMÉN, N.L. (1984). Viscoelastic properties of in situ lignin under water-saturated conditions. *Journal of Material Science, 19*(9):3090-3096.

SALMÉN, N.L., & HAGEN, R. (2001). Viscoelastic properties. In: *Handbook of physical testing of paper*. Mark, R.E., Habeger, C.C. Jr., Borch, J. & Lyne, B.M. (eds.). Marcell Dekker Inc. New York, Basel, pp. 77-113. ISBN 9780824707866.

SCHAPERY, R.A. (1966). A theory of non-linear thermoviscoelasticity based on irreversible thermodynamics. In: *Proceedings of the Fifth U.S. National Congress of Applied Mechanics*. University of Minnesota, USA, pp. 511-530.

SCHNIEWIND, A.P. (1968). Recent progress in the study of the rheology of wood. *Wood Science and Technology, 2*(3):188-206.

SCHNIEWIND, A.P. & BARRETT, J.D. (1972). Wood as a linear orthotropic viscoelastic material. *Wood Science and Technology, 6*(1):43-57.

STEVENS, W.C. & TURNER, N. (1970). *Wood bending handbook*. Woodcraft supply corp. Woburn, Massachusetts.

SULSKY, D., CHEN, Z. & SCHREYER, H.L. (1994). A particle method for history-dependent materials. *Computer Methods in Applied Mechanics and Engineering, 118*(1/2):179-196.

SULSKY, D. & SCHREYER, H.L. (1996). Axisymmetric form of the material point with application to upsetting and Taylor impact problems. *Computer Methods in Applied Mechanics and Engineering, 139*(1/4):409-429.

TORATTI, T. (1992). *Creep of timber beams in variable environment*. PhD. Thesis No. 31, Helsinki University of Technology, Laboratory of Structural Engineering and Building Physics, Finland.

TRELOAR, L.R.G. (1975). *The physics of rubber elasticity*. Oxford University Press, Oxford.

TSCHOEGL, N.W. (1997). Time dependence in material properties: An overview. *Mechanics of time dependent materials, 1*(1):3-31.

VINCENT, P. (2006). Comportement viscoélastique thermoactivé en torsion du bois de peuplier à l'état vert. (Thermoactivated viscoelastic behaviour in torsion of poplar wood in a green state.). PhD. Thesis No. 152. Monpollier, Université de Monpollier 2.

VALANIS, K.C. & LANDEL, R.F. (1978). The strain-energy function for a hyperelastic material in terms of extension ratios. *Journal of Applied Physics 11*:1969–1978.

WARD, I.M. (1985). *Mechanical properties of solid polymers*. (2nd edition) John Willy and Sons, New York.

WHALE, L.R.J. (1988). *Deformation characteristics of nail or bolted timber joints subjected to irregular short or medium term lateral loading*. PhD. Thesis, South Bank Polytechnic, London.

CHAPTER 5

INFLUENCE OF THE THM PROCESSING PARAMETERS ON THE MECHANICAL AND CHEMICAL DEGRADATION OF WOOD

5.1 INTRODUCTION

The term "plastic" is given to macromolecular organic materials that can be softened and melted by heating, as in thermoforming, extrusion or a molding process. Like plastics, wood consists of organic macromolecules but it nevertheless exhibits a different rheological behavior under thermo-hydrous conditions because of its composite cellular wall structure. Wood consists primarily of cellulose, hemicelluloses and lignin with small quantities of minerals and extractive substances. The principal components are organized in a multi-layer fibrous composite forming the cellular walls and middle lamella. The fundamental properties of wood are related to its principal components. The complex properties of wood result not only from the simple summation of the properties of each component taken individually but also from their interaction. The apparent plasticity of wood under thermo-hydro-mechanical (THM) treatments is one revealing example.

When wood is subjected to thermo-hydrous conditions, its amorphous components soften and become easy to deform which opens the way to numerous industrial processes such as molding, densification, large bending, shaping, surface densification etc. Unsuitable high-temperature THM conditions can mechanically damage and chemically degrade the wood. It is thus essential to understand the elasto-viscoplastic behavior of wood during THM processing as well as its chemical degradation. It is demonstrated in the next sections that thermo-hydrous processes at higher temperatures may lead to severe degradation in the wood and reduce the mechanical and fractural strength. This chapter describes the effect of THM actions and processing time on both the mechanical and the chemical degradations of wood at temperatures up to 200 °C. A case study presents the mechanical behavior and chemical degradation of small wood samples during THM densification in a closed system. This case study was chosen for two reasons:

- Wood densification in a closed system is a process in which all the parameters involved in THM treatment, such as time, temperature, humidity and force, are present, and each of them can be controlled accurately during the treatment in a closed system.
- Samples of small size make it possible to guarantee the homogeneity and reproducibility of the wood specimens, and the uniformity of temperature and moisture content inside the specimens during the densification and post-treatment processes.

The properties of polymeric materials depend on the major thermal transitions, the primary transition being melting and secondary transitions being the glass transition and other transitions in the glassy state below the glass transition temperature. The amorphous wood constituents, hemicelluloses, lignin and regions of semi-crystalline cellulose, soften during the THM process. It is important to understand the softening of the wood constituents, which is related to their glass transition temperature (T_g), a topic well developed in polymer science.

5.2 GLASS TRANSITION TEMPERATURE OF AMORPHOUS AND SEMI-CRYSTALLINE POLYMERS

The transition temperature of an amorphous polymer marks the border between two fundamentally different states: the glassy, hard and brittle state and the soft, rubbery state, Figure 5.1. The physical and mechanical properties of materials in these states can be characterized as follows:

- In the glassy state, the physical properties of an amorphous material are very similar to those of a solid phase whose values are not strongly sensitive to the chemical nature of the material, such as the topology of the polymeric network. The molecular movements that occur in this state are of low amplitude (movements of side groups or co-operative movements of a few monomeric units).
- In the rubbery state, the modulus of elasticity is 3000 to 4000 times lower than that of the glassy state, and the elongation at break is about 100 times greater. This high extensibility is due to the fact that thermal action decreases inter- and intra-molecular cohesion (Van der Waals forces, hydrogen bonds). Molecular movements of large amplitude (macromolecular movements) and the complete extension of the segments of the macromolecular chain are then possible, thanks to rotation around the covalent carbon-carbon (–C–C–) and carbon-oxygen (–C–O–) bonds.

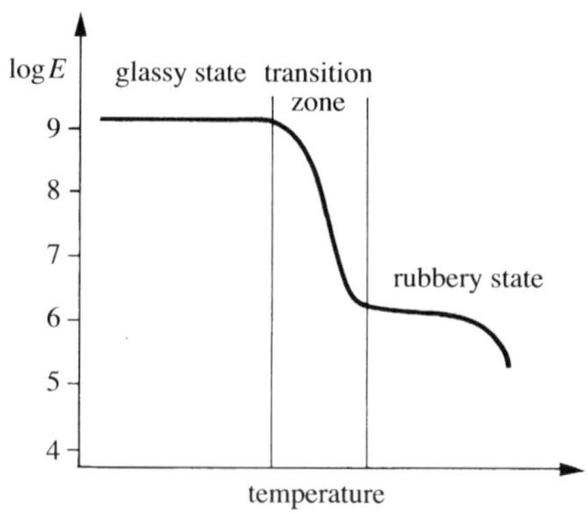

Fig. 5.1 Variation of the modulus of elasticity with the temperature for an amorphous polymer.

The ratio between the energy of the secondary forces responsible for the cohesion of the molecules in the glassy state and the thermal energy necessary for the displacement of a segment determines whether a whole macromolecule is a molten mass, or in a rubbery or glassy state. The glass transition is a characteristic of almost all organic polymers, but it is more apparent if the material is amorphous. Semi-crystalline polymers also exhibit such properties, but the amplitude of the phenomenon is often very low, as shown in Figure 5.2.

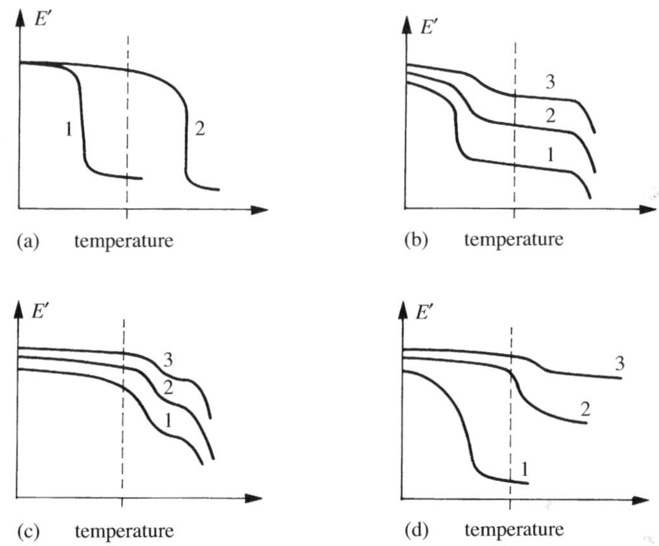

Fig. 5.2 Storage modulus as a function of temperature for (a) amorphous polymers (1: rubber, 2: organic glass), (b) semi-crystalline polymers such as PE or PP (increasing crystallinity from 1 to 3), (c) semi-crystalline polymers such as PA or PET (increasing crystallinity from 1 to 3), (d) amorphous polymers, three-dimensional (increasing cross-linking from 1 to 3). The dotted vertical lines indicate room temperature.

5.2.1 Glass transition temperature (T_g) of wood components

Investigations have been carried out to determine the glass transition temperatures of lignin, hemicelluloses and semi-crystalline cellulose by for instance Goring (1963), Baldwin and Goring (1968), Takamura (1968), Back and Didriksson (1969), Hatakeyama *et al.* (1972), Alfthan *et al.* (1973), Salmén (1979), Back and Salmén (1982) and Lapierre and Monties (1986). Baldwin and Goring (1968) have shown that the T_g values of the isolated components of wood differ from those in native wood. Back and Salmén (1982) studied the relationship between the glass transition temperatures of lignin, hemicelluloses and semi-crystalline cellulose and their moisture contents. It is often difficult to determine the T_g of wood components and the results differ, according to Irvine (1984), the results differ for several reasons:

- The chemically extracted components of wood may be degraded chemically during isolation or separation.

- The T_g of the components is determined by various techniques such as TGA (Thermo-Gravimetric Analysis), TDA (Thermo-Differential Analysis) and DSC (Differential Scanning Calorimetry).
- Some of the techniques are incapable of determining T_g of wood components in for example the saturated state. Others cannot be used to estimate the glass transition temperatures of the polymeric components of wood in anhydrous (dried) conditions where the transition occurs at relatively high temperatures and their decomposition rate is usually rapid. To limit the decomposition, it is thus important to use rapid techniques to determine T_g.

For anhydrous cellulose, the T_g lies between 200 and 250 °C. This interval is due to parameters such as the degree of crystallinity, the method used for the extraction of the cellulose, the time of measurement and different criteria defining the transition temperature. For anhydrous hemicelluloses, the glass transition temperature lies between 150 and 220 °C. This even larger interval is due partly to variations in the chemical composition and conformation of the hemicelluloses but more importantly to differences in their functional groups. The destruction of the functional groups and the reduction of the molecular weight during isolation can also influence T_g. According to Alfthan *et al.* (1973) and Yano *et al.* (1976), the average value of T_g of native hemicelluloses containing side groups in an anhydrous condition should be about 180 °C.

The variations observed in the T_g of lignin are probably due to the difficulties of extraction, leading to degradation of its chemical and physical structure. The results compiled for un-sulphonated and un-oxidized lignin show a variation between 140 and 190 °C. The molecular weight of this isolated lignin is in most cases significantly lower that of native lignin, and this difference has a notorious effect on T_g. Salmén (1979) has suggested a transition temperature of approximately 210 °C for native anhydrous lignin.

At present, thanks to the experimental results obtained by many researchers, the T_g values of wood components as a function of the relative humidity are fairly well known and presented in Figure 5.3.

Figure 5.3(a) shows the glass transition temperatures of the isolated wood polymers as functions of moisture content, and figure 5.3(b) provides the glass transition temperature of the native matrix of hemicelluloses and lignin as a function of the relative humidity. It is thus essential to consider wood as a cellulose – hemicelluloses – lignin material whose three principal components are chemically linked together. Figure 5.3(a) shows that, at relatively high moisture contents, the lignin has the highest glass transition temperature, apart from crystalline cellulose. Therefore, the T_g of lignin determines the limiting lowest temperature of the THM process for wood.

The choice of the temperature of forming must be selected according to two criteria. Firstly, the minimum temperature under which the wood can be formed. This is usually considered to be at least 25 °C higher than the T_g of the lignin, i.e., approximately 110 °C under moisture-saturated conditions and approximately 140 °C at 80% relative humidity. Secondly, the maximum temperature, usually considered to be 200 °C when air is saturated, limit the thermal degradation of the wood components. The thermo-hydrous window for the process of forming of wood is thus restricted to temperatures and relative humidity varying from 110 to 140 °C and from 80 to 100%,

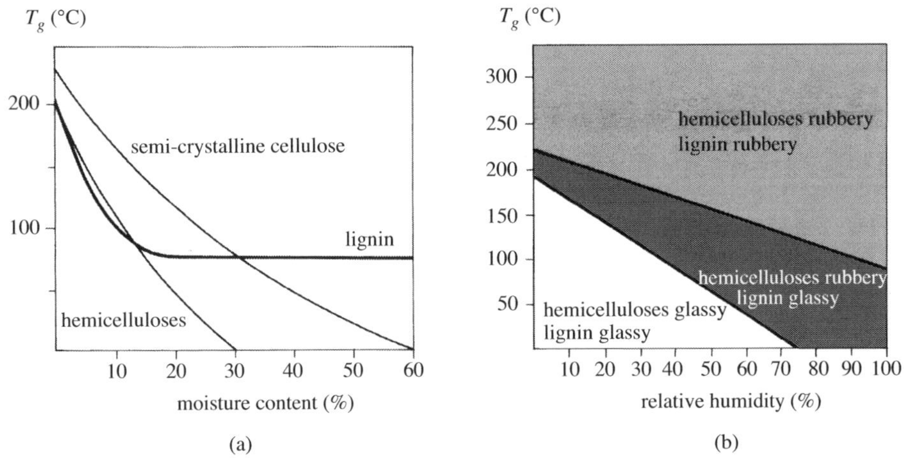

Fig. 5.3 Glass transition temperature for; (a) isolated components as a function of moisture content (Salmén, 1982), (b) matrix of native hemicelluloses-lignin as a function of the ambient relative humidity (Salmén *et al.*, 1986).

respectively. Under these conditions, lignin, hemicelluloses and the semi-crystalline cellulose are relatively mobile and can be deformed easily thanks to two molecular phenomena:

- The inter- and intra-molecular interactions of Van der Waals type in an amorphous or semi-crystalline polymer decrease strongly when the temperature is higher than the T_g.
- The inter- and intra-molecular interactions of the hydrogen bond in an amorphous polymer which has functional hydroxyl (–OH) groups decrease significantly when its moisture content increases. The adsorbed water molecules are placed between the molecules by the hydroxyl groups. Consequently the average inter-molecular distance and the mobility of the molecules increase, and this lowers the T_g.

The water acts as a plasticizer by decreasing the interactions between the hemicelluloses and lignin macromolecules and the amorphous regions of the cellulose. The semi-crystalline cellulose and hemicelluloses have many hydroxyl groups and are highly hygroscopic so that, when they are saturated with water, their T_g drops to a temperature lower than room temperature (Figure 5.3).

On the other hand, lignin contains fewer hydroxyl groups and water therefore does not have a great impact on its T_g which is approximately 85 °C from a moisture content of about 15% to saturation conditions. The large number of hydrogen bonds in crystalline cellulose means that it has a compact and stable structure and that water is unable to penetrate into the crystal lattice. Only the hydroxyl groups on the surface of crystallites can adsorb water. Apparently the percentage of semi-crystallinity of cellulose is not really important. Consequently, the fibrils remain crystalline and very rigid at high temperatures and elevated moisture contents.

5.3 INFLUENCE OF TEMPERATURE AND HUMIDITY ON THE HYGRO-PLASTICITY OF WOOD

Under saturated conditions at room temperature, the semi-crystalline cellulose and hemicelluloses are in their rubbery state. Lignin on the other hand is in its glassy state and a higher temperature is thus required if wood is to be deformed easily. Uhmeier *et al.* (1998) have shown that the plasticity threshold of spruce decreases with an increasing temperature, as shown in Figure 5.4, but that the derivative of the plasticity threshold with respect to temperature shows a maximum temperature at 85 °C, see Figure 5.5(b). This is probably related to the glass transition of lignin in this species under saturated conditions. The glass transition temperature (T_g) of the in-situ lignin under saturated conditions has been determined by Salmén (1984) and by Olsson and

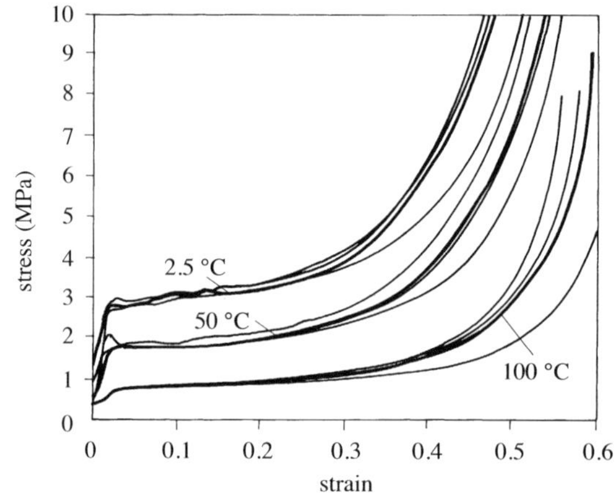

Fig. 5.4 Stress-strain curves at various temperatures. The "typical" curve for each temperature is indicated by a thicker line (Uhmeier *et al.*, 1998).

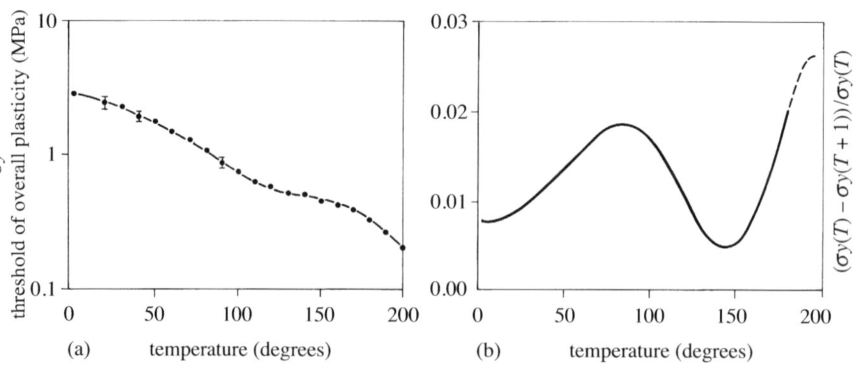

Fig. 5.5 (a) Threshold of plasticity in radial compression (logarithmic scale) as a function of temperature. (b) Derivative of the threshold of plasticity (tan(δ))with respect to temperature as a function of temperature (Uhmeier *et al.*, 1998).

Salmén (1992) using Dynamic Mechanical Analysis, DMA. They have shown that tan (δ), (Figure 5.5b) passes through a maximum at 83 °C. It is usually possible to mold polymers at a temperature of T_g + 25 °C, and one can therefore reasonably expect that wood can easily be deformed at temperatures above 110 °C under saturated condition.

The term "plasticity threshold" of wood is often employed improperly. It would be more appropriate to talk of the "buckling threshold" of the cellular walls in the direction of loading, since the material is viscoelastic and this deformation is partially or even totally reversible, so that the plasticity is only apparent. Only materials like metals, for example, present a truly plastic behavior (Figure 5.6). When metals are loaded above their elastic limit, they undergo an irreversible plastic deformation, which corresponds to dislocations due to local shear stresses, but their modulus is not modified during loading-unloading.

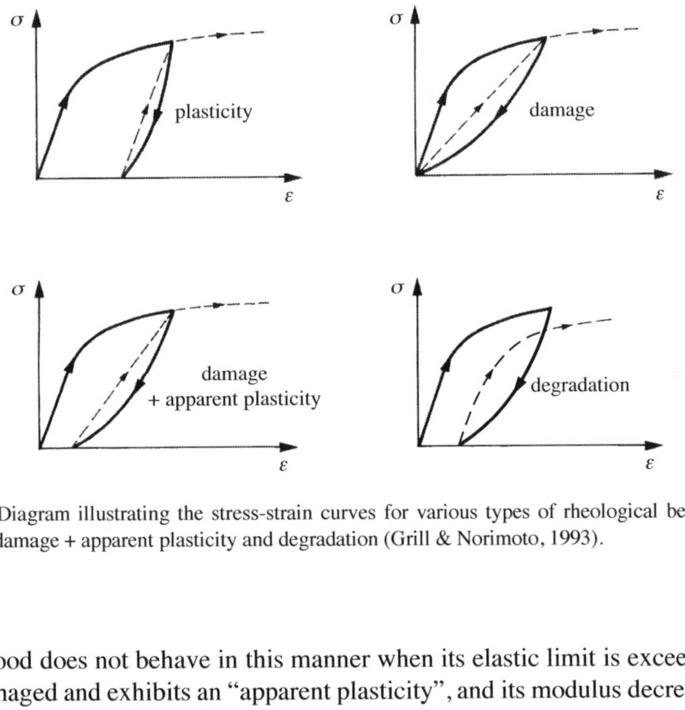

Fig. 5.6 Diagram illustrating the stress-strain curves for various types of rheological behavior: plastic, damage, damage + apparent plasticity and degradation (Grill & Norimoto, 1993).

Wood does not behave in this manner when its elastic limit is exceeded. Rather, it is damaged and exhibits an "apparent plasticity", and its modulus decreases in loading-unloading. The energy used for the deformation is partly stored by wood and the curve regains its initial shape if the wood is soaked in water or heated (Figure 5.7). Nevertheless, a permanent deformation remains, which is a sign of structural damage or chemical degradation of the wood components during the experiment. At temperatures below the T_g of lignin, the damage is significant and probably involves the structural damage of rays (when the densification is in the radial direction) and the cellular zones are largely deformed. On the other hand, above T_g, the permanent deformation is primarily due to thermal degradation of the wood components. In this case, the mechanical and physical characteristics of the wood are usually altered.

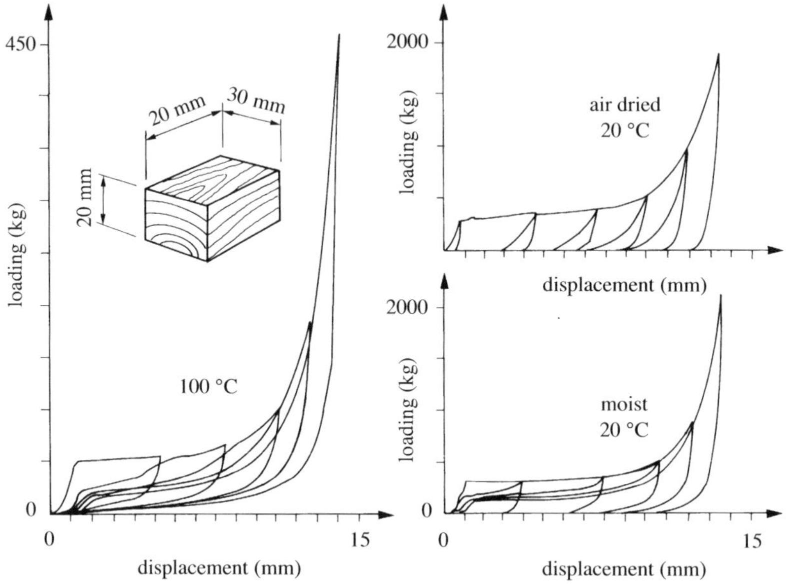

Fig. 5.7 Stress-strain curves of compression of a coniferous tree (Sugi) with restraints (Liu *et al.*, 1993).

5.4 INFLUENCE OF TEMPERATURE ON THE THERMAL DEGRADATION OF WOOD

The thermal degradation of wood can be revealed for example by studying the recovery-set of densified wood samples prepared at several temperatures under saturated

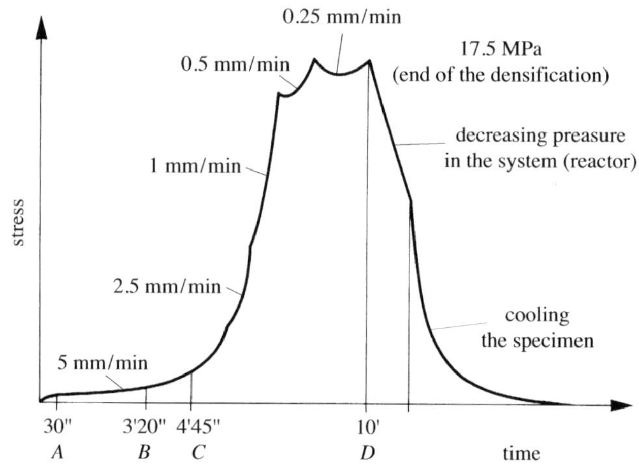

Fig. 5.8 The stress as a function of time during compression in the radial direction of $40 \times 40 \times 30$ mm^3 ($l \times t \times r$) spruce samples (Heger, 2004). The letters *A*, *B*, *C* and *D* refer to the photomicrographs in Figures 5.11 and 5.12.

steam conditions. Figure 5.8 shows the densification of spruce samples of $40 \times 40 \times 30$ mm^3 ($l \times t \times r$) densified in the radial direction in a closed system.

The specimens were compressed to a pre-determined level of $C = 65\%$ under the same conditions of loading and time, but at different temperatures. The total time of processing was 18 minutes and 20 seconds for all specimens. The compression-set C is defined as

$$C = \frac{X_0 - X_c}{X_0} \tag{5.1}$$

where X_0 is the initial thickness of the specimen in the compression direction and X_c is the thickness after compression.

The recovery test consisted in immersing the compressed samples in distilled water at 60 °C for 5 days and drying them at 30 °C for 5 days. After repeating this procedure 5 times, the samples were carefully dried to determine the recovery R as a percentage of total compression-set. The value of R is expressed as

$$R = \frac{X_c' - X_c}{X_0 - X_c} X\ 100(\%) \tag{5.2}$$

where X'_c is the thickness of the specimen after recovery. This test reveals any mechanical damage and thermal degradation due to the compressive forces, moisture vapor and heat. Figure 5.9 presents the recovery curves of the densified samples at different temperatures and Figure 5.10 indicates the recovery as a function of the temperature during compression.

Figure 5.10 shows that the recovery starts to diminish strongly above 150 °C and becomes negligible at 200 °C. The strong reduction observed above 150 °C may be related

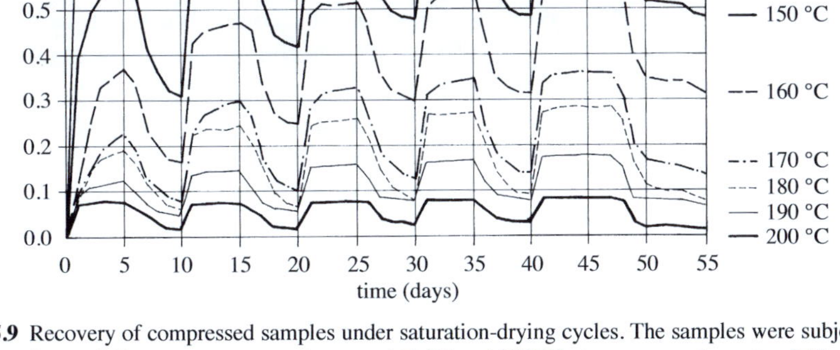

Fig. 5.9 Recovery of compressed samples under saturation-drying cycles. The samples were subjected to compression at different temperatures between 110 °C and 200 °C at a compression rate of 10 mm/min (Heger, 2004).

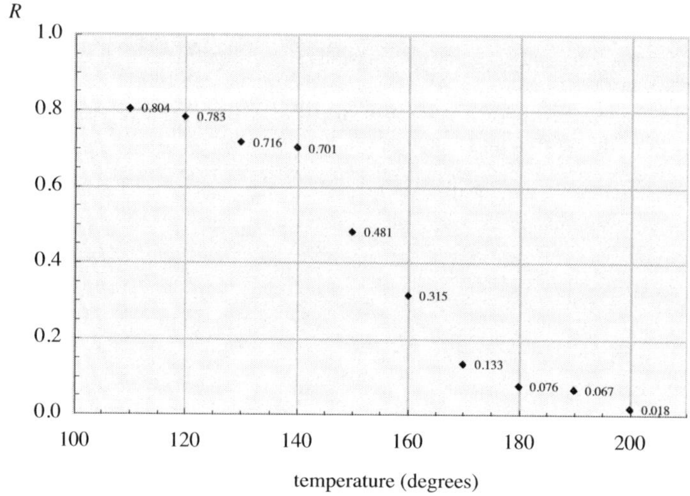

Fig. 5.10 Recovery (*R*) of spruce samples compressed at different temperatures between 110 °C and 200 °C at a compression rate of 10 mm/min (Heger, 2004).

to the increase in the derivative of the threshold of plasticity at about 150 °C observed by Uhmeier *et al.* (1998), as shown in Figure 5.5. They suggested that this increase was due to some phenomenon other than thermal softening. Takahashi *et al.* (1998) have observed extensive thermal degradation after wet thermal treatment for 4 minutes at 180 °C. The nature of this thermal degradation is further discussed in the section devoted to post-treatment of THM processing. The optimal compression temperature is therefore approximately 110 °C, which is the limiting temperature still permitting a good plastic deformation of the cellular structure of wood without any extensive thermal degradation.

5.4.1 Deformation of the cellular structure of spruce during transverse compression

When subjected to transverse compression, the cellular structure of wood undergoes a large deformation in precise stages (Schrepfer & Schweingruber, 1998). Figures 5.11 and 5.12 present confocal photomicrographs at the cellular level of specimens compressed to 5, 35, 50 and 65%. These compression ratios correspond respectively to the letters *A*, *B*, *C* and *D* in Figure 5.8. In these photomicrographs, several stages during the deformation of the fibers can be observed:

- When a compressive stress is applied, the cells deform elastically until the yield point (or buckling threshold) is reached, which corresponds to a stress level of about 0.5 MPa at 140 °C under saturated moisture conditions. After this point, while the stress level is increasing, the cell walls undergo a greater deformation and the lumen of the earlywood cells starts to close. The deformation occurs as though the cells were subjected to shearing[1]. The earlywood

1 In Figure 5.13(b), even under a low magnification, it is possible to see the mode of deformation by cellular shearing and the zigzagging of the woody rays.

cells are the first to be deformed, since their cell wall thickness (~2 μm) is much less than that of the latewood cells (~10 μm). In general, failure of cells starts near the interface between the last row of latewood cells and the first row of earlywood cells[2]. After the buckling threshold has been passed (*A* in Figure 5.8), the stress increases only slightly.

- When the first cellular row yields, the phenomenon is gradually propagated as shown in Figure 5.11(b). The stress increases almost linearly up to a compression of 35%. This increase is due to a closing of the lumens, which multiplies the points of contact between the walls of the earlywood cells, and makes it

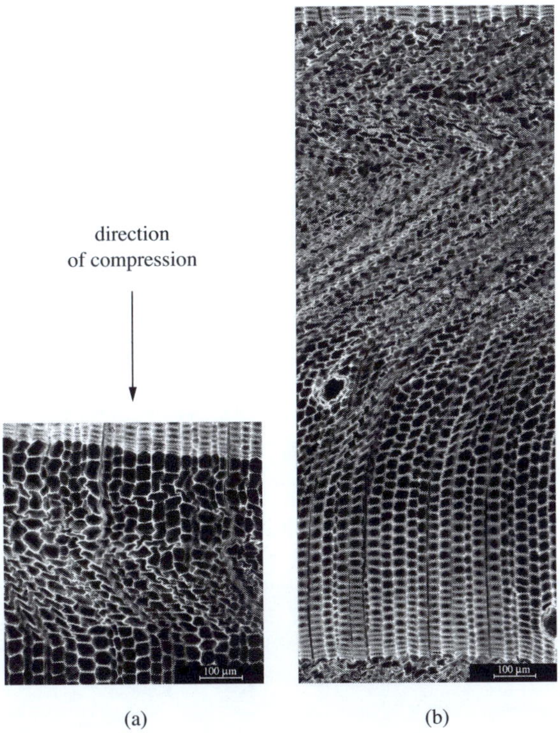

Fig. 5.11 Confocal photomicrographs of compressed specimens of spruce with a compression of: (a) 5% (*A* in Figure 5.8); a cellular shear band at earlywood zone can be seen, (b) 35% (*B* in Figure 5.8); most of the earlywood cells are buckled.

2 During compression, one can distinguish in the photomicrograph in Figure 5.14a two types of walls: the walls between two red points or two green points are tangential walls and the walls between a red point and a green point are radial walls. The latter are the first to be deformed. Moreover, they contain many pits which decrease their resistance. Figure 5.13b shows the sinusoidal deformation after compression.

possible to transmit a slightly higher stress to the closed cellular rows.[3] As the compression increases, so does the thickness of these buckled bands and other cell rows start to yield.

- When the multiplication of contact points has spread throughout the earlywood zone and the compressive stress has increased sufficiently, the walls of the latewood cells start to yield (Figure 5.12a). Above a compression of 35%, there is a more marked increase in the applied stress. The latewood cells do not deform in the same manner as the earlywood cells. Since their walls are thicker, they may not buckle but rather undergo large deformation. Moreover, the latewood zone is less porous than the earlywood zone.
- When all the cell walls have yielded, including those of the latewood cells, the stress necessary to continue the compression increases exponentially because the areas of the cell walls in contact becomes greater (Figure 5.12b). The compression is total when all the lumens are entirely closed. The compression cannot proceed further without risking degradation of the microstructure of the cell walls, since the cell wall is then compressed.

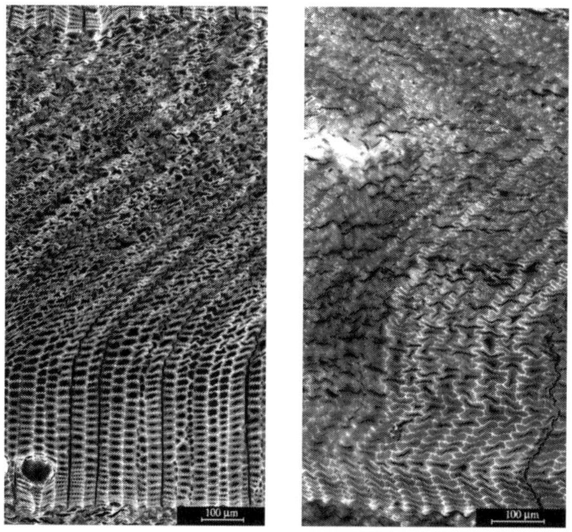

(a) (b)

Fig. 5.12 Confocal photomicrographs of compressed specimens of spruce with a compression of: (a) 50% (*C* in Figure 5.8); all the earlywood cells and some of the latewood cells are crushed, (b) 65% (*D* in Figure 5.8); all the spring and latewood cells are crushed.

---

3 Each cell has its own yield stress beyond which it is becomes crushed (buckled). This yield stress depends on the thickness of the walls. The cells with the thinnest walls in the direction of the applied compression, yield first. To deform cells with thicker walls, it is necessary to increase the applied pressure and for this pressure to be transmitted through the points of contact between cell walls. These points are visible in the photomicrograph in Figure 5.14b. When each row of cells buckles, this contributes to the multiplication of the points of contact and consequently a slight increase in its failure stress, but enough for the buckling of the next row of cells.

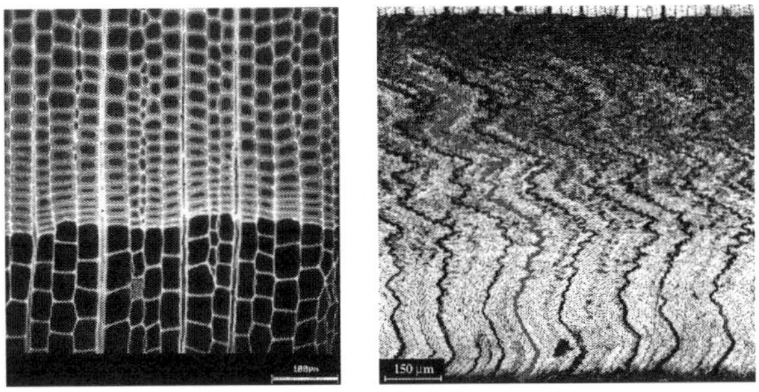

(a) (b)

Fig. 5.13 Confocal photomicrographs of a spruce sample (a) before compression and (b) totally compressed. In red: zigzag deformation of wood rays.

Under compression, the cell walls are deformed with a certain degree of regularity. When a cell row is crushed, the tangential walls that are in opposition shear and lead to a systematic deformation of the S-shaped or Z-shaped radial walls following the direction of shear (see Figs. 5.14 and 5.15). The middle lamella plays a dominant part in the compressive deformation of wood, since, thanks to pectin, it allows each cell to rotate relative to its neighbors in order to minimize the stress on the structure (Figure 5.16).

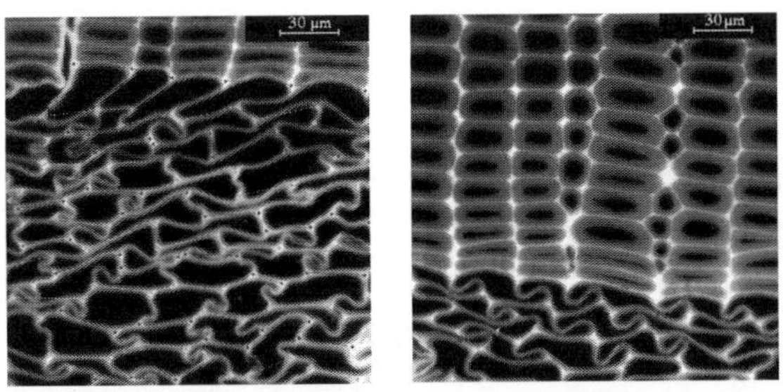

(a) (b)

Fig. 5.14 Confocal photomicrographs of a sample of densified spruce compressed to 5%; detail of a partially crushed earlywood zone. (a) The corners of each cell are indicated by red and green points. The red and green points represent horizontal faces in opposition. (b) Points of contact between cell walls are indicated by red points.

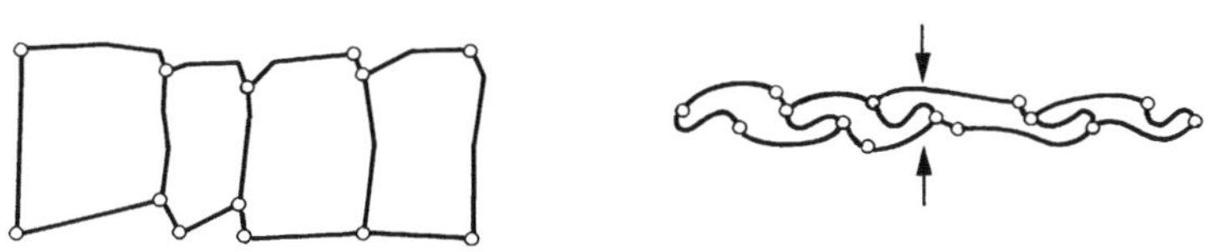

(a) (b)

Fig. 5.15 Diagram of the S-shaped deformation (or Z if shearing is exerted from left to right) of the radial cell walls of spruce; (a) cells before compression, (b) cells after compression.

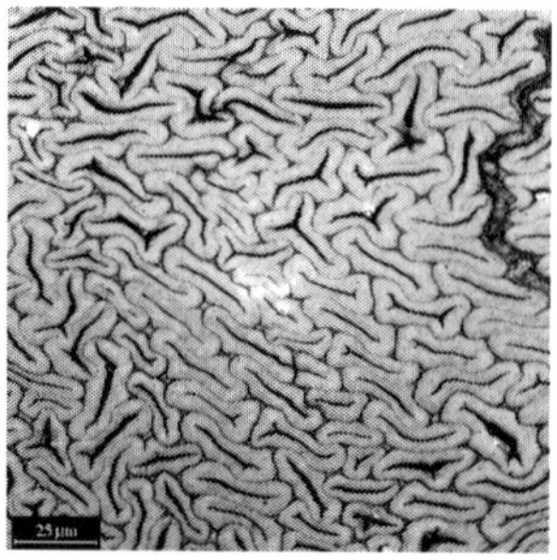

Fig. 5.16 Confocal photomicrograph showing the rotation of some cells relative to others after compression at 140 °C under saturated moisture conditions.

5.5 CHEMICAL DEGRADATION OF WOOD CONSTITUENTS UNDER THERMO-HYDRO-MECHANICAL ACTION AT TEMPERATURES UP TO 200 °C

5.5.1 Reactivity of the wood components in acid conditions

In previous sections, it was pointed out that high temperature steam, i.e., both a high humidity and a high temperature, may chemically degrade wood constituents. In this section, the degradation of wet wood components at a temperature up to 200 °C is discussed. Hemicelluloses contain many carboxyl groups, rendering wood a naturally acid material. The reactivity of the wood components under acid conditions is special. Indeed, the components that can be hydrolyzed participate in the acidification of the medium. The acid medium can differ significantly according to the conditions under which the THM treatment is carried out. The temperature, the moisture content distribution, the time of treatment and the pressure can strongly affect the reactivity of the medium.

Effects of temperature and moisture content

At high temperatures, the dissociation of water molecules is greater than at 20 °C. The concentration of hydroxide and hydronium ions (products of water dissociation) is 10^{-7} mol/l at 20 °C whereas, at a temperature of about 200 °C, the concentration is 10^{-6} mol/l. In other words, at this temperature, the pH of water is 6 and not 7. In most chemical reactions, the increase in temperature also increases the speed of the reaction. The effect of temperature may be increased in the presence of water.

In wood at high temperatures, water thus plays three important roles: it makes hydrolysis possible, it ensures the mobility of the protons and it participates in the acidification of the reactive medium. However, it is the quantity of water present which most strongly influences the acid medium. If there is only a very small quantity of water, it will take part only in the hydrolysis, and will not be sufficient to dissolve the reaction products. The reactive medium is thus solid and strongly localized around the glycosidic links. On the other hand, if the amount of present water is large, the reaction products will be dissolved and take part in the hydrolysis. Their mobility increases with an increasing moisture content.

5.5.2 Reactivity of polysaccharides

The principal functional groups that are potentially reactive in the polysaccharides are:

- reducing terminal groups;
- glucosidic links; and
- hydroxyl groups.

The reactivity of these functional groups varies strongly, however, between cellulose and hemicelluloses because of their chemical structure and their molecular arrangement within the cell wall and in the middle lamella.

Reducing terminal groups
All natural polysaccharides contain a reducing terminal group. This group is a hemi-acid, and it is partially converted into an aldehyde with an open chain in solution. This functional group can be reduced or oxidized to an alditol or an acid aldonic group, respectively. Also, since the hydroxyl group on the C1 carbon of the reducing terminal group is most strongly acid, it can be selectively esterified. The reducing terminal groups of cellulose are naturally less accessible than those of hemicelluloses because of the semi-crystalline nature of the microfibrils.

Glucosidic links
The glucosidic links can be hydrolyzed under acidic, alkaline and oxidizing conditions. Acidic hydrolysis is the basis of the process of saccharification, whereas alkaline hydrolysis requires much more complex conditions. The hydrolytic reaction of polysaccharides in general reduces the mechanical strength of wood.

The hydroxyl groups
The constitutive units of cellulose and hemicelluloses, except for those at the end of the chain, include a primary hydroxyl group (OH-6) for each anhydrous hexose and two secondary hydroxyl groups (OH-2 and OH-3) for each anhydrous hexose and pentose. These hydroxyl groups are sensitive to oxidation, and the aldehyde groups and resulting ketone groups can initiate other degradation reactions. Among the three hydroxyls groups, the OH-2 group is the most acidic because of the anomeric electrophilic influence of the center. The acidity of the OH-2 group is also influenced by other factors such as the intra- and inter-chains of the hydrogen bonds. The reactivity of the hydroxyl groups in a heterogeneous system is also affected by their accessibility. The hydroxyl group is in an area inaccessible to the reagent and is bound by a hydrogen bridge. For this reason, the semi-crystalline cellulose microfibrils are much less reactive than the amorphous hemicelluloses.

5.5.3 The hydrolysis of polysaccharides

Acidic hydrolysis or acidolysis is an important reaction of polysaccharides. The mechanism involved at the molecular level is well-known since acidolysis forms part of the process of chemical analysis. It was confirmed by several studies using models and tracers (Shafizadeh, 1963; Timell, 1964). Figure 5.17 presents the three principal stages of the reaction mechanism in the hydrolysis of polysaccharides. The upper reaction mechanism is the most probable, since it generates a tautomeric carbonium-oxonium ionic pair (Figure 5.18), which is more stable than the non-cyclic cation, i.e., carbonium, in the lower reaction mechanism. It should be noted that hydrolysis can take place only in the presence of water since water is consumed at the end of the reaction process, which is catalysed by the presence of acid. On the other hand, the absence of water does not prevent the formation of potentially reactive carbonium ions and thus a chemical decomposition (degradation) of polysaccharides.

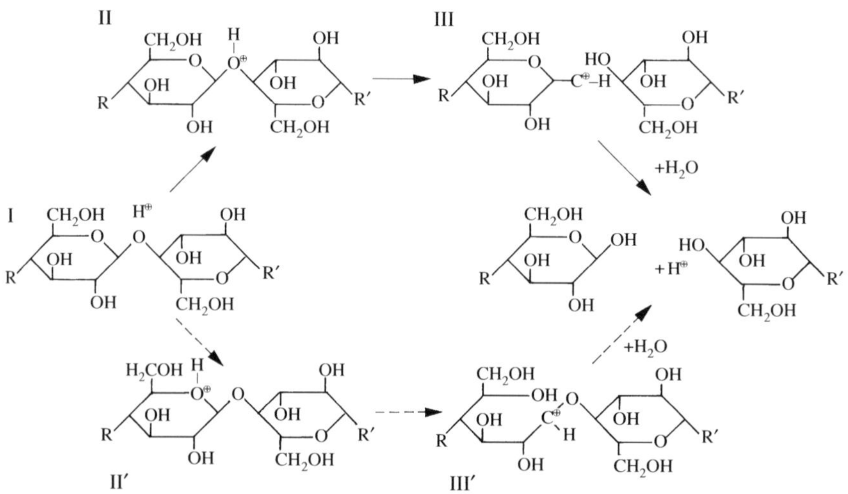

Fig. 5.17 Mechanisms involved in the acidic hydrolysis of polysaccharides (Fengel & Wegener 1983): (I) fast interaction of the proton of the catalyzing acid with the glycosidic oxygen linking two sugar units, (II) formation of a so-called conjugated acid or (II′) protonation may also occur at the ring oxygen, (III) slow cleavage of the C-O bond and formation of a cyclic carbonium cation or (III′) opening of the ring and formation of a non-cyclic carbonium cation, (IV) addition of a water molecule on the cation carbonium to release a proton and form a stable ring.

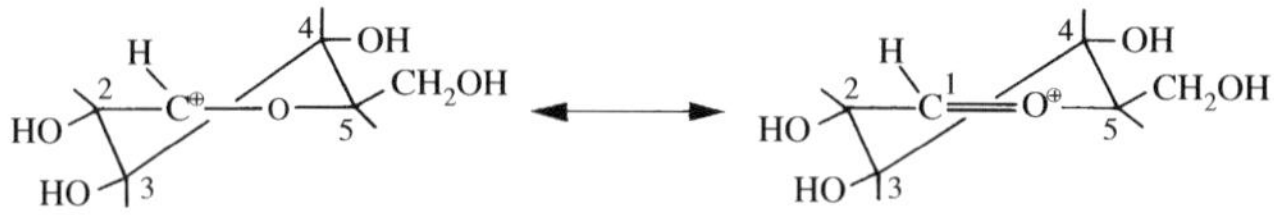

Fig. 5.18 Semi-chair conformation of the tautomeric carbonium-oxonium ionic pair (Fengel & Wegener, 1983).

Factors influencing the hydrolysis

The kinetics and the evolution of the hydrolysis reaction are influenced by the acidity of the medium and the state of the sample, Table 5.1. The most frequent case is that of cellulose or polysaccharides in a solid state in a more or less acidic solution. If the sample dissolves easily and completely in the acid, the hydrolysis is homogeneous in the liquid state. More rarely, hydrolysis can occur where the solid or dissolved sample is in contact with a solid acidic component, such as a resin able to exchange cautions, used for example in problems of selective acidic hydrolysis.

The hydrolyzing medium can be characterized by the type of acid, its concentration, the strength of the acid (pKa), its temperature and its pressure. The value of the hydrolytic activity expressed by the pH and the coefficient of activation of the proton (Zecchina *et al.*, 1998) (expressed by the acidity function of Hammett, H_0) also influence the kinetics of the hydrolysis. In addition, the sample can be characterized by its phase state, its physical structure and accessibility in the case of heterogeneous hydrolysis, the conformation of its polysaccharides, the structure and the constituents of the rings.

Table 5.1 Possible conditions during hydrolysis reactions (Fengel & Wegener, 1983)

State of the sample	State of the acid	Examples
Solid	Liquid	Cellulose in the presence of a dilute acid
Dissolved	Concentrated liquid	Cellulose in the presence of a concentrated acid
Dissolved	Solid	Sucrose in the presence of a resin that can exchange ions
Solid	Solid	Cellulose in the presence of a resin that can exchange ions

The hydrolysis of the glucosidal link follows a first order reaction with respect to the number of inter-monomer links (Daruwalla & Shet, 1962; Springer, 1966; Toussaint, 1990). The hydrolytic kinetics also depends on the parameters mentioned above. The rate of heterogeneous hydrolysis of cellulose is for example one to two orders of magnitude lower than the rate of homogeneous hydrolysis (Visapää, 1971, 1972). It is mainly influenced by the degree of crystallinity and the moisture content. The hydrolytic behavior of the glycosidic link is also influenced by the conformation of the constitutive sugars in the polysaccharides and by the inductive effect of some constituents (Fengel & Wegener, 1983). In theory, the influences of the conformation of sugars and their constituents are superimposed.

Dehydration and condensation reactions

Although the direct products of the hydrolysis (mono and oligosaccharides) are relatively stable under acid conditions, they can undergo other reactions such as dehydration (for instance during heat treatment of wood when the temperature and acid concentration are high), fragmentation or condensation. These reactions generate products such as phenolic furans and other compounds by elimination of water (Figures 5.19 and 5.20).

Hemicelluloses are more sensitive to hydrolysis than the cellulose microfibrils because of their amorphous and branched structure and the diversity of the non-glucosidic units forming them. These differences are at the level of their chemical structure which causes the dissimilarities in the configuration of the hydroxyl groups. Due to their high susceptibility to hydrolysis, hemicelluloses can be selectively extracted from lignocellulosic substrates. As shown in Table 5.2, the rate of homogeneous

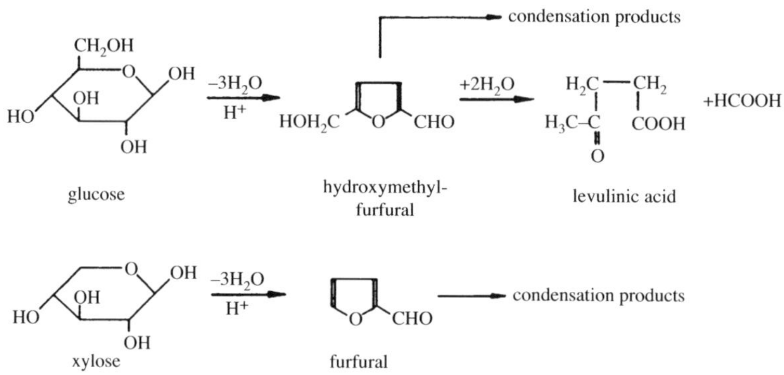

Fig. 5.19 Formation of furfural, hydroxy-methyl-furfural, levulinic and formic acids starting from monosaccharides in an acidic medium (Fengel & Wegener, 1983).

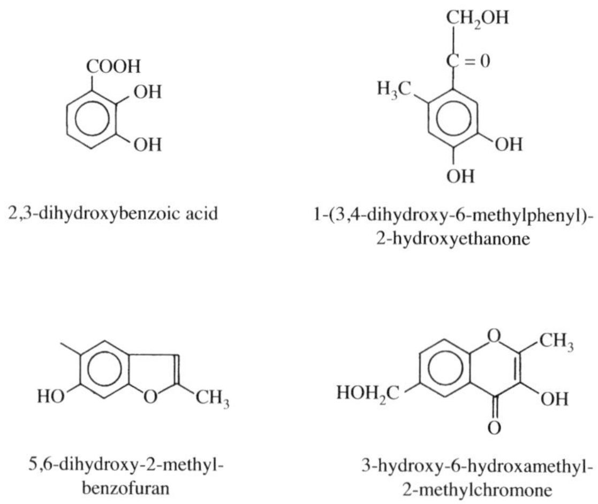

Fig. 5.20 Aromatic compounds formed by sugar dehydration in an acidic medium.

hydrolysis of polysaccharides with β-(1- 4) bonds increases in the order of cellulose (1) < mannan (2-2.5) < xylan (3.5) < galactan (4-5).

The heterogeneous hydrolysis of these polysaccharides reveals an even more marked difference in the order of cellulose (1) < mannan (60) < xylan (60-80) < galactan (300), which shows the dominant role of accessibility in the acidic degradation of polysaccharides.

The acetyl groups in the xylans of hardwood trees are, like the galacto-glucomannans, also hydrolyzable, particularly at high temperatures. The acetic acid thus released contributes significantly to the acidity of the reaction medium.

Table 5.2 Comparison of the kinetics of hydrolysis of the methylated pentopyranosides (Shafizadeh, 1963)

Hexose/Pentose	Rate of hydrolysis[a]		Conformation of the substituents[d]	
	k/k'[b]	k/k'[c]	Axial sub.	Equatorial sub.
α-D-Glucose	0.42	0.34	1	2, 3, 4, 5
β-D-Glucose	0.80	0.60	–	1, 2, 3, 4, 5, 6
α-D-Mannose	1.0	1.0	1, 2	3, 4, 5
β-D-Mannose	2.4	–	2	1, 3, 4, 5
α-D-Galactose	2.2	1.7	1, 4	2, 3, 5
β-D-Galactose	3.9	2.5	4	1, 2, 3, 5
α-D-Xylose	1.9	1.3	1	2, 3, 4
β-D-Xylose	3.8	2.8	–	1, 2, 3, 4
α-L-Arabinose	5.5	6.7	1, 2, 3	4
β-L-Arabinose	3.8	4.6	2, 3	1, 4

[a] Ratio of the constant speed k of Glycoside to that of the α-D-Mannopyranoside methylated k'
[b] Ratio based on hydrolysis with HCl (0.5 N) at 75 °C
[c] Ratio based on hydrolysis with HCl (2 N) at 60 °C
[d] Numbers of carbons comprising axial or equatorial substitutes

Xylans

The presence of uronic acid strongly reduces the rate of hydrolysis of xylans by an inductive effect (Table 5.2). The initial rate of the relatively weak hydrolysis of xylans of hardwoods by water at 170 °C is related to the uronic acid groups that they contain. For example, the xylans in softwood, which have a larger proportion of uronic acid groups, are more stable in the sulphite pulping processes.

Glucomannans

The α-D-(1 → 6)- galactosidic link of the galactoglucomannans shows a very low resistance to hydrolysis and can be broken selectively through treatment by dilute oxalic acid (0.05 M) at 100 °C. This low resistance is due to the presence of acetyl groups on the chain.

5.5.4 Reactivity of the cellulose microfibrils

The considerations presented above concerning the factors influencing the hydrolysis make it possible to understand why the cellulose microfibrils are chemically more resistant to the hydrolysis.

The chemical stability of the cellulose microfibrils is due mainly to their semi-crystalline nature. Hydrogen bonds present in the (020) plane of cellulose I_β ensure this stability. These bonds consist of two intra-chain hydrogen bonds between HO3- and O5′ and HO6- and HO2′-, and one inter-chain hydrogen bond between HO6- and HO3- of the close chain (Gardner & Blackwell, 1974) (Figure 5.21). The energy of this bond is two orders of magnitude higher than the energy of the Van der Waals bonds: 15 kJ/mol between a water molecule and cellulose and 0.155 kJ/mol between liquid water molecules. Consequently, the water molecules cannot be adsorbed in this crystalline structure; only the surface of the crystallites is accessible to water molecules. On the other hand, the amorphous parts of the cellulose microfibrils are able to adsorb water, since their structure is irregular and they have much fewer inter-chain hydrogen bonds.

If water is able to enter only into the amorphous (para-crystalline) parts of the microfibrils, the hydrolysis can thus be highly heterogeneous, because the reaction consumes water. The results obtained by Toussaint (1990) on the acidic hydrolysis of the cellulose of poplar (*Populus Tremula*) show that at high temperature (225 °C) and in the presence of sulphuric acid (1%), the degree of cellulose polymerization (measured by viscometry), decreased very quickly to a value of about 150 after only 60 seconds, after which the rate of hydrolysis was highly reduced. This limit corresponds approximately to the length of the crystallites present in the micro-fibrils and is known as the LODP (Leveling-Off Degree of Polymerization). The amorphous parts are thus preferentially hydrolyzed, at a rate that is two orders of magnitude higher than that of the crystalline parts.

Although the hydroxyl group is the principal functional group in the "ideal" cellulose chain, the cellulose extracted from wood or other plants has a more or less significant number of other functional groups, mainly carboxyl groups, distributed randomly along the chains. By an inductive effect, the carboxyl groups make the β-(1- 4) bonds, of which the electronic density is increased, more sensitive to hydrolysis.

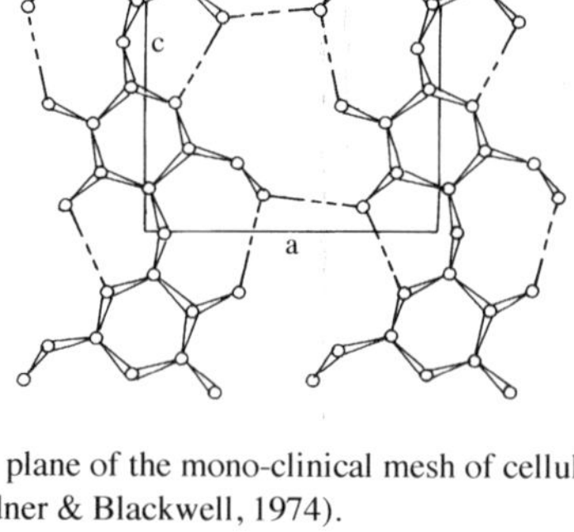

Fig. 5.21 Projection of the (020) plane of the mono-clinical mesh of cellulose I_β. The hydrogen bonds are indicated with dashed lines (Gardner & Blackwell, 1974).

5.5.5 Reactivity of lignin

Reactivity of the ether bonds in the classical model of lignin

The carbon-carbon link is chemically very resistant. The decomposition of lignin is mainly restricted to the ether bonds on the level of carbons α and β and the hydroxyl groups (Nimz, 1965a,b; Nimz & Gaber, 1965; Nimz, 1966a,b; Nimz, 1967a,b). The hydrolyzable ether functions are the β-aryl, α-aryl and α-alkyl bonds. The β-aryl (~50%), non-cyclic α-aryl (< 3%) and cyclic α-αaryl (~10%) bonds are also the most frequent in lignin (Figure 5.22). Adler *et al.* (1968) have measured a loss of 6-9% on lignin of ground spruce by a fairly acidic hydrolytic reaction. This loss is due mainly to the hydrolysis of the α-aryl ether bonds.

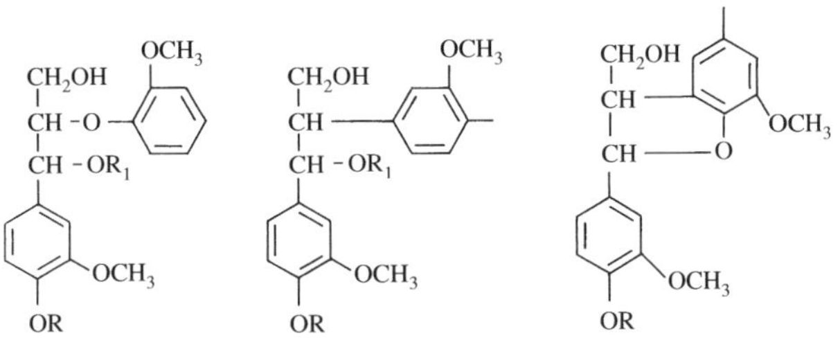

Fig. 5.22 Hydrolyzable bonds in lignin (according to Hon and Shiraishi, 1991): (a) β-aryl bond or ($R_1 = \text{H}$), non-cyclic α-aryl bond (R_1 = aryl), (b) non-cyclic α-aryl bond (R_1 = aryl), (c) α-aryl cyclic bond.

Reactivity of the hydroxyl groups

The phenolic hydroxyl groups are very important for the physical and chemical properties of lignin and thus play a dominating part in the process of delignification. Lignin also contains two types of aliphatic hydroxyl groups localized at the α- and γ- positions. The hydroxyl groups on the α-carbon are very reactive and play a dominant role in the reactions of lignin.

Reactivity of the lignin-carbohydrate complex

The concept of lignin-cellulose complexes or lignin-carbohydrate complexes (LCC) have arisen mainly from the difficulty encountered in the separation of the final residues of lignin during the cooking of wood pulp (a process used to separate of cellulose

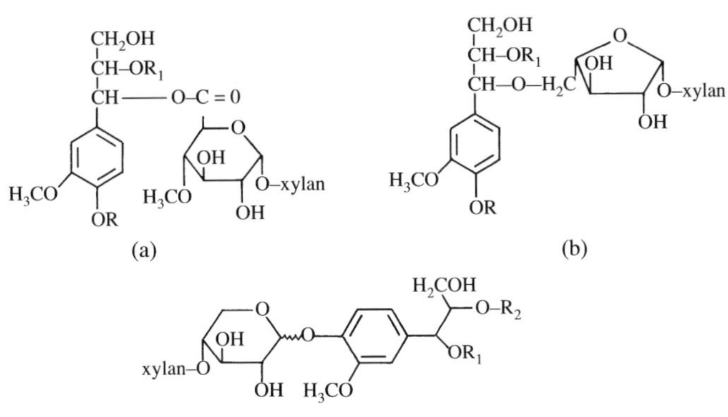

Fig. 5.23 Possible lignin-carbohydrate links: (a) benzyl ester, (b) benzyl ether, (c) glucosidic link (Hon & Shiraishi, 1991).

from hemicelluloses and lignin for paper making). Three types of LCC- linkages were detected: benzyl esters, benzyl ethers and glucosidic links shown respectively in Figures 5.23 (a), (b) and (c). Among these LCCs, the benzyl esters and the glucosidic links, are probably more susceptible to acidic hydrolysis than benzyl ethers. On the other hand, the reactivity of the benzyl ether is increased in the presence of a phenolic hydroxyl group.

5.6 PHYSICO-CHEMICAL MODIFICATIONS OF WOOD COMPONENTS DURING THM POST-PROCESSING

Numerous experiments have been conducted on densified wood with or without THM post treatment to better understand the physicochemical modification of wood components during THM post treatments. Some of these results on spruce wood are presented and discussed below. These experiments have been carried out at the Swiss Federal Institute of Technology in Lausanne where the densified samples have been subjected to heat treatment for different periods of time.

5.6.1 Quantitative analysis of hemicelluloses by GC

Neutral sugars of wood such as glucose, galactose, mannose, xylose and arabinose can be identified and quantified by gas chromatography, GC of the corresponding acetates on alditols (type of acidicpolyol derived from aldose). In order to determine the quantities of hemicelluloses present in THM post-treated wood after washing with water, three samples were prepared and analyzed by GC. Figure 5.24 shows their GC spectra and Table 5.3 gives the percentages of the different neutral sugars. The peaks were referenced and quantified according to an internal inositol standard.

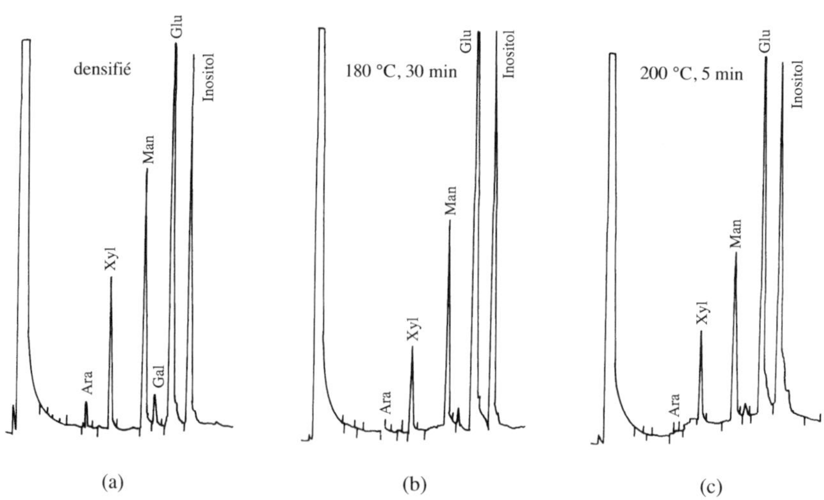

Fig. 5.24 GC spectra of three samples of spruce: (a) densified, (b) densified and post-treated at 180 °C for 30 min, and (c) densified and post-treated at 200 °C for 5 min.

Table 5.3 Percentages (of the initial mass) of neutral sugars in GC spectra, (Heger, 2004).

Treatments	Mannose	Xylose	Galactose	Arabinose	Total
Reference (natural)	13.6	5.6	2.8	1.2	23.2
Densified (140° C, 10 min)	9.7	4.7	1.2	0.7	16.3
Densified and post-treated (180 °C, 30 min)	5.6	2.0	0.5	no-traces	8.1
Densified and post-treated (200 °C, 5 min)	5.4	2.8	0.4	no-traces	8.6

The results in Table 5.3 show that 6.9% of the initial mass of the sample, consisting of cellulose, lignin and hemicelluloses (mannose, xylose, galactose and arabinose), was dissolved during the densification (140 °C, 10 min) and washing with water. The mass loss increased considerably when the sample was subjected to a post-treatment. The total lost was 15.1% during treatment at 180 °C for 30 minutes and 14.6% at 200 °C for 5 minutes. After the densification, the quantity of water-soluble hemicelluloses was already relatively high, signifying that the hydrolysis of hemicelluloses was not extensive, since hemicelluloses are soluble even at its relatively high degree of polymerization. On the other hand, during the post-treatment, the hydrolysis was relatively high since one third of the hemicellulose material was dissolved.

5.6.2 Determination of the degree of polymerization of cellulose by capillary viscometry

In order to determine the hydrolytic effect on the cellulose of the post-treated densified samples of spruce, the degree of polymerization of cellulose was determined according to the AFNOR (NF G06-037) standard with a capillary viscometer of Ubbelohde type (model SCHOTT, Type 501 10/I). The average degree of polymerization $\overline{DP_V}$ of the cellulose was determined from the intrinsic viscosity measured in a solution of cupri-ethylene-diamine (CED), according to the equation

$$\left(\overline{DP_V}\right)^{\alpha} = \frac{\eta}{K} \tag{5.3}$$

The parameters α and K in (5.3) are constants depending on the temperature, the polymer and the solvent. In this experiment, $\alpha = 1$ and $K = 7.5 \cdot 10^{-3}$. According to the AFNOR standard, the reproducibility is of the order of ± 3%, η is the limiting index of viscosity, calculated with the help of a table given in the AFNOR (NF G 06-037) standard. The relative increase in viscosity η_a can be calculated directly from the flow time using the equation

$$\eta_a = \frac{t - t_0}{t} \tag{5.4}$$

where t_0 is the average flow time of the solution of the CED solvent and t is the average flow time in seconds of the cellulose solution.

The samples tested were

- densified spruce;
- densified spruce post-treated at 180 °C for 20 minutes; and
- densified spruce post-treated at 200 °C for 4 minutes.

Two solutions of each sample were prepared and the viscosity of each solution was measured at least six times to assess the reproducibility of the measurements (the mode of extraction of cellulose is given in Heger 2004). Results are presented in Table 5.4.

Table 5.4 Degree of polymerization $\overline{DP_V}$ of three spruce samples.

Treatments	Densification	Densification post-treatment at180 °C, 20 min	Densification post-treatment at 200 °C, 4 min
$\overline{DP_V}$	1533	947	837

During the initial densification, cellulose microfibrils undergo deformation and store the elastic energy, which is at the origin of the shape memory of wood. The hydrolytic decomposition of the cellulose is thus probable to be an important phenomenon, which is likely to dissipate part of the stored energy. In practice, microfibrils at certain places are more susceptible to hydrolysis than others. Only the amorphous parts of the microfibrils and the surfaces of cellulose crystallites can be hydrolyzed. It is also known that the zones under stress can hydrolyze more quickly.

A comparison of the $\overline{DP_V}$ values of cellulose before and after post-treatment renders it possible to quantify the effect of this reaction. As shown in Table 5.4, the $\overline{DP_V}$ drops from 1533 to 837 during post-treatment at 200 °C for 4 min. At the molecular level, this means that on average each cellulose macromolecule was hydrolyzed at only one location. The hydrolysis of the cellulose microfibrils was thus not extensive and should not be a dominant factor in the elimination of the shape memory of the material. It is also possible to envisage a dissipation of the compression energy by a rearrangement of the semi-crystalline zones of the cellulose macromolecules and the fibrils as a result of the induced constraints. This rearrangement can result either in a relative increase in the proportion of crystalline cellulose or in an increase in the diameter of the crystallites. The samples were therefore studied by x-ray diffraction (XRD) in order to quantify the crystallinity changes resulting from the post-treatment.

5.6.3 Determination of the degree of crystallinity of the cellulose microfibrils by XRD

X-ray diffraction (XRD) makes it possible to determine both the index of crystallinity (C_rI) of cellulose and the diameter of the crystallites, see for instance Segal *et al.* (1959) and Wellwood *et al.* (1975). The diffraction was carried out on solid wood samples. Since wood is anisotropic, the samples were consistently oriented in the same manner so that the incident beam struck the transverse plan. The index of crystallinity is given by the empirical formula of Segal *et al.* (1959)

$$C_rI = \frac{I_{002} - I_{AM}}{I_{002}} \tag{5.5}$$

where I_{002} is the maximum intensity of the diffraction peak of the (002) plane and I_{AM} is the intensity diffused by the amorphous phase of the sample. The peak of the (002) plane was located at an angle of diffraction 2θ of 22.8° and the intensity diffused by the amorphous phase was measured of an angle of diffraction 2θ of 18.0°.

The diameter (L) of the cellulose crystallites was calculated from the width at half height of the (002) peak using the formula suggested by Wellwood *et al.* (1975)

$$L = \frac{0.9\lambda}{H \cos\theta} \tag{5.6}$$

where λ is the wavelength of the diffracting beam (line $K_{\alpha 1}$), H is the width at half height of the (002) peak and θ the angle of diffraction of the (002) peak, according to Figure 5.25.

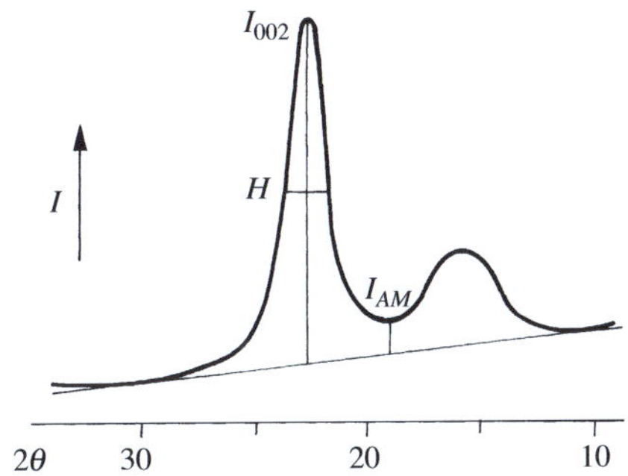

Fig. 5.25 Typical diffractogram of cellulose showing the position of the peak of the diffracting (002) plane and the position of the intensity diffused by the amorphous phase used for the calculation of the index of crystallinity.

The numerical values were $\lambda = 1.54060$ Å and $\theta = 0.3979$ radian. The results obtained are compiled in Table 5.5.

It is important to note that an increase in the index of crystallinity is due not only to an increase in the diameter of the cellulose crystallites produced by rearrangement of the para-crystalline parts, but also, as shown by the definition (5.5), the index of crystallinity is a ratio, considering the intensity diffused by the amorphous phase, i.e., the lignin, hemicelluloses and amorphous part of the cellulose. The increase in CrI of densified and post-treated wood can thus also be caused by a reduction in the amorphous phase by loss of hydrolyzed hemicelluloses and lignin from the sample.

The increase in diameter (L) of the crystallites suggests that the proportion of crystalline cellulose really did increase. However, it can also be caused by a rearrangement of hemicelluloses on the crystalline cellulose surface. Tanahashi *et al.* (1989) have shown by transmission microscopy on wood exploded by steam that the fibrils melted together forming fibrils with a diameter larger than the initial diameter. They reported that the diameter before treatment was approximately 25 Å and that it increased to about 65 Å in white birch and Japanese larch. A rearrangement of the semi-crystalline zones is thus not the only phenomenon leading to an increase in the diameter of crystallites, although this is very significant at 200 °C. These results

Table 5.5 Index of crystallinity (CrI) of the cellulose and diameter (L) of cellulose crystallites (Heger, 2004).

Treatment	Temperature (°C)	Time (minutes)	CrI	Diameter, L (Å)	Recovery R on % of total densification
No treatment	–	–	0.710	25.0	–
Densified	140	20	0.778	28.7	74.5
Densified and post-treated under saturated moisture conditions	140	60	0.809	31.3	18.7
	140	120	0.804	31.3	2.2
	140	180	0.830	32.5	0.9
Densified at 140 °C and post-treated under saturated moisture conditions	160	30	0.818	31.9	2.1
	160	60	0.811	32.5	0.7
	160	90	0.844	34.5	–1.9
	190	8	0.831	33.8	1.7
	190	16	0.843	35.2	–1.9
	190	24	0.853	36.7	–2.4
	210	2	0.844	35.2	0.7
	210	4	0.846	36.7	–1.4
	210	8	0.868	40.1	–2.3
	210	16	0.875	43.1	–0.3

show that the diameter of spruce crystallites was increased from 28.7 to 43.1 Å after a post-treatment at 200 °C for 16 minutes. The increase in diameter of the crystallites is of particular importance when the temperature is high and the duration of the post-treatment is long. The results of Tanahashi *et al.* (1989), for the wood explosion by steam, indicate that the increase in crystallite diameter and in the cellulose crystallinity is solely caused by the elevated temperature and the steam pressure. The analysis of the experiment is as follows:

- First, the steam, whose ionic activity is greater at high temperature and high pressure, reacts with polysaccharides, which are hydrolyzed to sugars of a lower molecular weight. Moreover, the acetic acid formed from the acetyl groups of hemicelluloses and the levulinic and formic acids partially formed by the degradation of hemicelluloses catalyze the hydrolysis of polysaccharides. The lignin is mainly degraded by rupture of the aryl-ether bonds.
- The wood components, partially degraded by these reactions, become more mobile allowing the internal stresses in the crystalline cellulose zone (in the densified wood) to be partially or even totally dissipated.
- Under these conditions, the crystallinity of cellulose can increase by the rearrangement and reorientation of the cellulose molecules in the para-crystalline zones during the treatment by steam under pressure.

- Moreover, the fibrils can fuse together to form micro-fibrils of larger dimensions. This phenomenon is possible due to the solubility and mobility of the lignin as a result of the steam treatment, which leaves more inter-fibril spaces. In addition, parts of the hemicelluloses are quickly hydrolyzed, breaking their connections with the cellulose and lignin, and the hydrolyzed hemicelluloses are almost soluble in the water. The mobility of these components thus leaves the cellulose free from the other wood components and renders it possible for the fibrils to fuse together and form fibrils and microfibrils of larger diameters.

In the tests carried out by Tanahashi *et al.* (1989), the temperature was higher than the post-treatment temperatures used in the present study. Nevertheless, the significant increase in the crystallinity index and crystallite diameter observed during the post-treatment suggests that the phenomena described by Tanahashi *et al.* (1989) may also occur in densified wood post-treatment, although to a lesser extent because of the lower thermal action. Two conclusions can be drawn from the results presented at Table 5.5:

- The increase in CrI and *L* (see the Table 5.5) after compression alone indicates that there is already a potential to dissipate the energy stored during the compression of the sample by rearrangement of the molecular structure of the cellulose microfibrils. This phenomenon would partially explain the fact that a sample compressed at 140 °C under saturated moisture conditions recovers to only 70% and not to 100%.
- The increase in CrI and *L* during post-treatment indicates that a rearrangement of the semi-crystalline parts of cellulose as well as a fusion between fibrils can be responsible for the elimination of the shape memory of densified wood, only although to a small extent. In fact, there is no correlation between the reduction in shape recovery and the increase in the diameter of the crystallites; for example, the recovery decreased from 18.7 to 0.9 between the first hour and the third hour of post-treatment at 160 °C, whereas the diameter of the crystallites increased only very slightly. It is thus possible that the main stress relaxation of the cellulose microfibrils is due to a reduction in their connections with hemicelluloses due to hydrolysis and not to any molecular rearrangement of the fibrils.

5.6.4 Determination of the glass transition temperature (T_g) of lignin by DSC

Lignin is a three-dimensional, cross-linked and amorphous polymer. It can be compared with a thermo-hardening polymer. Thus, its degree of cross-linking and the glass transition temperature (T_g) are likely to vary during THM post-treatment. Since the T_g of the in-situ lignin of water-saturated wood is at approximately 85 °C (Olsson & Salmén, 1992), thermal degradation is expected during post-treatment at a temperature of 140 °C, such as breaking of the α- and β-aryl-alkyl-ether links. It is in fact well known that during the shaping process, thermo-hardening polymers should not remain for long periods of time at high temperatures (compared to their T_g) because of the risk of degradation of the molecular structure.

The T_g of the lignin of several wood samples was measured by DSC (differential scanning calorimetry with a single scan) using the point of inflection criterion.

Measurements were made on samples kept at a constant relative humidity of 76% with a temperature increase of +20 °C/min. The results are presented in Figure 5.26. There was a peak T_g for almost all the post-treatment temperatures followed by a strong reduction, which suggests that the lignin first cross-linked before its thermal degradation started. In general, an increase in the degree of cross-linking is accompanied by an increase in T_g and thermal degradation (breaking of covalent bonds by hydrolysis) by a reduction in T_g.

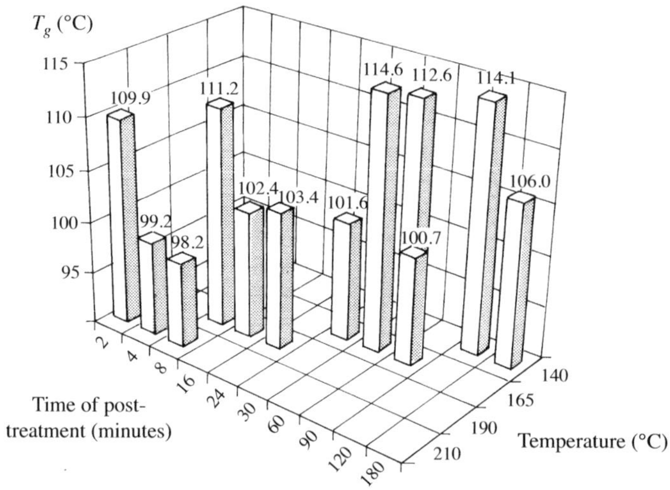

Fig. 5.26 T_g of the lignin of samples densified and post-treated in saturated steam as a function of the temperature and the time of processing (Heger, 2004).

For a given temperature and relative humidity after the T_g peak, the longer post-treatment results in a greater thermal degradation. It is interesting to note that at 165 °C, the peak of T_g occurs at roughly the same time as the densified spruce completely loses its shape memory (ca. 60 min).

Figure 5.27 presents micrographs obtained with confocal microscopy showing thermal degradation of the middle lamella, rich in lignin. Micro-cracks were due to two phenomena specific to the post-treatment process:

- An abrupt decrease in the steam pressure at the end of the post-treatment (purging of the engine) creating a large pressure increase inside the sample due to the steam being unable to leave the sample instantaneously. The pressure increase is important since the temperature and the relative humidity are high. The process where the pressure is abruptly released is used to disintegrate wood and is known as the Masonite process. The difference between wood densification followed by post-treatment and the Masonite process is that the densified sample remains under the mechanical pressure of the piston at the time of purging and the disintegration takes longer. The mechanical pressure of the piston can thus partially prevent intercellular debonding and micro-cracking.

- Cooling and drying of the sample after post-treatment creating a thermo-hydrous gradient, which induces mechanical constraints, which in turn can cause intercellular micro-cracking. When the densified samples are post-treated at too high a temperature (> 180 °C), the thermal decomposition of the lignin thus leads to a relatively extensive micro-cracking. These micro-cracks decrease the shape recovery, since this phenomenon dissipates the energy stored at the time of the densification, but it also greatly reduces the mechanical strength of the wood. It is thus preferable to post-treat the samples at temperatures lower than 180 °C. One can then to a certain extent limit the thermal decomposition of lignin, which reduces intercellular cohesion.

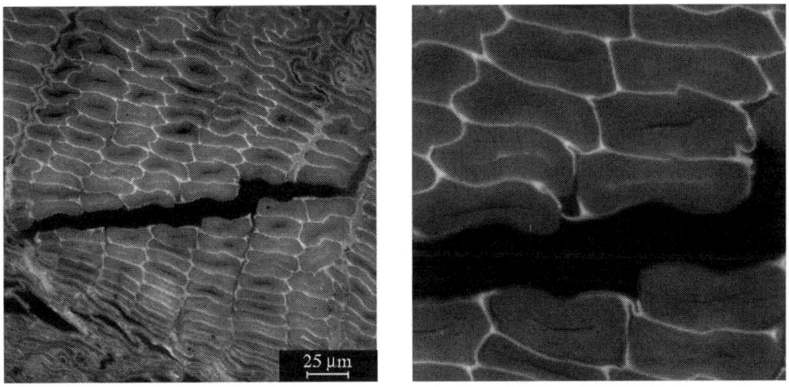

(a) (b)

Fig. 5.27 (a) Photomicrograph of a sample of spruce compressed and post-treated at 200 °C for 4 minutes. (b) Enlargement clearly showing the intercellular cracking due to the thermal degradation of the middle lamella as well as disintegration of the middle lamella between the cell walls.

5.7 INFLUENCE OF TEMPERATURE, HUMIDITY AND PROCESSING TIME ON THE LONGITUDINAL TENSILE STRENGTH AND THE YOUNG'S MODULUS OF DENSIFIED WOOD

Heger (2004) has measured the longitudinal tensile strength as well as Young's modulus of small samples with a cross section of 3×1 mm^2 and a length of 10 mm (Figure 5.28). The results were satisfactory although the dimensions of the samples were smaller than standard. Indeed, as the conditions of cooling and drying were not well adapted, the specimens treated at a temperature higher than 170 °C cracked. It was therefore necessary to decrease the size of the samples to avoid the influence of the cracking on the results. This measurement was carried out on specimens prepared from densified spruce treated under saturated condition at different temperatures and processing times, i.e.,

- natural wood;
- wood densified at 140 °C during 20 minutes;

- wood densified at 140 °C during 20 minutes and post-treated at 140 °C for 3 hours;
- wood densified at 140 °C during 20 minutes and post-treated at 160 °C for 1 hour;
- wood densified at 140 °C during 20 minutes and post-treated at 180 °C for 20 minutes; and
- wood densified at 140 °C during 20 minutes and post-treated at 200 °C for 4 minutes.

At least 10 specimens of each sample type were tested in tension mode. The results are given in Figure 5.29.

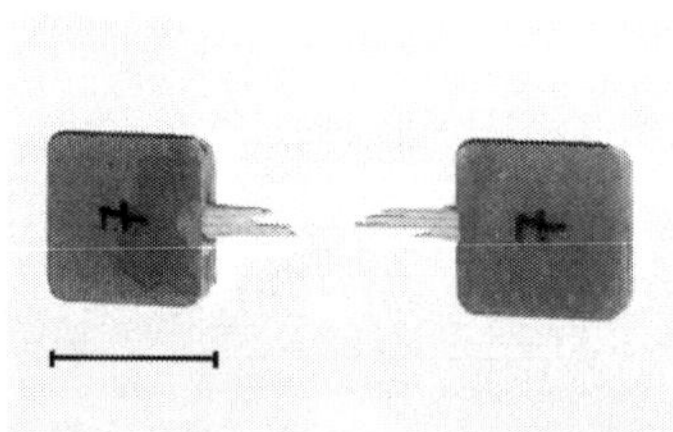

Fig. 5.28 Photograph of a specimen after fracture under tensile testing. The two ends of the specimens were reinforced with resin before testing.

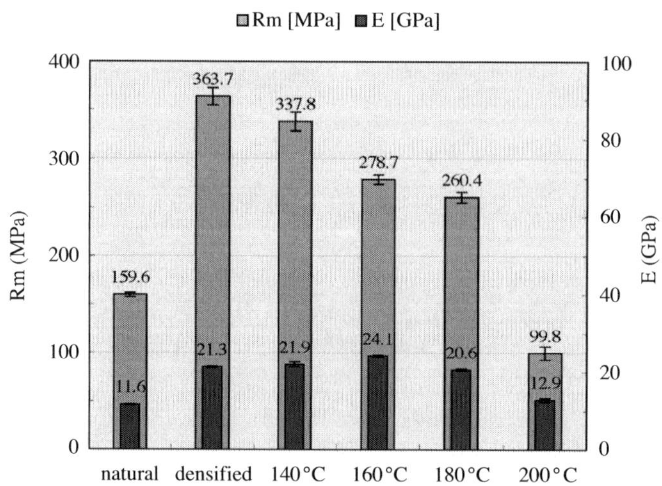

Fig. 5.29 Longitudinal tensile strength[4] and Young's modulus of densified samples post-treated at 140, 160, 180 and 200 °C in comparison with the same type of natural wood sample.

4 The average value of longitudinal tensile strength given by the "Swiss Union in Favour of Wood" (LIGNUM) is 69 MPa for natural spruce with a moisture content between 10 and 15% and for a standard test. The value varies according to the section of the specimen because of defects in the wood. For example, for a section 60.4 and 1.4 cm^2, the strength is respectively 69.85 and 115 MPa. It can even reach 300 to 500 MPa when the tensile load is applied to some isolated wood fibers. The value given for tensile longitudinal modulus by (LIGNUM) is 10.5 to 12 GPa for natural spruce with a moisture content between 10 and 15%.

The results given in Figure 5.29 show that the tensile strengths of the specimens of natural spruce are in agreement with the values given by the Swiss Union in Favour of Wood[4] (LIGNUM). The average compression ratio "C" of the densified samples is 67.6%, which means that the density of the wood has increased by a factor of approximately 3. Theoretically the longitudinal tensile strength should also be increased by a factor of 3, but the densification process has mechanically and thermally degraded the samples. The tensile strength of the densified spruce must thus be less than 3 times the strength of the natural spruce. The value obtained was in fact 363.7 MPa, which is approximately 2.3 times greater than that of the natural spruce. Post-treatment at 140 °C leads to a loss of about 7% of the wood strength, but the reduction is greater at 160, 180 and 200 °C where losses of respectively 23, 28 and 73% have been estimated from the values given in Figure 5.29, compared to natural spruce.

The strong reduction in strength at temperatures higher than 180 °C can be due to two phenomena. On the one hand, the rapid pressure drop at the end of the post-treatment leads to micro-cracks in the sample. We have observed that this type of damage is significant when the temperature approaches 200 °C. In fact, the pressure of saturated steam increases exponentially with temperature. The pressure of the saturated steam at the post-treatment at 200 °C is about 1.6 MPa, whereas it is only 0.4 MPa at 140 °C. On the other hand, the thermal degradation of lignin becomes significant above 190 °C (Figure 5.27).

Theoretically, Young's modulus of the densified spruce should also increase threefold compared to the natural spruce, but this was not the case: Its value increased by a factor of only 1.9. This relatively small increase compared to expectations may be explained in the following way. During the densification, the middle lamella fractured and even underwear displacements compared to the cells, which can be significant (Figures 5.16 and 5.27). The middle lamella, made up mainly of lignin, constitutes a matrix that connects the various wood cells. These movements thus decrease the cellular cohesion. The Young modulus increased by 3% and 13% respectively at 140 and 160 °C. Above 160 °C, the modulus started to decrease, probably due to the evolution of the characteristics of the components of wood. Figure 5.26 shows that the T_g of lignin had a maximum for a post-processing of 165 °C at 1 hour, and that the time decreased to 8 min at 190 °C and to 2 minutes at 210 °C. As the increase in T_g leads to an increase in the rigidity of the lignin, this could partly explain the variation in the modulus.

It has been noted that from 80 to 100% RH and 110 to 140 °C the densification considerably improved the mechanical properties of wood. On the other hand, the form obtained by densification was not stable. It has also been observed that THM post-processing constitutes an effective means of fixing the shape of the densified samples. The temperature should however not exceed 160 or 170 °C since the mechanical properties of new material may then greatly deteriorate.

5.8 REFERENCES

Adler, E. von, Miksche, G.E. & Johanson, B. (1968). Über die Benzyl-aryllätherbindung im Lignin I. freilegung von phenolischem hydroxyl in Ligninpräparaten durch Spaltung leicht hydrolysierbarer Alkyl-aryllätherstrukturen. (On the benzyl aryl ether linkage in lignin. I. Liberation of phenolic hydroxyl in lignin preparations by cleavage of easily hydrolizable alkyl aryl ether structures.) *Holzforschung, 22*(6):171–174.

ALFTHAN, E., DE RUVO, A. & BROWN, W. (1973). Glass transition temperatures of oligosaccharides. *Polymer, 14*(7):329-330.

BACK, E.L. & DIDRIKSSON, E.I.E. (1969). Four secondary and the glass transition temperatures of cellulose, evaluated by sonic technique. *Svensk Papperstidning, 72*(21):687-694.

BACK, E.L. & SALMÉN, N.L. (1982). Glass transitions of wood components hold implications for molding and pulping processes. *Tappi, 65*(7):107-110.

BALDWIN, S.H. & GORING, D.A.I. (1968). The thermoplastic and adhesive behaviour of thermomechanical pulps from steamed wood. *Svensk Papperstidning, 71*(18):646-650.

DARUWALLA, E.H. & SHET, R.T. (1962). Heterogeneous acid hydrolysis of alpha-cellulose from Sudanese cotton. *Textile Research Journal, 32*(11):942-954.

FENGEL, D. & WEGENER, G. (1983). *Wood: chemistry, ultrastructure, reactions*. Walter de Gruyter, Berlin.

GARDNER, K.H. & BLACKWELL, J. (1974). The structure of native cellulose. *Biopolymers, 13*(10):1975-2001.

GORING, D.A.I. (1963). Thermal softening of lignin, hemicellulose and cellulose. *Pulp and Paper Magazine of Canada, 64*(12):517-527.

GRIL, J. & NORIMOTO, M. (1993). Compression of wood at high temperature. In: *Proceedings of workshop on wood plasticity and damage. In: COST 508 Wood Mechanics, Workshop on Wood: plasticity and damage,* Birkinshaw, C. (ed.). University of Limerick, Ireland, pp. 135-144.

HATAKEYAMA, H., KUBOTA, K. & NAKANO, J. (1972). Thermal analysis of lignin by differential scanning. *Cellulose Chemistry and Technology, 6*:521-529.

HEGER, F. (2004). *Etude du phénomène de l'élimination de la mémoire de forme du bois densifié par post-traitement thermo-hydro-mécanique*. (Study of the mechanisms of elimination of the memory form of densified wood by post-processing thermo-hydro-mechanics.) PhD. Thesis No. 3004, EPFL, Lausanne, Switzerland.

HON, D.N.-S. & SHIRAISHI, N. (eds.) (1991). *Wood and cellulosic chemistry.* Marcel Dekker, Inc., New York, Basel, ISBN 0-8247-8304-2.

IRVINE, G.M. (1984). The glass transitions of lignin and hemicellulose and their measurements by differential thermal analysis. *Tappi Journal, 67*(5):118-121.

LAPIERRE, C. & MONTIES, B. (1986). Differential calorimetric study of pine and poplar lignins between 300 and 525 K. *Journal of Applied Polymer Science, 32*(4):4561-4572.

LIU, Y., NORIMOTO, M. & NOROOKA, T. (1993). The large compressive deformation of wood in the transverse direction I. Relationship between stress-strain diagrams and specific gravities of wood. *Journal of the Japan Wood Research Society (Nippon Mokuzai Gakki), 39*(10):1040-1145.

NIMZ, H. (1965a). Über die milde Hydrolyse des Buchenlignins, I. Isolierung die Dimethyl-pyrogallylglycerins. (About the mild hydrolysis of the lignin from beech, I. Isolation of Dimethyl pyrogallylglycerine.) *Chemical Berichte, 98*:3153-3159.

NIMZ, H. (1965b). Über die milde Hydrolyse des Buchenlignins, II. Isolierung eines 1.2-Diaryl-propan-Derivates und seine Überführung in ein Hydroxystilben. (About the mild hydrolysis of the lignin from beech, II. Isolation of a 1.2-diaryl-propan-derivative and its transfer to a hydroxystilben.) *Chemical Berichte, 98*:3160-3164.

NIMZ, H. (1966a). Über die milde Hydrolyse des Buchenlignins, III. Isolierung von zwei weiteren Abbauphenolen mit einer 1.2-Diaryl-propan-Struktur. (About the mild hydrolysis of the lignin from beech, III. Isolation of two other degradation of phenols with 1,2-diaryl-propan-structure.) *Chemical Berichte, 99*:469-474.

NIMZ, H. (1966b). Oligomere Abbauphenole des Fichtenlignins. (Oligomeric degradation of phenols in lignins from spruce.) *Chemical Berichte, 99*:2638-2651.

NIMZ, H. (1967a). Isolierung des Guajacylglycerins und seines dimeren β-Aryläthers aus Fichtenlignin. (Isolation of dimeric β-arylethers from of Guajacylglycerins in lignins from spruce.) *Chemical Berichte, 100*:181-186.

NIMZ, H. (1967b). Über zwei aldehydische Dilignole aus Fichtenlignin. (About two aldehyde Dilignole from Fichtenlignin.) *Chemical Berichte, 100*:2633-2639.

NIMZ, H., & Gaber, H. (1965). Isolierung von DL-Syringaresinol aus Buchenholz. (Isolation of DL-syringaresinol beech.) *Chemical Berichte, 98*:538-539.

OLSSON, A.-M. & SALMÉN, N.L. (1992). Viscoelasticity of in situ lignin affected by structure – softwood vs. hardwood. In: *Viscoelasticity of biomaterials*. Glasser, W.G. (ed.). American Chemical Society, Washington, DC, pp.133-143.

SALMÉN, N.L. (1979). Thermal softening of the components of paper: its effect on mechanical properties. Canadian *Pulp and Paper Canadian Association, Transaction of the Technical Section, 5*(3):45-50.

SALMÉN, N.L. (1982). *Temperature and water induced softening behaviour of wood fiber based materials*. PhD Thesis, Royal Institute of Technology, KTH, Stockholm.

SALMÉN, N.L. (1984). Viscoelastic properties of in situ lignin under water-saturated conditions. *Journal of Material Science, 19*(9):3090-3096.

SALMÉN, N.L., KOLSETH, P. & RIGDAHL, M. (1986). Modelling of small-strain properties and environmental effects on paper and cellulose fiber. In: *Composite system from natural and synthetic polymers*. Salmén, N.L., de Ruvo, A., Seferis, J.C. & Stark, E.B. (eds.). Elsevier Science Publisher B.V., Amsterdam, pp. 211-223.

SCHREPFER, V. & SCHWEINGRUBER, F.H. (1998). Anatomical structures in reshaped pre-dried wood. *Holzforschung, 52*(6):615-622.

SEGAL, L., CREELY, J.J., MARTIN, Jr. A.E., & CONRAD, C.M. (1959). An empirical method for estimating the degree of crystallinity of native cellulose using the X-ray diffractometer. *Textile Research Journal, 29*(10):786-794.

SHAFIZADEH, F. (1963). Acidic hydrolysis of glycosidic bonds. *Tappi, 46*(6):381-383.

SPRINGER, E.L. (1966). Hydrolysis of aspenwood xylan with aqueous solutions of hydrochloric acid. *Tappi, 49*(3):102-106.

TAKAMURA, N. (1968). Studies on hot pressing and drying process in the production of fibreboard. III. Softening of fibre components in hot pressing of fibre mat. *Journal of the Japan Wood Research Society (Mokuzai Gakkaishi), 14*(2):75-79.

TANAHASHI, M., GOTO, T., HORII, F., HIRAI, A. & HIGUCHI, T. (1989). Characterization of steam-exploded wood. III. Transformation of cellulose crystals and changes of crystallinity. *Journal of the Japan Wood Research Society (Mokuzai Gakkaishi), 35*(7):654-662.

TAKAHASHI, K., MOROOKA, T. & NORIMOTO, M. (1998). Thermal softening of wet wood in the temperature range of 0 to 200 °C. *Wood Research, 85*(1):79-80.

TIMELL, T.E. (1964). The acid hydrolysis of glycosides. I. General conditions and the effect of the nature of the aglycone. *Canadian Journal of Chemistry, 42*(6):1456-1472.

TOUSSAINT, B. (1990). *Autohydrolyse rapide du bois de peuplier (Populus tremula), modifications ultrastructurales, hydrolyse enzymatique, inhibitions et reversions*. (Rapid autohydrolysis of poplar (*Populus tremula*), ultrastructural modifications, enzymatic hydrolysis, inhibition and reversions.) PhD. Thesis, Centre de Recherches sur les Macromolécules Végétales (CERMAV) – CNRS, Université J. Fourier, Grenoble, France.

UHMEIER, A., MOROOKA, T. & NORIMOTO, M. (1998). Influence of thermal softening and degradation on the radial compression behaviour of wet spruce. *Holzforschung, 52*(1):77-81.

VISAPÄÄ, A. (1971). Heterogeneous acid hydrolysis of cellulose, Part V. The effect of ball-milling on the hydrolysability. *Paperi ja puu, 53*(7):397-408.

VISAPÄÄ, A. (1972). Heterogeneous acid hydrolysis of cellulose. Part VI. The interdependence of the rate of hydrolysis upon the conditions of hydrolysis and the state of order of cellulose with ten different qualities of cellulose. *Paperi ja puu, 54*(6):353-364.

WELLWOOD, R.W., SASTRY, C.B.R., MICKO, M.M. & PASZNER, L. (1975). X-ray diffraction studies: A statistical evaluation of crystallinity index of Ramie by different method. *Journal of the Japan Wood Research Society (Mokuzai Gakkaishi), 21*(5):272-277.

YANO, S., HATAKEYAMA, H. & HATAKEYAMA, T. (1976). Effect of hydrogen bond formation on dynamic mechanical properties of amorphous cellulose. *Journal of Applied Polymer Science, 20*(12):3221-3231.

ZECCHINA, A., LAMBERTI, C. & BORDIGA, S. (1998). Surface acidity and basicity: general concepts. *Catalysis Today, 41*(1/3):169-177.

CHAPTER 6

WOOD DENSIFICATION AND FIXATION OF THE COMPRESSION-SET BY THM TREATMENT

6.1 FORMING OF SOLID WOOD BY THM TREATMENT

The process of imposing a specific form on a wood element is called ***forming of wood***. It includes bending, twisting, compressive molding and densification. The forming of wood is achieved through the plasticization of the polymeric components of wood using hot water, steam, or other plasticizers such as ammonia (NH_3) or urea.

Usually, the THM process for forming wood consists of four distinct stages. These are

- the softening or plasticization of wood;
- the forming of wood;
- hardening; and
- blocking.

Softening of wood

The plasticization of wood is a treatment aiming at reducing the glass transition temperatures (T_g) of the polymeric constituents. The process of plasticization decreases the internal cohesive forces in the wood. Wood is a composite material made up of various polymers: cellulose, hemicelluloses and lignin. Chemical and physical bonds bind these macromolecules together, and this confers on wood its strength. The most important of the physical bonds is the hydrogen bond, which binds oxygen and hydrogen atoms of the polar OH-groups of the constituents. The action of plasticizers on the hydrogen bonds can be either structural or molecular. Structural plasticization involves using a substance that blocks the polar groups of wood and weakens the cohesive forces between these groups. Molecular plasticization, in contrast, deactivates hydrogen bonds and has a lubricating effect.

Plasticization of wood by the hydro-thermal process has been known for a long time. This method involves replacing the hydrogen bonds by less energetic bonds or simply reducing the number of hydrogen bonds through the reaction of water molecules with the OH-sites. An increase in temperature increases the molecular thermal agitation and thus decreases the action of physical forces such as the Van der Waal's force. Other plasticization treatments use various chemical substances such as ammonia, urea etc.

Ammonia, like water, is used to plasticize wood, and it plays the same role as water by lowering the glass transition temperature of wood. This is due to the NH_3

molecule having small dimensions and a strong polarity. It is also speculated that the ammonia molecules are able to enter the crystalline parts of the fibril, forming a cellulose-ammonium structure. Softening with the aid of ammonia is described in detail in Chapter 10.

During the first stage of forming, the wood element is heated while its moisture content is increased by wetting, by soaking in warm water or by conditioning in saturated steam at a high temperature. The action of water lowers the glass transition temperature of the wood constituents. In industry, one seeks to lower the temperature for wood softening by increasing its moisture content, and it is possible to form wood within a relatively low temperature range of 90 °C to 120 °C thanks to the reduction of the glass transition temperature brought about by humidification. Figure 6.1 shows a schematic illustration of a simple system consisting of a chamber, a steam generator and pressure controllers used for the softening of the pieces of wood before bending or molding.

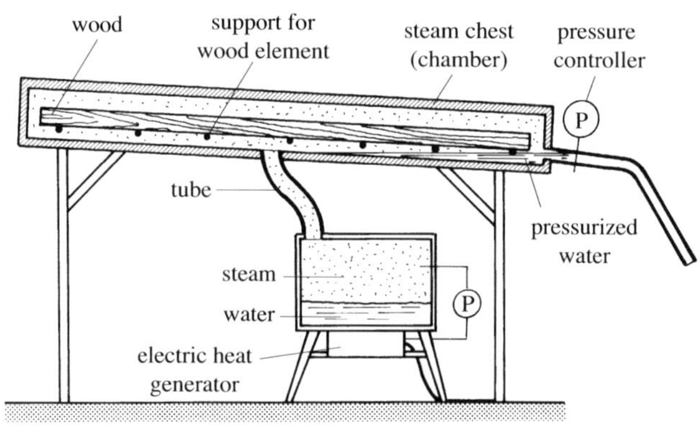

Fig. 6.1 Steam generator used for softening the wood.

The forming of wood

During the forming of a wet wood element, the temperature is higher than the respective glass transition temperatures of the lignin, hemicelluloses and the semi-crystalline cellulose. The wood element can thus be deformed easily under the action of mechanical loads in the radial, tangential and longitudinal directions. The application of a compressive load in the transverse direction leads to a flattening of the cells and reduces the volume of the lumen (Figures 5.11-5.16). The total or partial pressing can be unilateral or multilateral. On the other hand, the application of a compressive load in the longitudinal direction of softened wood causes

a) plastic deformation of the cell wall with the formation of slip plans;
b) cell wall buckling;
c) localized shear-band deformation (in macro-level); and
d) diffuse shear-band deformation (homogeneous).

These deformation types are shown diagrammatically in Figure 6.2. Photographs illustrating the formation of slip and the slip planes are presented in Figure 6.3.

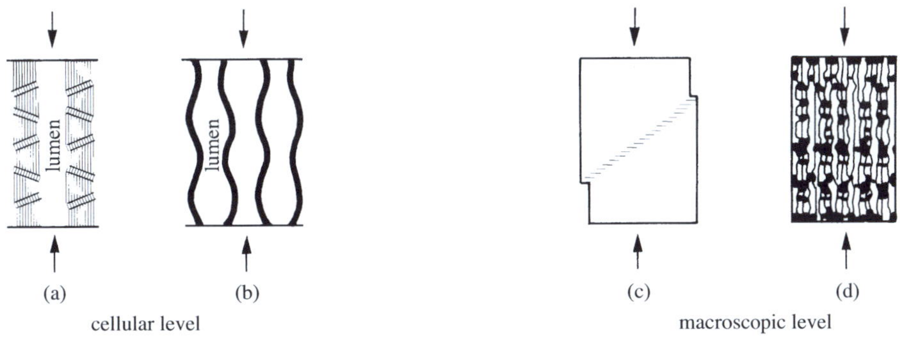

Fig. 6.2 Various modes of large deformation of the cell walls of wood under compression in the longitudinal direction; (a) plastic deformation of the cell wall (formation of slip planes), (b) lateral buckling of the cell walls, (c) localized shearing band, (d) diffuse buckling of the cell walls.

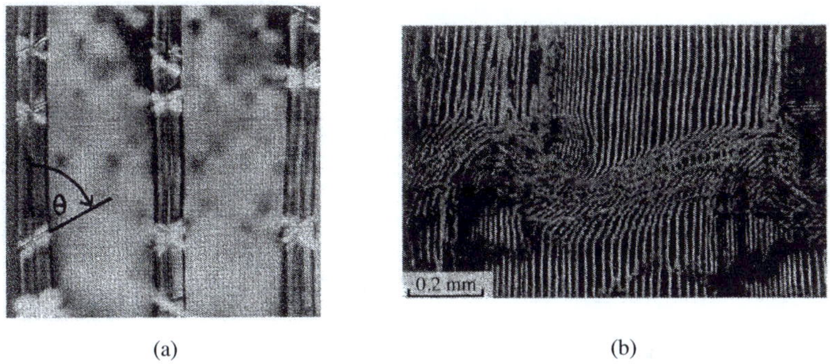

Fig. 6.3 Photomicrographs showing the deformation of the cell walls; (a) observation by polarized microscope of the slip plane on the level of the cell wall, (b) shear bands (Hoffmeyer, 1990).

Hardening and blocking

After being shaped, the wooden element is cooled and dried. When the temperature drops below the respective glass transition temperatures of lignin, hemicelluloses and semi-crystalline celluloses, they return to the glassy state, and the reduction in the molecular thermal activation energy and of the moisture content allow Van der Waal's and hydrogen bonds to be created at the molecular level. The molecular conformation of wood constituents in the cell walls and the middle lamella is considered to be "frozen".

The form fixed by the drying and cooling operation is unstable. If the sample is again humidified and heated in the absence of any external force, it reverts to its original dimensions. The purpose of a blocking stage is to permanently fix the form of the sample. To prevent the formed wood from returning to its initial form, one can employ

- mechanical fixing by gluing, nailing, screwing, etc.;
- chemical modification by deactivation of the OH-sites through acetylation (substitution of the -OH by CH_3COO-groups), or formaldehydation (fixing of H_2CO between two hydroxyls to obtain a strong chemical bond), etc.;
- a thermo-hydro-mechanical treatment. It is necessary to keep the samples in a dryer for one or two weeks at a temperature between 40 °C and 80 °C, for

30 hours at 140 °C, or for 12 hours at 180 °C. During drying, the mechano-sorptive effect plays a large role to reducing the internal stresses in the wood; or

- hydrolysis of hemicelluloses under the effect of water and high temperature in order to relax the internal stresses.

Figure 6.4 illustrates the four stages in the densification of wood at the cellular level by compression in the radial direction and fixation of the compression set.

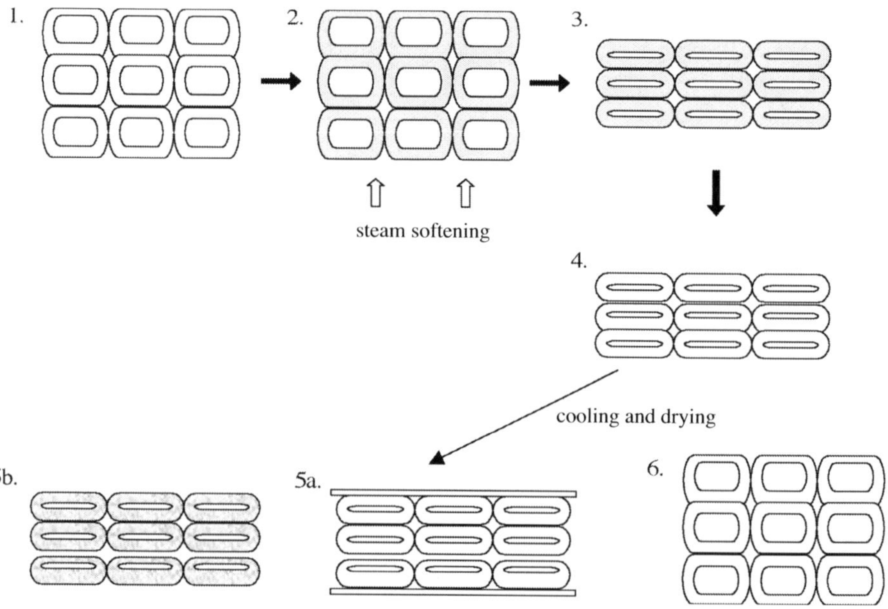

Fig. 6.4 Schematic illustration of wood densification in the transverse direction at the cell level and fixation of the compression set: 1. initial state, 2. softening, 3. densification (or forming), 4. hardening, 5a. mechanical fixation, 5b. chemical fixation or fixation by THM treatment, 6. total set-recovery.

6.2 WOOD DENSIFICATION

This section presents the various THM processes (open or closed systems), after which the fixation of the compression-set by THM treatment is discussed.

6.2.1 Densification inside an open system

The densification of wood by thermo-hydrous and mechanical actions was developed and industrialized in the 1940s. The product was marketed under the name "Staypack" (Seborg *et al.*, 1945). This process consists in applying a compressive force in the radial direction of the wood under high temperature and at a sufficiently high moisture content so that plasticization of the wood occurs. Densification leads to a reduction in the porosity of wood and to an improvement of its mechanical characteristics, but this process cannot adequately eliminate the shape memory.

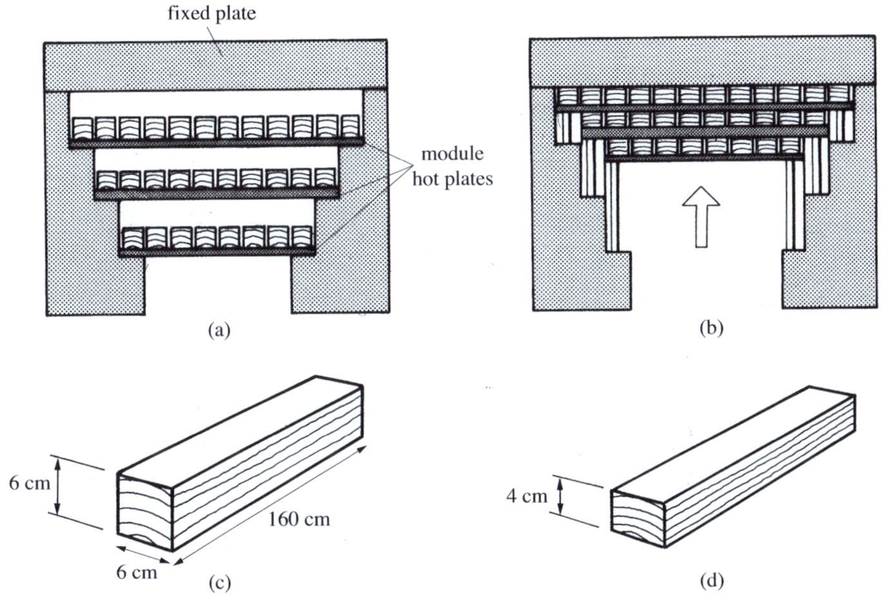

Fig. 6.5 Diagram showing the different stages of wood densification in an open system by heating and pressing in a flat press; (a) the hot press with three rows of wood pieces, (b) the press during wood densification, (c) a piece of wood to be heated, (d) a wood piece densified by 33%.

The process of densification can be implemented in an open or a closed system. It is not possible to adequately control the moisture content of the elements during densification in an open system. The process of densification in an open system is thus can be named Thermo-Mechanical (TM). On the other hand, it is possible to control the moisture content as well as the temperature in a closed system, and the process is called Thermo-Hydro-Mechanical (THM). Staypack or TM-densified wood is produced by a simple process using a hot press (open system). The different stages in this open system process are shown in Figure 6.5 and consist of

1. preparing the elements of wood in the longitudinal direction;
2. conditioning the wood to a moisture content of 13-15%; and
3. compressing the wood in the radial direction after softening. A press can densify about 20 wood elements $60 \times 60 \times 1600$ mm^3 in size. The parameters of densification are a pressure of 25 MPa, a temperature of 130 °C and a duration of at least 2.5 hours. During this treatment, the elements undergo a large displacement from 30% to 40%. Figure 6.6 shows the temperature variation within the press during the process.

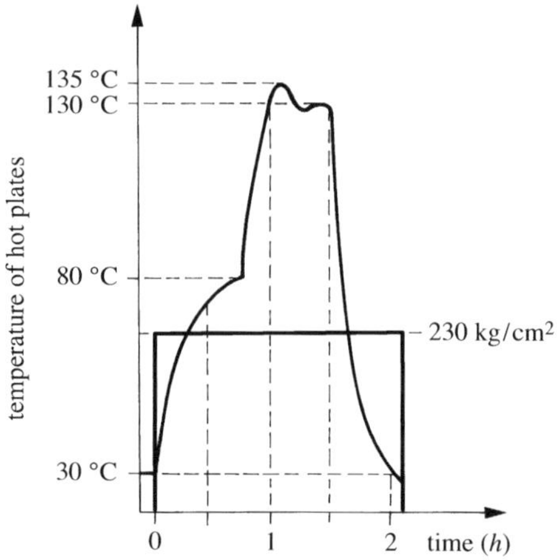

Fig. 6.6 The variation of temperature in a heating press (by company Stahel Köng AG, Hinwil, Zürich).

6.2.2 Mechanical and moisture characteristics of wood densified in an open system

Table 6.1 presents values of the apparent density measured on three types of beech.

Table 6.1 Dried apparent density (Huguenin & Navi, 1995).

Type of wood	Density (g/cm^3)
Beech for joinery	0.63
Beech to be densified	0.82
Densified Beech	1.13

Two types of mechanical tests were carried out: an impact test and a three-point bending test. The first test determines the capacity to resist shocks by indicating the energy per unit of area necessary to cause fracture, under a radial or tangential dynamic loading. The second type of test measures the deflection as a function of the force applied and calculates the maximum elastic tensile stress, the maximum stress to rupture, and the modulus of elasticity. The results presented are from tests carried out with a radial loading.

Table 6.2 shows the impact strength in the radial and tangential directions. The values are the averages of five tests.

Table 6.2 Impact resistance (Huguenin & Navi, 1995).

Type of wood	Tangential (J/cm^2)	Radial (J/cm^2)
Beech for joinery	3.8	8.5
Beech to be densified	8.2	10
Densified beech	14	15

The results presented in Table 6.2 show an increase in impact strength after densification of more than 70% in the tangential direction and 50% in the radial direction, these values being similar to the increase in density.

The results of the bending tests are given in Table 6.3, and the values are averages of 3 tests.

Table 6.3 Mechanical properties in bending (Huguenin & Navi, 1995).

Type of wood	Crack opening stress in mode I, $\sigma_{Rup.}$ (MPa)	Linear elastic stress limit, σ_{max} (MPa)	Modulus of elasticity in longitudinal tension (GPa)
Beech for joinery	68	29	6.9
Beech to be densified	117	53	11.0
Densified beech	200	95	22.9

According to Table 6.3 the densification increased the crack-opening stress in mode I by 70%, the yield stress (maximum linear stress) by 80% and the modulus of elasticity by 110%. Nevertheless, the breaking strength data indicate that densification can partially weaken the intercellular middle lamella, even at a relatively low temperature of approximately 130 °C.

Tests of the directional swelling were carried out on densified beech and beech to be densified, in the three principal directions of wood, at relative humidities of 76% and 97%. Figure 6.7 illustrates the development of the directional swelling with

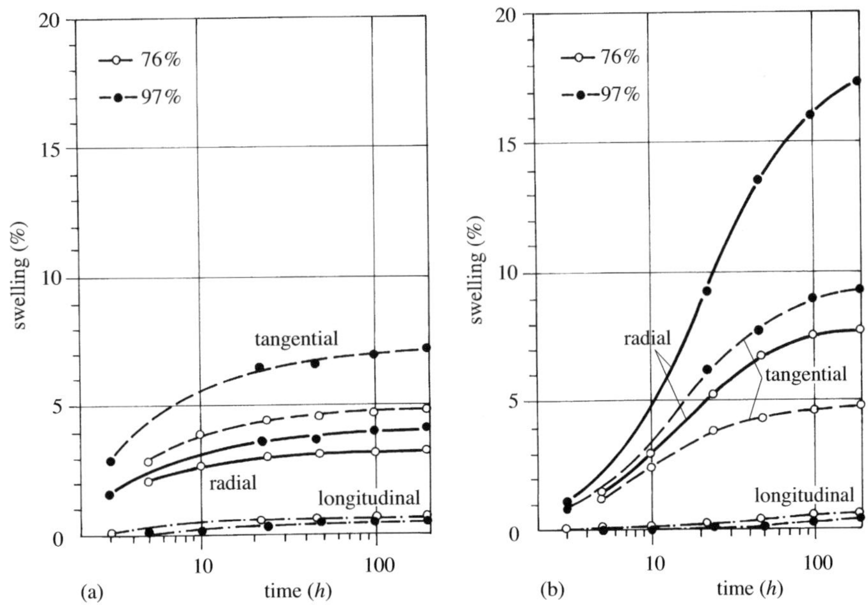

Fig. 6.7 The swelling of beech as a function of time (logarithmic scale) during humidification in an open system: (a) beech before densification, (b) beech after densification (Huguenin & Navi, 1995).

the time in a moist atmosphere. In the non-densified beech, the tangential swelling is greater than the radial swelling which in turn is greater than the longitudinal swelling. In the densified beech, on the other hand, the radial swelling is the greatest. This is in the direction of the compression and is thus mainly a recovery effect (shape memory).

The development of the swelling with time for the densified beech and the non-densified beech immersed in water and at relative humidities of 76% and 97% in the tangential and longitudinal directions is presented in Figure 6.8. The behavior in these two directions should be more and less identical for the two types of wood, since they were not largely deformed during the densification.

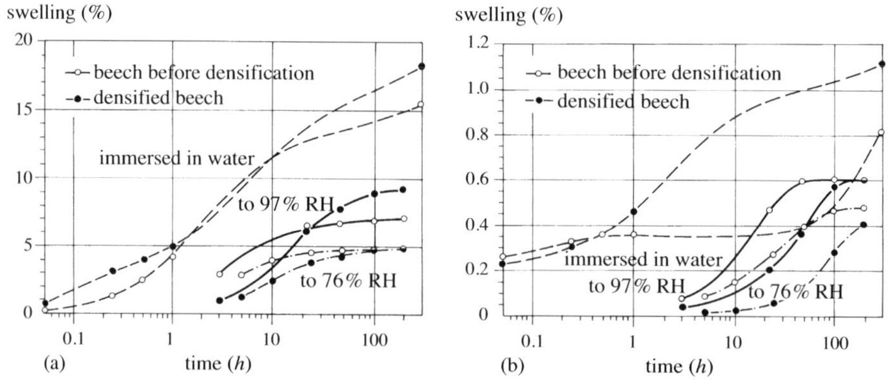

Fig. 6.8 The swelling of beech as a function of time (logarithmic scale) of beech under high relative humidity and immersed in water: (a) tangential swelling, (b) longitudinal swelling (Huguenin & Navi, 1995).

Figure 6.9 shows that the difference in radial swelling between densified and non-densified wood increases as the relative humidity is raised. It can also be seen that densified wood tends to return to its initial dimensions when immersed in water for a long period of time.

The densification of wood in an open system makes it possible to increase the mechanical strength of wood to a significant degree. Dimensional instability is the principal weakness of this type of densified wood, particularly at a high relative humidity and elevated temperatures for long exposure times of several days. At a relative humidity of 97%, the swelling in the radial direction (direction of densification) of densified beech is about 20%, whereas it is only 5% for the non-densified beech under equivalent conditions. Wood densified in an open system can swell in water by more than 50% depending on the porosity of wood. Consequently, wood densified by this process cannot be used under conditions of high moisture, and this greatly reduces its field of application in construction work. To fix the compression-set of the densified wood and to reduce its recovery and dimensional instability, impregnation with a synthetic resin, mechanical fixing or a thermo-hydro-mechanical treatment at high temperature and moisture is usually necessary.

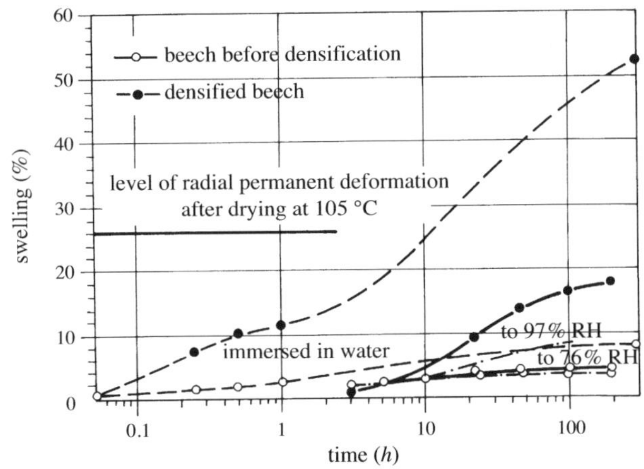

Fig. 6.9 The radial swelling as a function of time (logarithmic scale) of densified and non-densified beech at different RHs and immersed in water (Huguenin & Navi, 1995).

6.2.3 A new understanding of densified wood with shape memory

Wood behaves as a brittle material with a very small tensile strain of about 1-2%. The above-mentioned paragraph demonstrated that wood could be densified more than 50% on its transversal direction by closing up the cell lumens. If one recovers this compression-set completely, the tensile strain would be more than 100%.

Based on this idea in Institute für Stahl-und Holzbau at the, Technische Universität Dresden, Germany, glued laminated timber boards and rolled wooden tubes were made from densified spruce, Haller (2007). Figures 6.10, 6.11 and 6.12 show the steps of manufacturing of timber boards and the production of rolled tube in a mold.

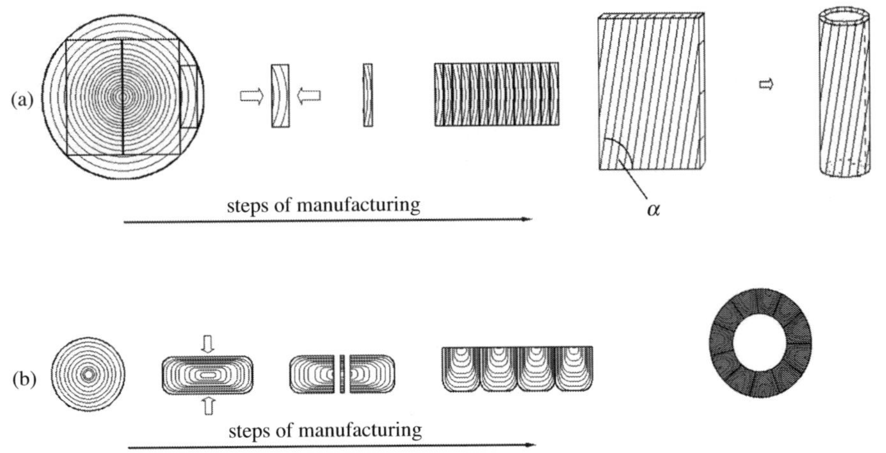

Fig. 6.10 Manufacturing of wooden tubes from densified sawn wood (a) and round wood (b), (courtesy Prof. P. Haller).

Fig. 6.11 Installation of a laminated compressed panel inside the mold (courtesy from Prof. P. Haller).

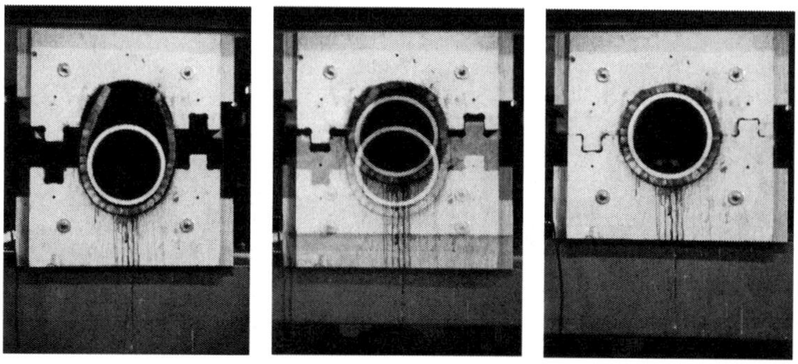

(a) charging (b) closing (c) forming

Fig. 6.12 Production of a rolled wooden tube in a mold (courtesy from Prof. P. Haller).

6.2.4 Densification in a closed system

The densification of wood in an open system under wet conditions and at a high temperature (about 130 °C) improves the mechanical and moisture-resistant properties of wood, but the compression deformation is not stable. Post-processing by a THM treatment has however been shown to constitute an effective means of eliminating the set-recovery.

Advanced investigations of wood densification and its treatment by THM in a closed system have been reported by Tanahashi (1990), Inoue *et al.* (1993), Norimoto *et al.* (1993), Ito *et al.* (1998a,b), Navi *et al.* (2000), Navi and Heger (2005), Kamke (2004), Kutnar *et al.* (2009) and many others. The purpose of recent research on the densification of wood and the development of new techniques based on THM post-treatment is to mold wood elements with larger dimensions with no set-recovery. This section presents different techniques have been recently developed mainly for research at the laboratory level.

Wood densification and post treatment in a closed chamber (a sealed cell) with a simple wall

To densify small wood elements in the transverse direction under saturated vapor and at various temperatures, a small thermo-hydrous chamber, a steam generator and a press are required. A photograph of a chamber (a sealed cell) with a simple wall for densifying small specimens approximately $40 \times 40 \times 30$ mm^3 in size is shown in Figure 6.13. The steam generator and the press are outside of the chamber.

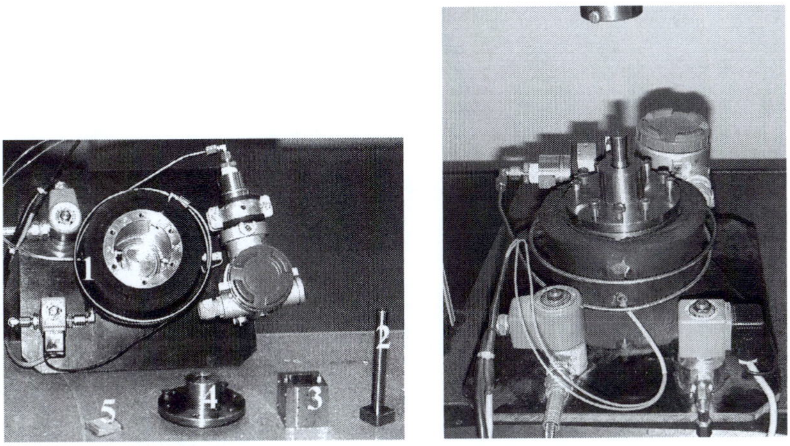

(a) (b)

Fig. 6.13 Apparatus for densifying small wood elements at temperatures up to 150 °C in saturated steam: (a) photograph of the apparatus under open conditions, 1) chamber, 2) piston, 3) mold, 4) cover, 5) densified wood, (b) photograph of the apparatus under closed conditions, (Navi *et al.*, 2000)

The physical and mechanical properties of the wood densified inside this chamber depend mainly on the type of wood and on the THM-treatment process. It seems that each wood species needs particular conditions to eliminate or reduce the shape memory (set-recovery). The THM treament process (with this apparatus) is divided into several stages summarized in Figure 6.14. The various stages of densification and post-processing (heating in steam) of the specimen are marked separately.

Figure 6.14 illustrates a diagram of the load as a function of time during a THM process. The process consists of five stages. Stage one, curve 0-a-b, represents the plasticization or softening of the specimen in saturated steam at about 150 °C. This stage takes about 10 minutes for small elements. The second densification stage, curve b-c, represents the application of a controlled displacement under constant temperature up to a prescribed pressure. During the third stage, curve c-d, the load and the temperature is kept constant. During the fourth stage, curve d-e, the steam flow is stopped and the specimen starts to dry. Lastly, during the fifth stage, curve e-f, the specimen is completely unloaded. The total process time is about 3 hours. Post-processing includes the third and fourth stages.

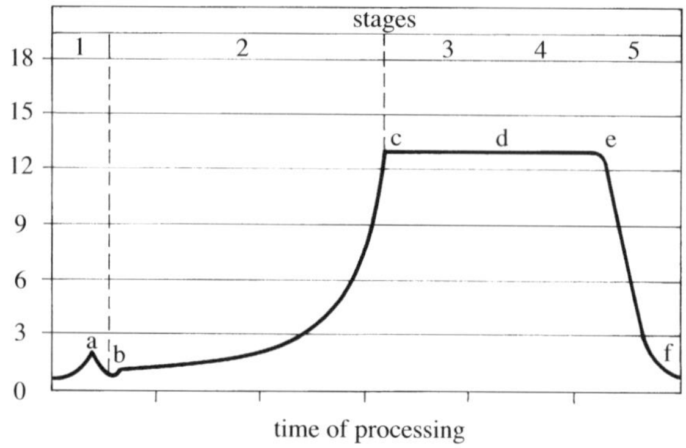

Fig. 6.14 Schematic diagram of the load as a function of time during a THM densification process.

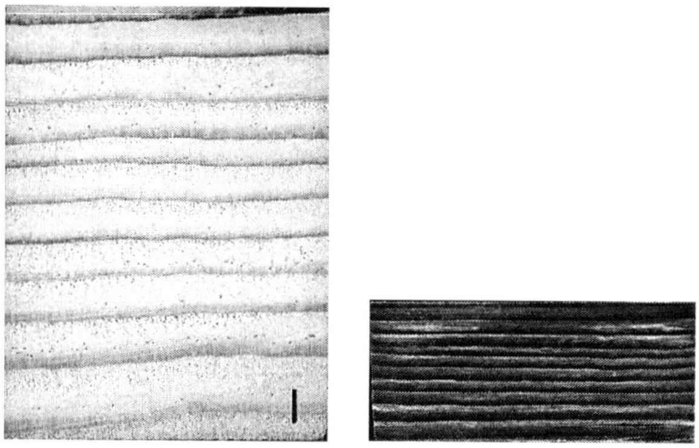

(a) (b)

Fig. 6.15 Spruce (a) before densification (density 430 kg/m^3), (b) after THM densification (density 1290 kg/m^3). The annual rings are perpendicular to the direction of densification, and the compression rate defined in equation 5.1 is 68% (Navi *et al.*, 2000).

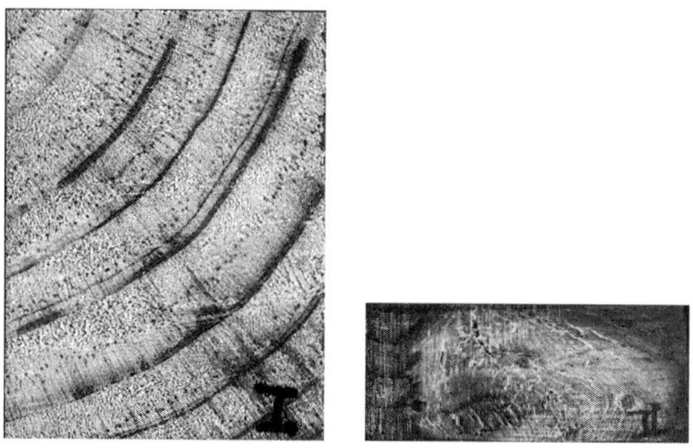

(a) (b)

Fig. 6.16 Pine: (a) before densification (density 490 kg/m^3), (b) after THM densification (density 1300 kg/m^3). The annual rings are 45 degrees to the direction of densification, and the compression set defined in equation 5.1 is 68% (Navi *et al.*, 2000).

It is also possible to densify wood in the transverse direction (parallel or not to the rings), as well as pieces of wood containing large knots. Photographs of specimens before and after densification, taken from the transverse plane are presented in Figures 6.15 to 6.17.

(a) (b) (c)

Fig. 6.17 Pine with a knot: (a) before densification (density 620 kg/m^3), (b) and (c) showing the knot at tangential and radial planes after THM densification (density 1280 kg/m^3). The annual rings are perpendicular to the direction of densification, and the compression rate defined in equation 5.1 is 67% (Navi *et al.*, 2000).

Effect of THM treatment on the mechanical properties of wood

Many different tests have been carried out to evaluate the mechanical properties of THM-densified wood in comparison with those of initial non-densified specimens. The shear strength has been determined in the LR plane (parallel to the fibers), and Brinel surface hardness tests have been carried out on both the tangential and radial planes. Figures 6.18 and 6.19 show respectively the shear strength and surface hardness of the THM densified wood and the initial wood.

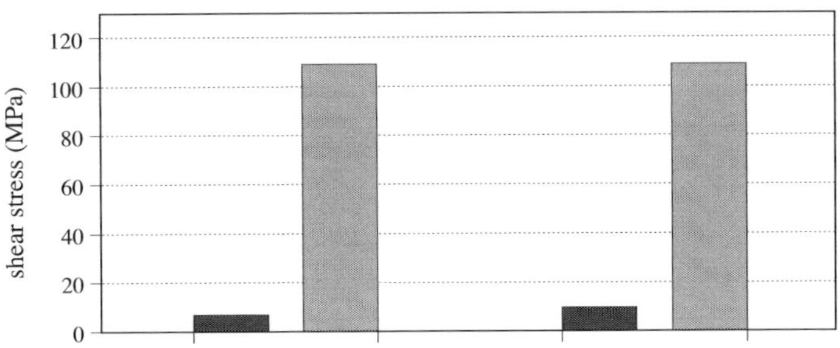

Fig. 6.18 Shear strength of the initial wood (black) and wood densified in a THM closed system (grey) for spruce and maritime pine (average values) (Navi *et al.*, 2000).

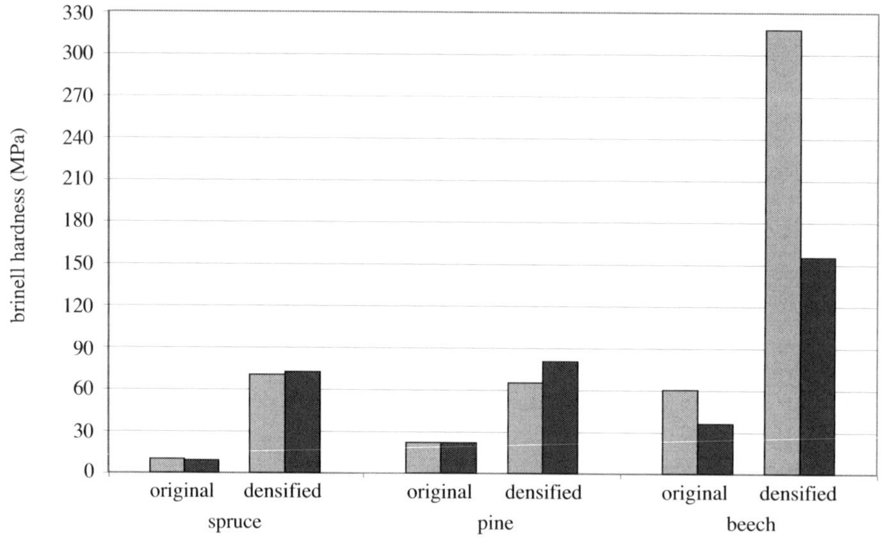

Fig. 6.19 Brinell hardness of initial wood and densified wood in a THM closed system for spruce, beech and maritime pine (average values) (Navi *et al.*, 2000).

Hygroscopic character and swelling of THM densified wood

To show the more hygroscopic character of THM-densified wood, sorption isotherm tests were carried out on initial, TM-densified wood (open system without post-treatment) and THM-densified wood (closed system with post-treatment). The results are given in Figure 6.20.

Swelling due to water sorption was measured at 18 °C. The samples of THM-, TM- and initial wood were immersed in water and their swelling in the radial direction was measured. The results are presented in Figure 6.21.

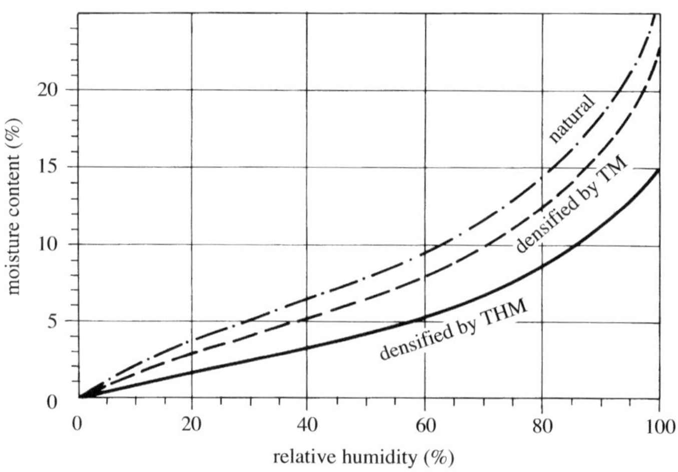

Fig. 6.20 Adsorption isotherms for initial, TM- and THM-treated beech (Navi *et al.*, 2000).

It is also possible to densify wood in the transverse direction (parallel or not to the rings), as well as pieces of wood containing large knots. Photographs of specimens before and after densification, taken from the transverse plane are presented in Figures 6.15 to 6.17.

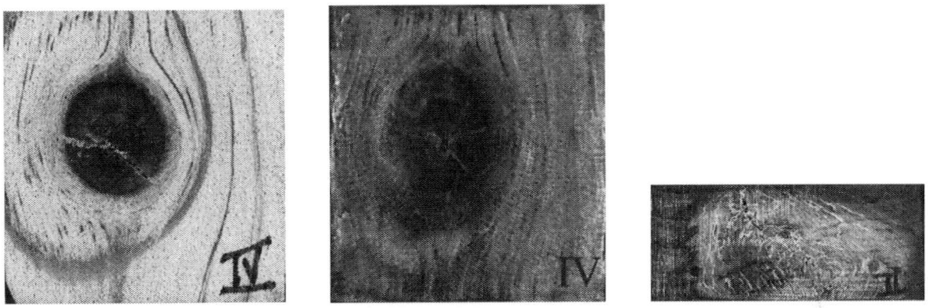

(a) (b) (c)

Fig. 6.17 Pine with a knot: (a) before densification (density 620 kg/m^3), (b) and (c) showing the knot at tangential and radial planes after THM densification (density 1280 kg/m^3). The annual rings are perpendicular to the direction of densification, and the compression rate defined in equation 5.1 is 67% (Navi *et al.*, 2000).

Effect of THM treatment on the mechanical properties of wood

Many different tests have been carried out to evaluate the mechanical properties of THM-densified wood in comparison with those of initial non-densified specimens. The shear strength has been determined in the LR plane (parallel to the fibers), and Brinel surface hardness tests have been carried out on both the tangential and radial planes. Figures 6.18 and 6.19 show respectively the shear strength and surface hardness of the THM densified wood and the initial wood.

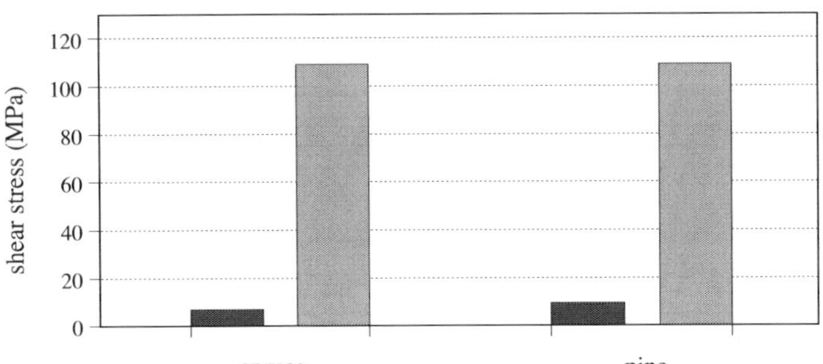

Fig. 6.18 Shear strength of the initial wood (black) and wood densified in a THM closed system (grey) for spruce and maritime pine (average values) (Navi *et al.*, 2000).

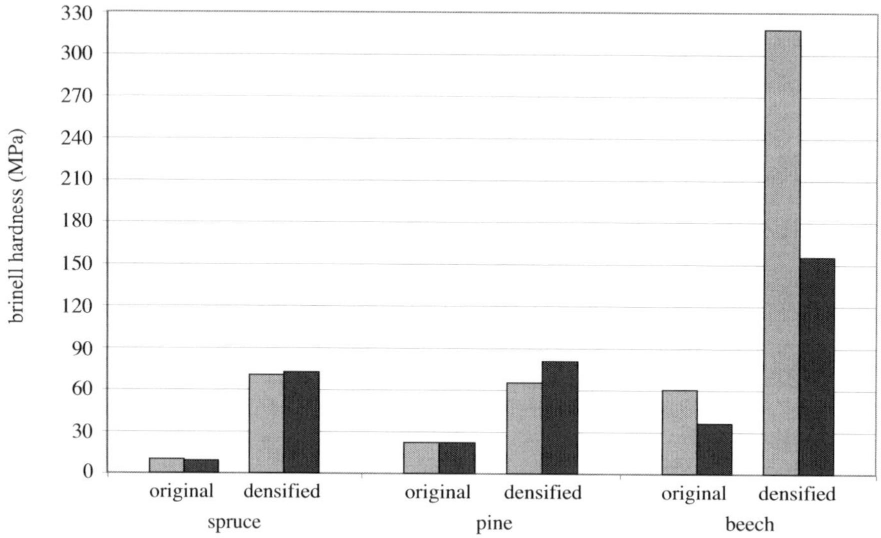

Fig. 6.19 Brinell hardness of initial wood and densified wood in a THM closed system for spruce, beech and maritime pine (average values) (Navi *et al.*, 2000).

Hygroscopic character and swelling of THM densified wood

To show the more hygroscopic character of THM-densified wood, sorption isotherm tests were carried out on initial, TM-densified wood (open system without post-treatment) and THM-densified wood (closed system with post-treatment). The results are given in Figure 6.20.

Swelling due to water sorption was measured at 18 °C. The samples of THM-, TM- and initial wood were immersed in water and their swelling in the radial direction was measured. The results are presented in Figure 6.21.

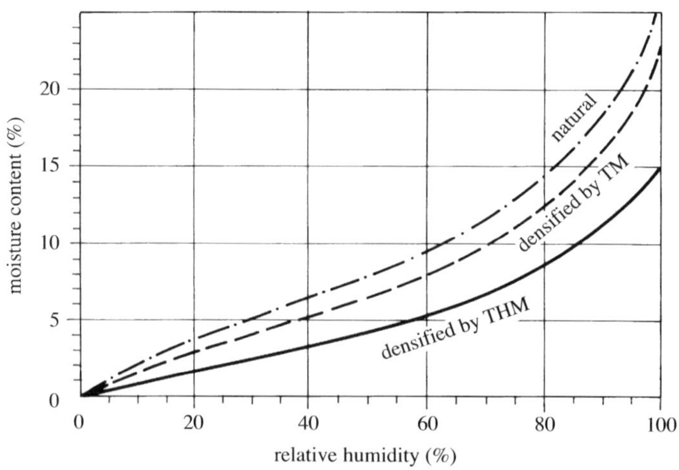

Fig. 6.20 Adsorption isotherms for initial, TM- and THM-treated beech (Navi *et al.*, 2000).

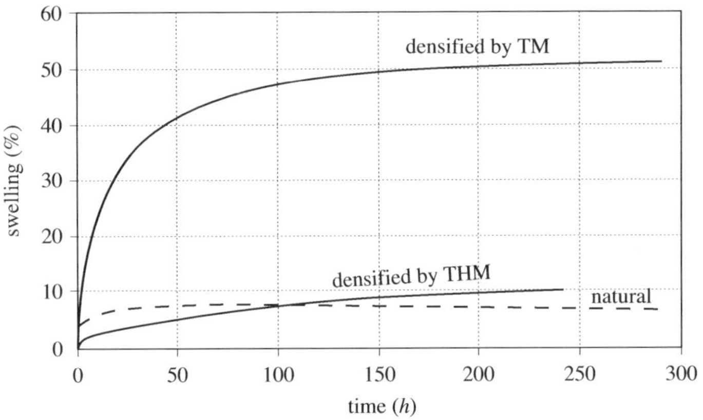

Fig. 6.21 Swelling in the radial direction of initial, TM- and THM-post-treated wood immersed in water at 18 °C (Navi *et al.*, 2000).

Wood densification and post treatment in a closed chamber with a double wall
In a chamber with a simple wall, the samples are directly heated by saturated steam, so that their temperature and moisture content are interdependent and remain unchanged during the process. On the other hand, in a reactor (composed of several chambers) with a double wall, it is possible to control these two parameters independently and the sample can be treated under various humidity conditions, not always a saturated one. In this system, the generation of high temperature steam and the press are external to the reactor, and the temperature can be raised up to 200 °C. The advantage of this system is that the moisture content of the sample can vary from the dry to the saturated condition during the treatment. Figure 6.22 shows a photograph of the control panels, the press and the THM reactor. In this system the watertight reactor consisted of 4 chambers. In Figure 6.23(c), area 6 represents the treatment chamber in which the mold and the specimen

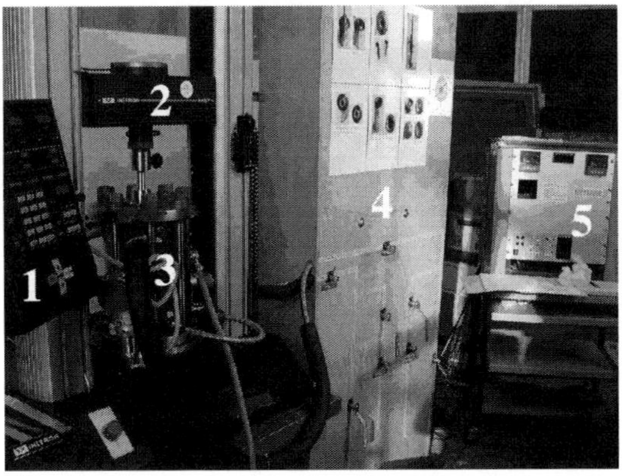

Fig. 6.22 Photograph of a THM chamber together with control panels and a press; (1) and (2) are the control panel for the press, (3) is the THM-chamber with thermal isolation, (4) and (5) are control panels for the high steam temperature (Navi & Heger, 2005).

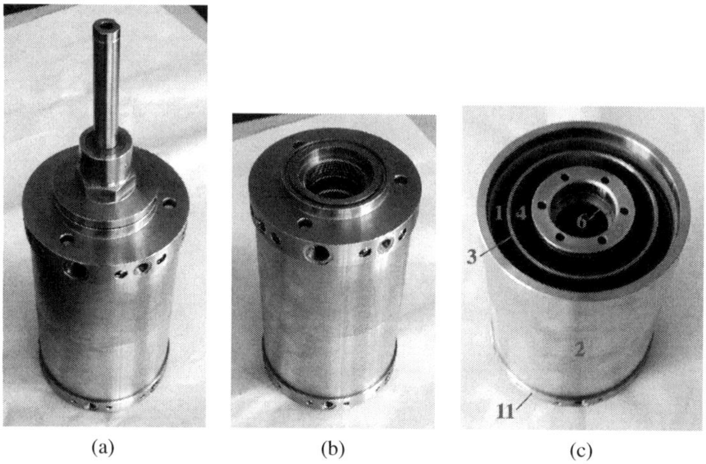

Fig. 6.23 Photographs showing the THM reactor; (a) – THM reactor with piston, (b) – THM reactor without piston (c) – THM reactor after removing the cover. Chamber 1 and cylinder 5 (cylindrical chamber) are connected to a heating circuit and are completely watertight. Chamber 4 is connected to chamber 6 (treatment chamber and mold) by the cover, (Navi & Heger, 2005).

are placed during the treatment. Chamber 1 and cylinder 5 (Figure 6.23c) are connected to a heating circuit and both are completely watertight. During THM-processing, chambers 1 and 5 are heated by saturated steam to reach the required temperature and the specimen placed in chamber 6 (treatment space) is heated by conduction.

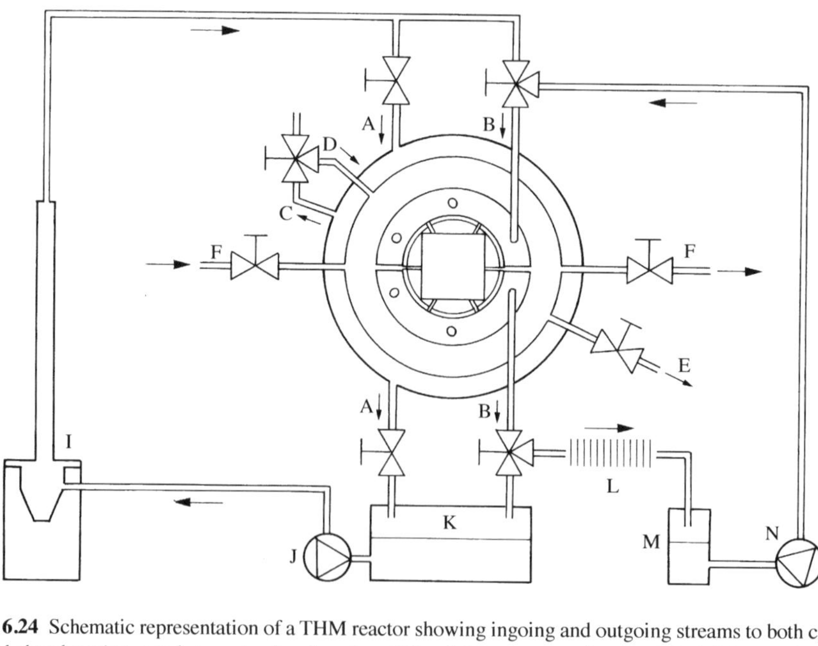

Fig. 6.24 Schematic representation of a THM reactor showing ingoing and outgoing streams to both cylindrical chambers (saturated steam to chambers 1 and 5) and the treatment chamber (Navi & Heger, 2005).

A diagram of the THM reactor is given in Figure 6.24. The temperature of the cylindrical confinements and the treatment chambers (or specimen) remains constant during the densification process and post-treatment, and the relative humidity inside the treatment chamber is set to the required level. A press is used to compress the specimen under a controlled displacement mode.

Loading during THM-densification and the post-treatment process

A cylindrical spruce specimen with a diameter of 40 mm and a length of 150 mm in the radial direction was densified and post treated in the reactor shown in Figure 6.23. A diagram of the loading variation during the densification and the post-treatment is given in Figure 6.25. The densification consisted of two steps: the sample was first heated to 140 °C under saturated conditions for up to 10 minutes (plasticization-step) and then densified under a controlled displacement mode. In this example, the maximum applied load was about 22 kN and the compression set was 66% (see Eq. 6.1).

After densification, the sample was post-treated at various temperatures and moisture contents. In this example, the temperature was 160 °C and the relative humidity was saturated. After densification, the specimen was maintained in the same position throughout the post-treatment. At the end of the post-treatment, the reactor was purged by letting the steam leave the cylindrical chamber. The sample was then immediately cooled with water at 60 °C and removed from the reactor.

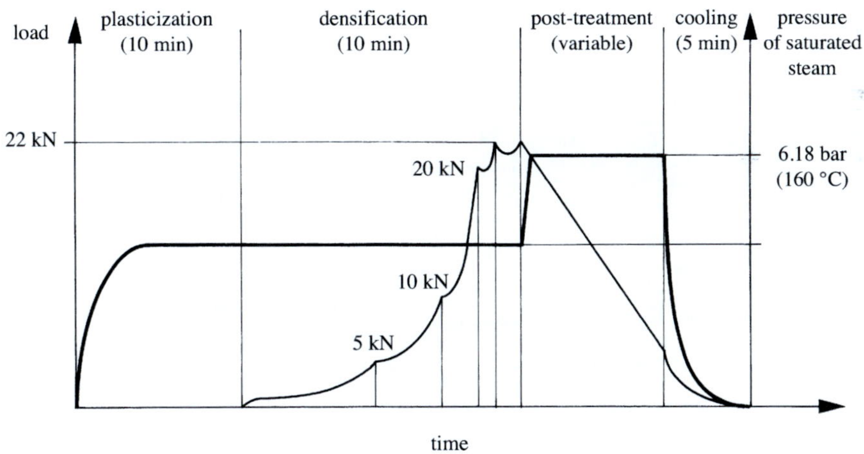

Fig. 6.25 Diagram of loading as a function of time during THM densification and post-treatment and steam pressure variations: The pressure is given for saturated steam (Navi & Heger, 2005).

Recovery

To examine the effect of THM post-treatment on the fixation of the compression set of densified wood, a recovery test was conducted on each specimens. This test consisted of subjecting samples to soaking-drying cycles in water at 60 °C and in an oven at 30 °C. The recovery R is defined as

$$R = \frac{R_c' - R_c}{R_0 - R_c} \tag{6.1}$$

where, R_0 is the length of the specimen before densification, R_c is the length of the specimen after densification and R_c' is the length of the densified sample after the recovery test. R can vary between 0% and 100%. When R = 0%, there is no shape memory or the compression-set is completely fixed, and when R is 100%, the recovery is total. When the specimen is not completely fixed, R will take on a value between 0% and 100%. Figure 6.26 presents the results of the recovery tests for densified spruce post-treated under saturated steam at 160 °C and 180 °C for different periods of times. When the recovery (R) of a sample remains at about zero after several soaking-drying cycles, the compression is permanent and the shape memory effect is completely eliminated. For example, Figure 6.26 shows that the compression of a small densified spruce element can be completely fixed by a THM treatment using saturated steam at temperatures of 160 °C and 180 °C for 60 and 20 minutes, respectively.

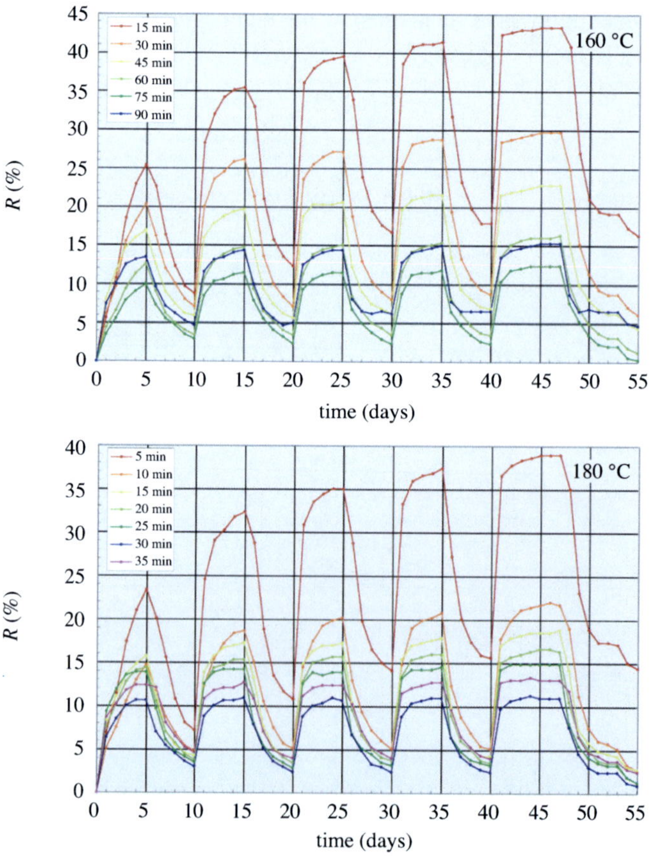

Fig. 6.26 Compression-recovery curves of densified spruce samples, post-treated under saturated steam for different times and at two temperatures; upper at 160 °C; lower at 180 °C (Navi & Heger, 2005).

In Japan, various investigations on the THM-molding of wood have been performed and various processes have been developed. Two of these are presented in the following.

1- Two-directional transverse densification of wood elements
This processes involves four stages: wood plasticization by high-temperature steam, compressive molding, THM post-treatment and cooling. In a technique developed by Ito *et al.* (1998a), small blocks of wood with a round section can be transformed into blocks with a square section. The THM apparatus, the molding block and a photograph of a transformed trunk are shown in Figures 6.27 and 6.28.

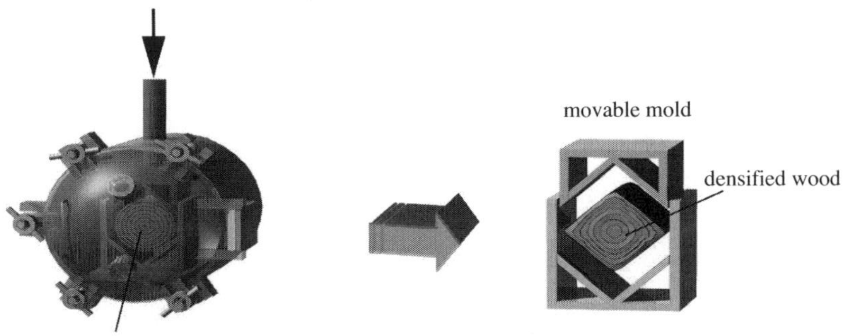

Fig. 6.27 Thermo-hydro-mechanical apparatus and the internal molding block.

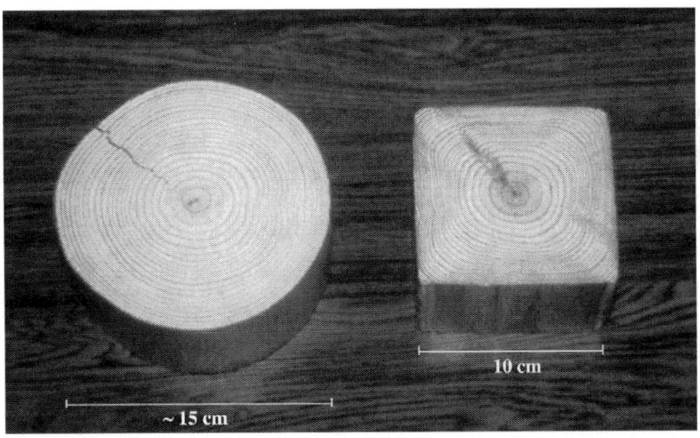

Fig. 6.28 Transformation of a circular trunk to a square-section block by two-dimensional densification (Ito *et al.*, 1998a).

2- Compressed Lumber Processing system
A second process was developed to obtain high-quality structural timber from low-density timber through THM-processing. This technique uses a "compressed lumber processing system" and consists of four stages.

1. Plasticization of low density lumber. The piece of wood is immersed in water at 80 °C for 60 minutes.

2. Forming of densified lumber (molding in compression). The low-density plasticized lumber is placed in an aluminum mold and compressed by a mechanical load.
3. Thermo-hydro-mechanical post-processing to fix the compression-set. The piece of wood with the mold is placed in a chamber with saturated vapor and a high pressure. The temperature variation within the chamber is shown in Figure 6.29.

During THM post-treatment, the temperature of the chamber is controlled in various stages.

- The temperature is raised to 180 °C.
- The temperature is maintained at 180 °C for 20 minutes.
- The steam is evacuated during 10 minutes.
- The material is vacuum-pumped for 20 minutes.
- The material is cooled with water for 15 minutes.

4. The mold is opened at room temperature. A photograph of pieces of densified lumber manufactured by the "compressed lumber processing system" technique is shown in Figure 1.7 in Chapter 1.

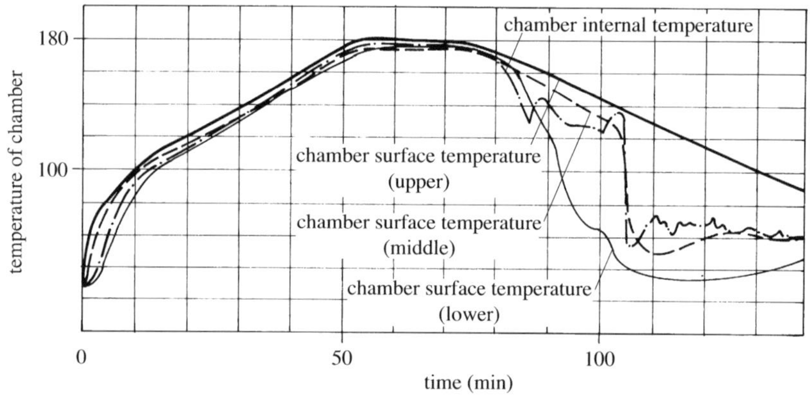

Fig. 6.29 The temperature in the THM chamber during the post-processing.

Since the 1990s, various investigations and research projects have been started in Europe, in the United States and recently in Canada to study densification of low-density veneers and massive wood molding, densification, fixation of the compression-set by THM treatment. In USA, due to the harvesting of fast-growing low-density wood, wood researchers have shown an interest in wood densification and the application of densified low-density veneers for the construction of composite materials (Kamke, 2004; Kutnar *et al.*, 2008). To densify small low-density hybrid poplar specimens, Kamke and Sizemore (2005) have developed a new closed system that is different from other THM-reactors. They called the process "viscoelastic thermal compression" (VTC).

Viscoelastic thermal compression

Kamke and Sizemore (2005) developed and built two VTC devices for wood (veneer) THM treatment; one "small VTC" and one "large VTC" device. Figure 6.30 and 6.31 present diagrams and photographs of respectively the small and large VTC. The small VTC device makes it possible to densify and post-treat up to 200 °C a veneer of 170×70 mm^2, and the large VTC device has a capacity for specimen up to 610×250 mm^2. The sample is heated by platens as in an open system, and the atmosphere surrounding the specimen during processing is controlled by steam coming in and out of the chamber.

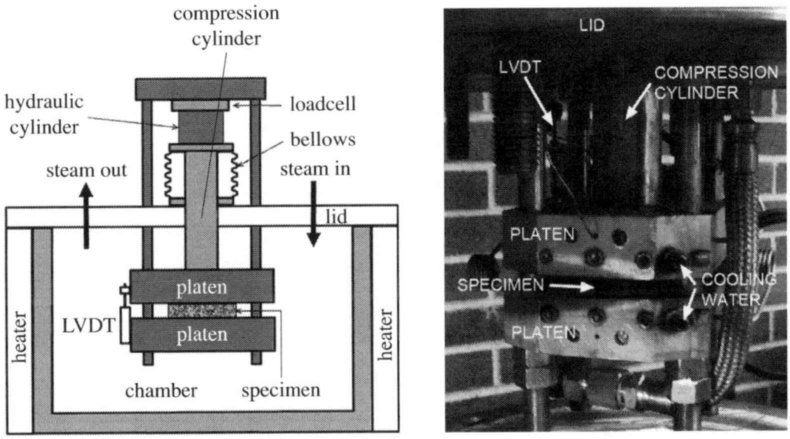

(a) (b)

Fig. 6.30 Small Viscoelastic Thermal Compression device: (a) diagram showing the hydraulic cylinder outside of the chamber, platens and the specimen inside the chamber with steam going out and in, (b) photograph of the interior of the chamber (Courtesy F. Kamke).

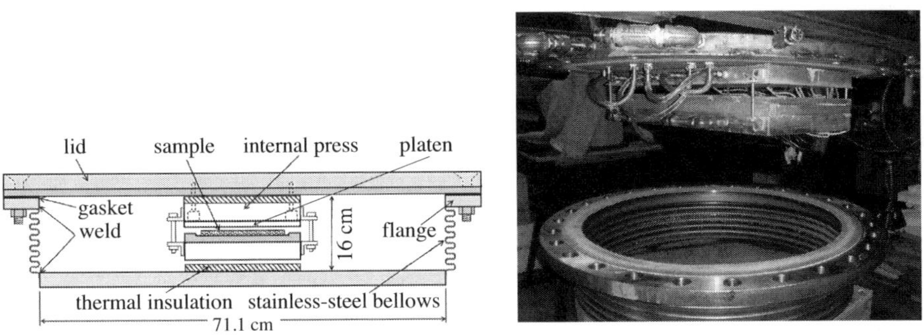

(a) (b)

Fig. 6.31 Large Viscoelastic Thermal Compression device: (a) diagram showing the press inside the chamber, platens and the specimen between the platens, (b) photograph of the interior of the chamber (Courtesy F. Kamke).

6.3 ORIGIN OF SHAPE MEMORY AND FIXATION OF COMPRESSION-SET BY THM-TREATMENT

Before discussing the origin of shape memory, the large transverse compressive deformation of wood during molding and the densification at the cell wall level will be described.

6.3.1 Wood deformation during densification at the cell-wall ultra-structural and molecular level

During plasticization of a wooden element by steaming at 120 °C, the temperature of the lignin, hemicelluloses and the zones of amorphous cellulose are above their respective glass transition temperatures T_g, and these components deform easily. However, the macromolecular movements of these non-crystalline regions are restricted. Indeed, on the one hand, hemicelluloses are strongly linked to lignin by covalent bonds (complex lignocarbohydrates) and, on the other hand, the cellulose microfibrils are surrounded by hemicelluloses (Figures 3.55, and 3.56) in a very coherent macromolecular system. This limits the slip of the microfibrils with respect to the hemicelluloses. Cohesion between the various components of wood is such that no separation between the middle lamella, the primary wall and the secondary walls was observed at a temperature lower than 160 °C, even after total recovery, as indicated by the micrographs presented in Figure 6.34. The molecular organization thus evolves or moves relatively little during the densification, if the temperature of densification is less than 160 °C and the processing time is short.

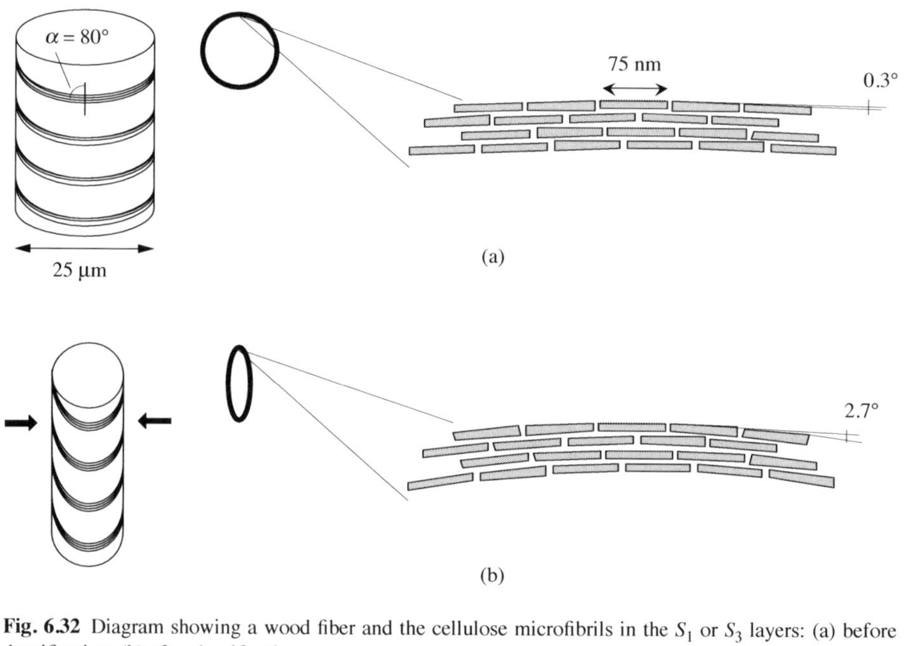

Fig. 6.32 Diagram showing a wood fiber and the cellulose microfibrils in the S_1 or S_3 layers: (a) before densification, (b) after densification.

With regard to the cellulose microfibrils, all the models of the macromolecular arrangement of cellulose proposed to date include semi-crystalline and crystalline zones (Figure 3.43) (Fengel & Wegener, 1983; Horii *et al.*, 1984). According to Heger (2004), the mechanism of the transverse compressive deformation of the microfibrils during densification is as follows: since the temperature of the quasi-amorphous areas of cellulose are higher than their T_g, these areas will be preferentially deformed during compression and play the role of articulations between the crystallites. The molecular configuration of the semi-amorphous zones will thus probably become modified. Figure 6.32 presents a diagram of such modifications. In an average wood cell with a diameter of 25 μm, an average length of crystallites of 75 nm and a microfibrillar angle of 80 degrees in the S_1 or S_3 layer, the number of end-to-end crystallites is approximately 1275. In that case, the average swing angle between each crystallite is approximately 0.3°. After densification, certain cells of earlywood may deform to a great extent so that each circular cell is flattened and becomes a lamella, leading to the microfibrils becoming curved through 180 degrees over a length of 5 μm (cell-wall thickness), which represents approximately 67 crystallites with an angle between each crystallite of 2.7 degrees. During the densification, the maximum variation of this angle is thus ca. 2.4 degrees. A small variation in the angle between each crystallite can therefore lead to a large deformation at the level of the cellular wall. As the microfibrillar angle is greater in the S_1 and S_3 layers[1], the variation in the angle between crystallites is higher and the deformation of the microfibrils is greater there. The zones of amorphous cellulose can be deformed during the densification thanks to the variation in the angles Φ (rotation around the connection C1-O1) and Ψ (rotation around the connection O1-C4′) glycosidic connection β- (1 → 4), Figure 6.33.

This model shows that the microfibrils undergo a relatively weak elastic deformation during the densifcation and that no irreversible plastic deformations are believed to be induced. However, this weak molecular deformation induces high internal stresses within the sample[2].

The sample thus stores elastic energy during densification. The internal stresses cannot be released, as long as sufficient physical Van der Waal's and hydrogen bonds maintain the deformation of the sample. Indeed, the cellulose microfibrils are surrounded by lignin and hemicelluloses in a matrix that prevents them from adopting another molecular configuration. The internal stresses are, of course, transmitted to all

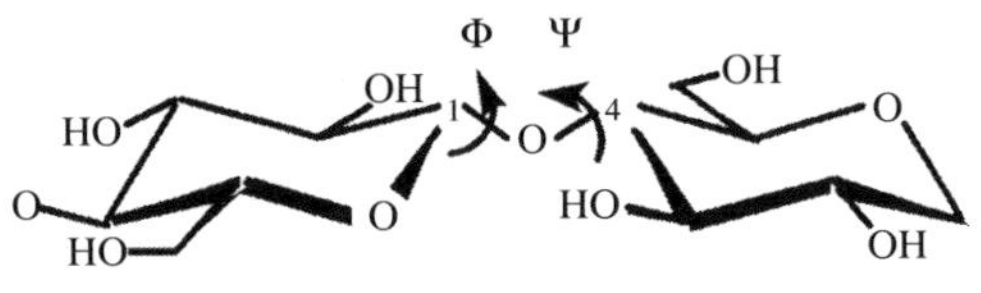

Fig. 6.33 Diagram showing the rotation of angles Φ and Ψ of the glycosidic connection *β*- (1 → 4) of the cellulose macromolecules in tension and compression (Horii *et al.*, 1984).

1 In the S_2 layer, where the microfibrillar angle is smaller than in the S_1 and S_2, the number of crystallites per whirl is definitely higher (~ 2500 for an angle microfibrillaire of 30°). The variation in the swing angle between each crystallite during the densification is smaller (~ 1.4°), since the deformation of the microfibrils is carried out on a longer length.

2 The compressive stress on the sample at the end of the densification reaches 17.5 MPa.

the components of the wood. When the densified sample is cooled to a temperature lower than the T_g of the wood constituents and dried, the reduction in the molecular thermal activation energy and in the moisture content allows of Van der Waal's and hydrogen bonds to reform. The molecular conformation of the cellular walls and the middle lamella of the compound becomes "frozen" as they are. This is known as the effect of cooling and drying on the densified wood.

6.3.2 Origin of compression recovery

During the densification process, wood is generally heated with high-temperature steam at 110 °C, mainly under saturated conditions to reach a softened state. After the wood has been plasticized, it is compressed in the transverse direction and undergoes elastic-visco-plastic deformation, before being cooled and dried. When the sample is moistened and heated in the absence of any external force, it generally returns to its original shape, depending on the temperature and moisture content of the sample. The thermo-hydro-plasticity of wood under these conditions of densification is thus only apparent. The phenomenon is known as "shape memory" or "compression-set recovery".

The origin of the shape memory is to be found at the level of the cell-wall ultrastructure and molecular structure. The assumptions formulated by Norimoto *et al.* (1993) are summarized as follows.

Moisture and the temperature act in various manners on the matrix and the microfibrils. A rise in temperature under wet conditions softens the matrix (amorphous hemicelluloses and lignin) and the semi-crystalline zones of the cellulose, and these components pass from the glassy to a quasi-rubbery state. However, the cellulose microfibrils (crystalline regions) remain in their glassy state because of their crystalline nature and they are almost unaffected by the moisture and heat. When a compressive load is applied to wood in the transverse direction, the load is almost totally supported by the microfibrils. The softening of the matrix allows a relative displacement of the microfibrils so that the framework of microfibrils becomes elastically deformed (Figure 6.32) to take up the local loads. As lignin is a polymer with slight cross-linking, its deformation can be regarded as viscoelastic rather than plastic. The elimination of the water molecules during drying and the reduction of thermal activation energy due to cooling lead to a reformation of the hydrogen bonds between the molecules of the matrix components. As a result of the reduction in temperature during cooling, the densified wood returns to a glassy state, where the elastic strain of the microfibrils and the matrix are frozen. Consequently, no recovery can occur until the matrix is again softened. However, as soon as the matrix is humidified and heated, the wood almost entirely recovers its initial form because of the release of the elastic energy stored in the microfibrils and the entropic and elastic molecular movements within the matrix.

Cell recovery of densified wood

The strain energy stored during the densification is proportional to the transverse deformation. The cell walls of earlywood, which are deformed to a larger extent as opposed to those of the latewood cells are the first to regain their initial shape under the effect of humidity and temperature. Normally the earlywood recovers first, before the latewood cells. However, the recovery of the sample never reaches 100%, since the components of the middle lamella of the compound, which were replaced (dislocated) during the densification, prevent the cells from finding exactly their initial forms (Figure 6.34). If the densification is carried out in the radial direction, the woody rays may undergo irreversible plastic deformation (Figure 5.13b and 6.34). In the best case scenario, when the reconditioning was carried out in the THM reactor at the same temperature (110 °C) and with the same relative humidity (RH = 100%) of the densification, the recovery was about 88.5%.

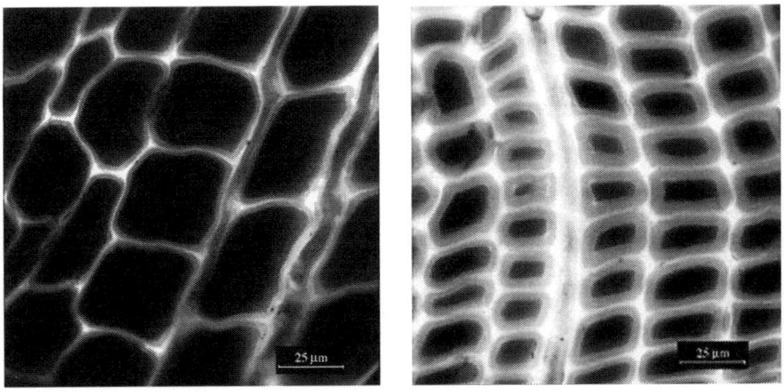

(a) (b)

Fig. 6.34 Confocal micrographs of densified spruce sample after a quasi-total recovery by distilled water at 60 °C; (a) cells of earlywood, (b) cells of latewood.

6.3.3 Elimination of shape memory of densified wood by THM-treatment

To obtain a stable densified wood (i.e., a densified wood with no shape memory), it is necessary for the densified wood to undergo a post-treatment. Therefore, THM-processes usually involve two steps. The transverse densification of the wood (first step) is followed by a second post-treatment step to eliminate the shape memory. The fixation of the compressive deformation of densified wood has been investigated using saturated steam (Inoue *et al.*, 1993; Dwianto *et al.*, 1997; Ito *et al.*, 1998a,b; Navi *et al.*, 2000; Navi & Heger, 2005) and also by dry heating (Dwianto *et al.*, 1997). These investigations have shown that increasing the temperature and the wood moisture content strongly reduces the post-treatment time necessary to eliminate the shape memory. The work carried out by Dwianto *et al.* (1997), illustrated in Figure 6.35, shows that heating under vacuum (totally dry conditions) at 180 °C should be carried out for more than 24 hours whereas heating at 200 °C under the same condition leads to the elimination of the shape memory after only 6 hours.

On the other hand, Inoue *et al.* (1993) have shown that the use of saturated steam during the post-treatments greatly reduces the time necessary for the heat treatment. Figure 6.36(b) illustrates that with saturated steam at temperatures of 180 °C and 200 °C, the shape memory was eliminated in respectively 8 and 2 minutes. Figure 6.36(a), on the other hands displays that under dry conditions the time required is 20 and 5 hours. The presence of water thus makes it possible to reduce the time by more than two orders of magnitude.

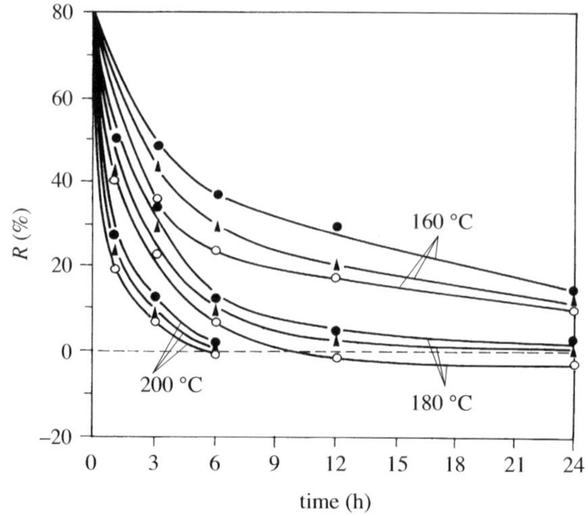

Fig. 6.35 Recovery of compressive deformation of densified pine samples (Pinus radiata) of $5 \times 30 \times 20$ mm^3 ($L \times R \times T$) as a function of the temperature and duration of the heat treatment: ○ heating by air, ▲ heating by molten metal, ● heating under vacuum, (Dwianto *et al.*, 1997).

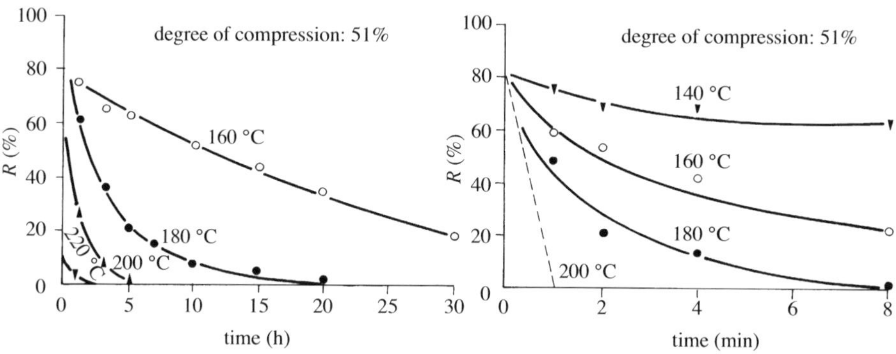

Fig. 6.36 Recovery of compressive deformation of samples of sugi (Cryptmeria japonica) of $20 \times 20 \times 30$ mm^3 ($L \times R \times T$) densified as a function of the temperature and time of the treatment. (a) heat treatment in a furnace, (b) treatment with saturated vapor (Inoue *et al.*, 1993).

Heger (2004) and Groux (2004) have studied the influence of the parameters of THM-treatment (moisture content, temperature and time) on the elimination of shape memory of densified wood using the THM apparatus, Figure 6.22-6.24. They showed that hydrolysis of the principal hemicelluloses of the softwoods, xylans and glucomannans, might be the main reason for the elimination of the shape memory.

In the following section, the results obtained using a multi-parametric THM reactor are presented together with a simple model that estimates the time necessary to eliminate the shape memory in a densified wood under given conditions of temperature and relative humidity.

Recovery test

To examine the effect of THM post-treatment on the fixation of the compressed-set of a densified wood sample, a hydral recovery test is necessary. This test consists of cyclic soaking-drying cycles of samples in water at 60 °C and in an oven at 30 °C. A complete description of the recovery test and definition of *R* is given by Equation 6.1 earlier in this chapter.

6.3.4 Experimental recovery results for densified wood THM-treated under saturated and unsaturated steam conditions

The recovery value (*R*) was measured by Heger (2004) and Groux (2004) for small densified spruce specimens, post-treated at several temperatures, under saturated and unsaturated RH for various times. Table 6.4 presents the values obtained for (*R*) defined by Equation (6.1). It is assumed that complete fixation of the compression is achieved when *R* is about 2%.

Table 6.4 Experimental results: time in boxes represents the minimum post-treatment time necessary to totally eliminate the shape memory (when *R* is about 2%). The uncertainty of the evaluation of *R* is about ±1%.

RH/T	140 °C	*R* (%)	160 °C	*R* (%)	180 °C	*R* (%)	200 °C	*R* (%)
100%	30 min	32.8	15 min	16.3	5 min	14.4	1 min	9.6
	60 min	19.7	30 min	6.1	10 min	2.7	2 min	12.0
	90 min	12.7	45 min	4.4	15 min	2.8	3 min	3.5
	120 min	5.6	60 min	1.3	20 min	1.3	4 min	1.5
	150 min	3.5	75 min	0.3	25 min	1.3	5 min	1.6
	210 min	4.0			30 min	0.9	6 min	1.3
80%	600 min	4.6	80 min	8.1	25 min	18.1	5 min	8.2
	1200 min	2.0	100 min	2.3	35 min	9.1	9 min	1.8
			120 min	2.6	40 min	1.9	13 min	1.1
			140 min	2.5	45 min	2.7	17 min	0.5
					50 min	1.9	20 min	0.6
60%	1200 min	5.7	330 min	2.3	80 min	5.7	10 min	2.1
	2000 min	6.6	420 min	2.7	90 min	4.0	25 min	0.1
			600 min	2.3	100 min	3.7	40 min	0.5
					120 min	2.7	55 min	1.1
					140 min	1.2	65 min	0.1
40%	1200 min	16.9	600 min	3.1	150 min	8.2	65 min	1.8
	2400 min	7.1	720 min	3.6	240 min	2.1	80 min	1.6
	3600 min	6.4			270 min	3.0	95 min	1.4
					300 min	1.5	110 min	1.4
					330 min	2.0	120 min	1.8
0	192 h	21.4	96 h	10.7	24 h	17.5		
			144 h	7.0	48 h	5.5	The sample burned	
			192 h	5.7	72 h	5.0		

Table 6.5 summarizes the conditions giving complete fixation.

Table 6.5 Post-treatment time (in minutes) for complete fixation of the compression ($R < 2\%$) for different temperatures and relative humidities.

Relative humidity	0	40%	60%	80%	100%
140 °C	not achieved*	6000	not achieved	1200	210
160 °C	24250	not achieved	not achieved	120	60
180 °C	5680	330	140	50	20
200 °C	not achieved	80	25	13	4

* For various reasons complete fixation of the compression ($R < 2\%$) at some temperatures and relative humidities was not achieved. These reasons were:

a) When the specimens were kept for a long time (more than 50 minutes) in the reactor at 200 °C with low moisture content, the specimens become burned.

b) When the specimens were treated at low temperatures with low relative humidity, it was probably difficult to achieve complete fixation, because these conditions were lower than the glass transition temperature of lignin.

6.3.5 Role of hemicellulose hydrolysis on the fixation of compression-set

The post-treatment time necessary to eliminate the shape memory increased nonlinearly when decreasing the relative humidity. To understand the mechanisms responsible for the elimination of shape memory, some chemical and physical analyses were carried out on densified wood before and after post-treatment, and the influence of THM parameters on the recovery value was examined through the development of a model. Figure 6.37 shows the post-treatment times (log scale) as a function of temperature for different relative humiditiesy. For a given relative humidity, the logarithm of time varied linearly with respect to temperature and all the lines were parallel. This

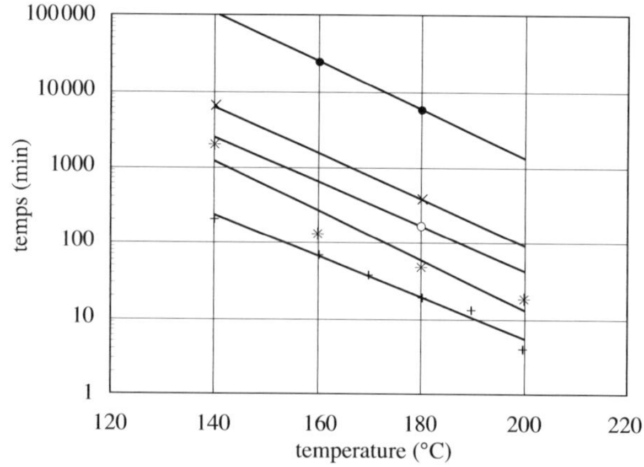

Fig. 6.37 The minimum post-treatment time (log scale) to achieve complete fixation of compression-set as a function of temperature and RH. RH: ● = 0%, × = 40%, ○ = 60%, ∗ = 80%, + = 100%.

regularity suggests a mathematical relation between the time necessary to eliminate shape memory or for relaxation of the strain energy accumulated during densification and the temperature and relative humidity. Figure 6.38 shows the same data plotted against $1/(RT)$, where $R = 8.317$ is the ideal gas constant and T is the absolute temperature in second.

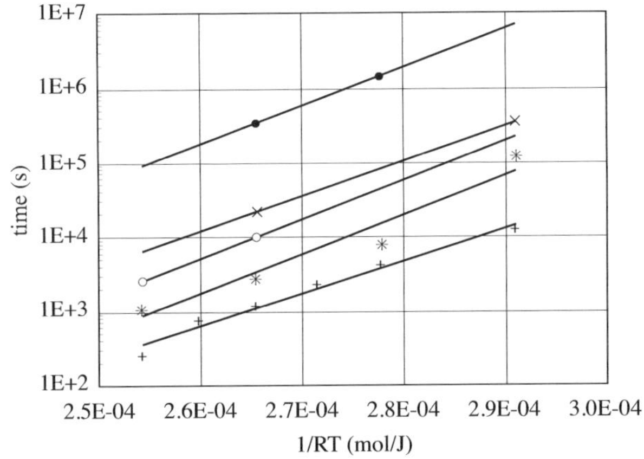

Fig. 6.38 The minimum post-treatment time (log scale) plotted against 1/(RT). RH: ● = 0%, × = 40%, ○ = 60%, ∗ = 80%, + = 100%.

This presentation makes it possible to calculate the activation energy (E_A) of the processes responsible for the elimination of shape memory or for eliminating the internal strain energy at each relative humidity according to the following equation

$$t(T, h) = \alpha e^{E_A/RT} \tag{6.2}$$

where α is a parameter depending on the relative humidity, and also on the pH of the solution during the post-treatment, and E_A is the activation energy. The value of α is given graphically as the intercept of the straight line on the time axe and E_A is the slope of the line. The values for E_A obtained from Figure 6.38 are given in Table 6.6.

Table 6.6 The activation energy obtained from the curves in Figure 6.35 in kJ/mol.

RH	100%	80%	60%	40%	0%	Average
Values of E_A (kJ/mol)	98.5	116.6	118.8	111.5	118.4	112.8

The values of the activation energy are more or less constant at 98.5-112.8 kJ/mol, regardless of the relative humidity, which means that the mechanisms responsible for the elimination of the shape memory are the same independently of the temperature and relative humidity during the post-treatment. These activation energy values are very close to the result obtained by Springer (1966) for xylan hydrolysis (E_A = 118 kJ/mol). The slightly lower value at 100% RH might be due to the fact that the hydrolysis was initiated by saturated steam.

Norimoto *et al*. (1992) have considered three mechanisms to be essential to prevent the compressive recovery:

- The formation of crosslinks between molecules of the matrix.
- Relaxation of the stresses stored in the microfibrils and the matrix during densification.
- Formation of polymers from hydrophilic materials inside the cell wall to avoid the softening of hemicelluloses by moisture.

The results presented in this section show that the hydrolysis of hemicelluloses during THM post-treatment plays an important role in the elimination of shape memory through the dissipation of the stresses stored in the microfibrils and matrix. It should nevertheless be noted that to achieve a total elimination of the shape memory, it is necessary for the lignin to be in its rubbery state during the THM-treatment.

6.4 REFERENCES

DWIANTO, W., INOUE, M. & NORIMOTO, M. (1997). Fixation of compressive deformation of wood by heat treatment. *Journal of the Japan Wood Research Society (Mokuzai Gakkaishi), 43*(4):303-309.

FENGEL, D. & WEGENER, G. (1983). *Wood: chemistry, ultrastructure, reactions*. Walter de Gruyter, Berlin.

GROUX, M. (2004). *Effet de la pression de vapeur lors des traitements thermo-hydro-mécaniques sur la stabilité de l'épicea densifi'e: influence de l'hydrolyse sur l'élimination de la forme*. (Effect of steam during thermo-hydro-mechanic treatment on the stability of densified spruce: influence of hydrolysis on the elimination of the form.) Master of Science Thesis, EPFL, Lausanne, Switzerland.

HALLER, P. (2007). Manufacturing of wooden profiles by forming., In: *Proceedings of Third International Symposium on Wood Machining – Fracture Mechanics and Micromechanics of Wood Composites with Regard to Wood Machining*, Navi, P. & Guidom, A. (eds.). PPUR, Lausanne, Switzerland, pp. 152-162.

HEGER, F. (2004). *Etude du phénomène de l'élimination de la mémoire de forme du bois densifié par post-traitement thermo-hydro-mécanique*. (Study of the mechanisms of elimination of the memory form of densified wood by post-processing thermo-hydro-mechanics.) PhD. Thesis No. 3004, EPFL, Lausanne, Switzerland.

HOFFMEYER, P. (1990). *Failure of wood as influenced by moisture and duration of load*. PhD Thesis, State University of New York, College of Environmental Science and Forestry, Syracuse, New York.

HORII, F., HIRAI, A. & KITAMARU, R. (1984). CP/MAS Carbon-13 NMR study of spin relaxation phenomena of cellulose containing crystalline and noncrystalline components. *Journal of Carbohydrate Chemistry, 3*(4):641-662.

HUGUENIN, P. & NAVI, P. (1995). *Bois densifié sans résine synthétique*. (Wood densification without use of synthetic resin.) Ingénieurs et Architectes Suisses, IAS, Switzerland, 13:262-268.

INOUE, M., NORIMOTO, M., TANAHASHI M. & ROWELL R.M. (1993). Steam or heat fixation of compressed wood. *Wood and Fiber Science, 25*(3):224-235.

ITO, Y., TANAHASHI, M., SHIGEMATSU, M., SHINODA, Y. & OTHA, C. (1998a). Compressive-molding of wood by high-pressure steam-treatment: Part 1. Development of compressively molded squares from thinnings. *Holzforschung, 52*(2):211-216.

ITO, Y., TANAHASHI, M., SHIGEMATSU, M. & SHINODA, Y. (1998b). Compressive-molding of wood by high-pressure steam-treatment: Part 2. Mechanism of permanent fixation. *Holzforschung, 52*(2):217-221.

KAMKE, F.A. (2004). A novel structure composite from low-density wood. In: *7th Pacific Rim Bio-Based Composites Symposium*, Nanying, China, pp. 176-185.

KAMKE, F.A. & SIZEMOORE, H. (2005). *Viscoelastic thermal compressed of wood*. U.S. Patent No. US2005/006004AI.

KUTNAR, A., KAMKE, F.A. & SERNEK, M. (2009). Density profile and morphology of viscoelastic thermal compressed wood. *Wood Science and Technology, 43*(1/2):57-68.

NAVI P., GIRARDET F. & HEGER, F. (2000). Thermo-Hydro-Mechanical post-treatment of densified wood. In: Proceedings of 5th Pacific Rim Bio-Based Composites Symposium Canberra, Australia, pp. 439-447.

NAVI, P. & HEGER, F. (2005). *Comportement thermo-hydromécanique du bois*. (Behaviour of thermo-hydro-mechanical processed wood.) Presses polytechniques et universitaires romandes, Lausanne, ISBN 2-88074-620-5.

NAVI, P., MEYLAN, B., PLUMMER, C.J.G., SPYCHER, M. & HEGER, F. (2005). Role of hemicelluloses in stress relaxation. In: *Proceedings of the hemicelluloses workshop,* Entwistle, K.M. & Walker, J.C.F., (eds.). Christchurch, New Zealand, pp. 121-136.

NORIMOTO, M., OTA, C., AKITSU, H. & YAMADA, T. (1993). Permanent fixation of bending deformation in wood by heat treatment. *Wood Research, 79*:23-33.

SEBORG, R.M., MILLET, M.A. & STAMM, A.J. (1945). Heat stablized compressed wood: Staypack. *Mechanical Engineering, 67*(1):25-31.

SPRINGER, E.L. (1966). Hydrolysis of aspenwood xylan with aqueous solutions of hydrochloric acid. *Tappi, 49*(3):102-106.

TANAHASHI, M. (1990). Characterization and degradation mechanisms of wood components by steam explosion and utilization of exploded wood. *Wood Research, 77*:49-117.

CHAPTER 7

WOOD WELDING

7.1 INTRODUCTION

The dimensions of wood are limited to the dimensions of the tree. Wood pieces are often joined together to produce larger sizes by mechanical fasteners such as nails, bolts and screws or by means of laminar fasteners (adhesives) like gelatine, glue and resins. Since 2002, a new type of laminar bonding of wood using frictional movement has been developed at the Bern University of Applied Sciences (BFH-AHB) in cooperation with the University of Nancy (France) and also at the Chair of Timber Construction of the Swiss Federal Institute of Technology in Lausanne. This wood bonding technique is called friction welding and has been shown to have several technical advantages over adhesive bonding. The time required for friction welding is short, less than a few minutes. Since the method is free from adhesives, it also promises environmentally friendly joints.

This chapter, presents a brief explanation of the welding process, defines the parameters of welding, their influences on the physical and mechanical behavior of the contact layer (interfacial zone), and reviews the microstructure of the contact zone and the chemical reactions of the wood constituents during welding processes. Two examples of wood joining, i.e., by linear welding to construct surfing board and by circular frictional welding to obtain a multilayer wood element are demonstrated.

7.2 JOINING OF WOOD BY FRICTION WELDING TECHNIQUES

Over the past 50 years, frictional welding has been developed and applied successfully to join pieces of metals and thermoplastics. The first work on the frictional welding of wood was documented and the process patented by Suthoff *et al.* (1996), in Germany. This patent and a subsequent one (Suthoff & Kutzer, 1997) reported that pieces of wood can be joined together through an oscillatory frictional movement. It also suggested that it may be possible to join pieces of timber with a wood dowel through rotary frictional welding. This patent also includes the concept of welding of wood under an inert atmosphere or vacuum.

In 2002, the first attempt at wood welding by a linear vibration movement was carried out in Switzerland. Some of the research results on the topic are given by Gfeller *et al.* (2003), Pizzi *et al.* (2004), Leaban *et al.* (2004, 2005), Ganne-Chédeville *et al.* (2005), Stamm (2005), Stamm *et al.* (2005a,b,c), Boonstra *et al.* (2006), Ganne-Chédeville *et al.*

(2006a,b), Stamm *et al.* (2006), Ganne-Chédeville *et al.* (2007), and Ganne-Chédeville *et al.* (2008a,b,c). The majority of the wood welded by friction is Norway spruce (*Picea abies* (L.) Karst.) and beech (*Fagus sylvatica* L.), and investigations have mostly been carried out on small samples with dimensions of about $150 \times 20 \times 15$ mm^3. The samples were welded on surfaces with dimensions of 20×15 mm^2. The maximum average tensile strength obtained was about 10 MPa for beech and 2 MPa for Norway spruce. Recently, in Switzerland at BFH-AHB, a multilayer of poplar was fabricated by linear friction welding, where the dimensions of each lamella were $160 \times 4 \times 2$ cm^3 and the welded surface had an area of 160×4 cm^2.

7.3 WELDING PROCESSES

The process of friction welding can be divided into several stages that are independent of the material of the pieces to be connected. Schlarp (1989) has described the process of friction welding of thermoplastics in seven stages, Figure 7.1. Usually the stages involved in wood friction welding are similar to those of thermoplastics.

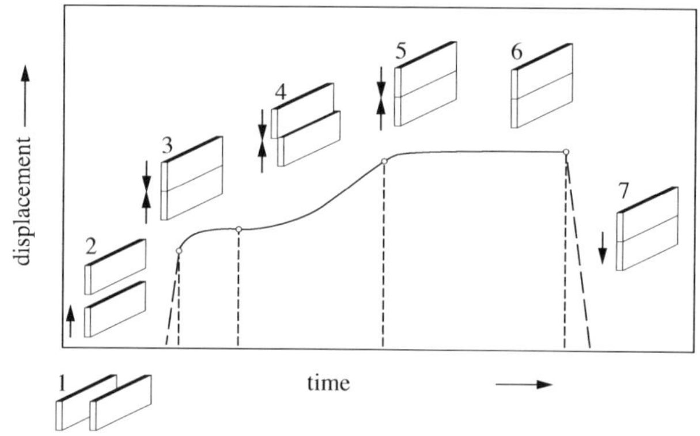

Fig. 7.1 Stages required during the linear friction welding of thermoplastics (Schlarp, 1989).

In this process, stages 1, 2 and 3 represent respectively the preparation of the pieces to be welded, machine feeding and the bringing into contact of the samples under a normal welding pressure. Stage 4 is the friction stage, where a frictional movement is applied. Stages 5 and 6 are curing stages. In stage 5, the frictional movement is stopped and a curing pressure is applied to the pieces to bring them closer. In stage 6, the applied normal pressure is removed and, in stage 7, the welded compound is removed from the welding machine.

The heat generated during the welding process depends on various parameters such as the settings of the welding machine (frequency, amplitude, welding pressure) and material properties (wood species, surface conditions, moisture content etc.). These parameters have important influences on the heat generation rate and consequently on the processing time and the bonding quality.

Usually, four different types of oscillatory friction movement are used for welding. These four types of welding are shown schematically in Figure 7.2, also illustrating the type of frictional movement, the variation of velocity and the force during the oscillatory movement.

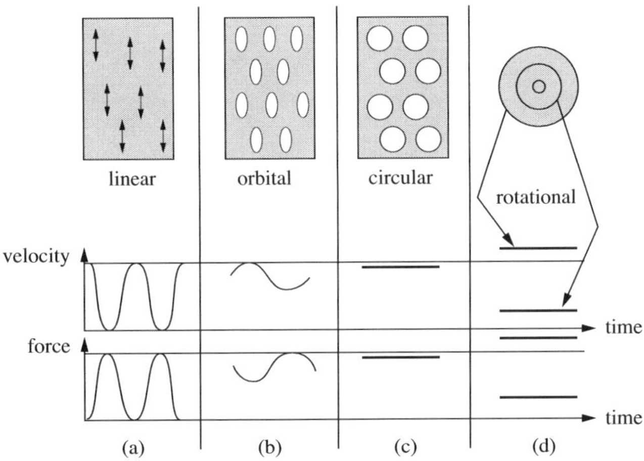

Fig. 7.2 Different movements, velocities and forces in 4 types of frictional welding; (a) linear friction, (b) orbital friction, (c) circular friction and (d) rotary friction welding (Stamm, 2005).

Linear welding is characterized by a linear oscillatory movement. The movement stops and changes direction when reaching the amplitude of the oscillation. In this movement, the velocity and its corresponding frictional force (parallel to the surface) are sinusoidal, as shown in Figure 7.2(a).

The frictional movements in orbital and circular welding are similar. In orbital friction welding, the movement is elliptical so that the force and velocity are sinusoidal, Figure 7.2(b). Whereas in circular welding, the movement is circular and the velocity and the frictional force are constant during the processing, Figure 7.2(c).

In rotary friction welding, the connection between the pieces is made by dowels and the method is usually called "dowel rotary friction welding", Figure 7.2(d). In this method, a cylindrical dowel is pressed into predrilled holes while rotating around its axis. A frictional force arise between the contact surfaces of the dowel and the hole as the consequence of the rotary movement of the dowel and the inward compressing force induced by the dowel due to the diameter of the dowel being slightly larger than the diameter of the hole. The first three types of frictional welding can be employed to join planar surfaces, whereas rotary friction welding can be employed to join axially symmetrical bodies. In rotary welding, the velocity and friction force are uniform (if the materials are considered to be homogeneous) for all points at a given distance from the rotation axis.

Rotary friction welding was first developed by Pizzi *et al.* (2004) to join two wood pieces by wooden dowels, and it was shown that the method allows an effective

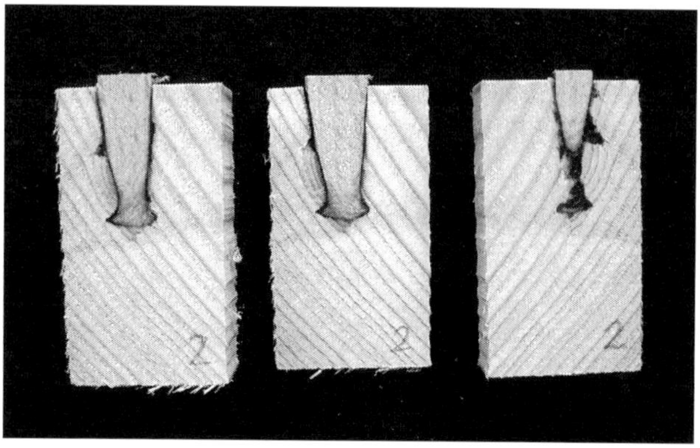

Fig. 7.3 Photograph of cross-sectional profiles of a ligno-bonded dowel by rotary friction welding (Ganne-Chédeville *et al.*, 2005).

ligno-bonding between the wooden dowel and the predrilled wood. Figure 7.3 displays a photograph of cross-sections of dowel rotary welding.

An X-ray inspection of dowel rotary friction has been carried out to investigate the ligno-bonding of the dowel and wooden pieces, Figure 7.4.

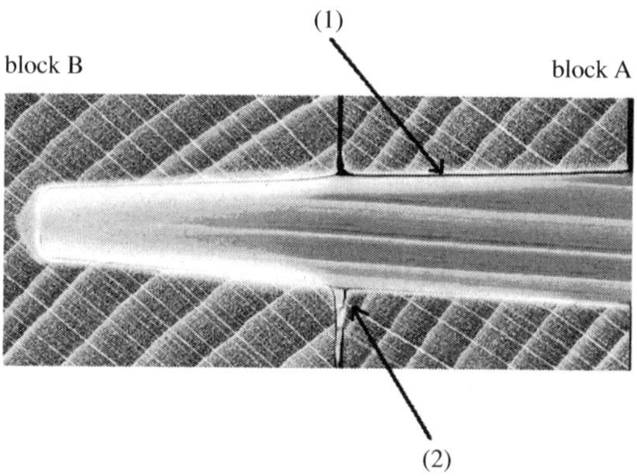

Fig. 7.4 X-ray micrograph of a beech dowel inserted and welded to two blocks of beech by rotary friction. The arrow (1) shows a section with no welding, and arrow (2) shows the welding with a microcrack (Ganne-Chédeville *et al.*, 2005).

In Switzerland, different types of wood welding were developed for different types of welded joints. Gfeller *et al.* (2003) used linear friction welding to join spruce and beech wood pieces at their edges, whereas Stamm *et al.* (2005a,b) employed circular friction welding to join two wood pieces of spruce or beech at their faces. Using frictional welding, Stamm studied the effect of wood machine parameters on wood-wood ligno-bonding. A schematic representation of a circular welding of two pieces with dimension of $110 \times 50 \times 10$ mm^3 is given in Figure 7.5. The two pieces were welded laterally by friction on a surface with an area of 50×50 mm^2.

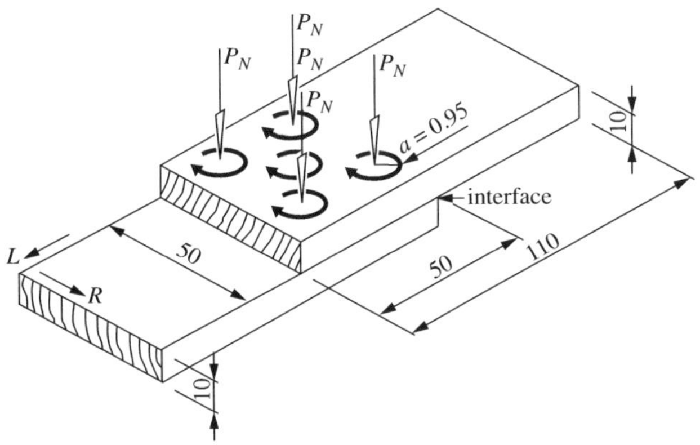

Fig. 7.5 Schematic representation of circular welding of two wood pieces. The dimensions of the pieces and the amplitude of the circular movement a are given in mm, normal pressure P_N, L = longitudinal direction, R = radial direction.

7.4 INFLUENCE OF MACHINE PARAMETERS AND SURFACE CONDITIONS ON THE HEAT GENERATION DURING FRICTIONAL MOVEMENT

To join wood pieces by frictional movement at the interface of two solid pieces of wood a thin layer on each surface should be converted into a "molten film", and a certain quantity of heat is required to make possible such a phase transition. The heat is generated by the sliding movement at the interface of two pieces subjected to a normal pressure P_N. The frictional thermal energy generated can be estimated as the product of the frictional shear stress τ_{fr} induced by the applied movement and the velocity v of the movement.

The frictional shear stress at the interface of two surfaces is defined as

$$\tau_{fr} = \mu P_N \qquad \text{[MPa]} \tag{7.1}$$

where μ is the coefficient of friction. According to Coulomb (1785), the frictional force (or frictional stress τ_{fr}) is independent of the velocity of the frictional movement. From relation (7.1) the calculated coefficient of friction is

$$\mu = \tau_{fr} / P_N \tag{7.2}$$

From this relation, it is evident that, for a given P_N, μ is also independent of the velocity of frictional movement.

Consider the frictionally generated power per surface area q [J/(s m^2] in a frictional welding, given by

$$q = \tau_{fr} \cdot v \tag{7.3}$$

The introduction of equation (7.1) into equation (7.3) gives:

$$q = \mu P_N \cdot v \tag{7.4}$$

In the following, the dependence of the value of q on μ, P_N and v is examined for two different friction welding processes.

Process 1: Linear frictional welding

A schematic representation of a linear frictional welding movement is given in Figure 7.6, where block *A* and block *B* are subjected to a compression force P_N and block *A* makes an oscillatory movement on block *B* with a velocity $v(t)$. The contact surfaces of blocks *A* and *B* are respectively s_A and s_B and the contact area is dA.

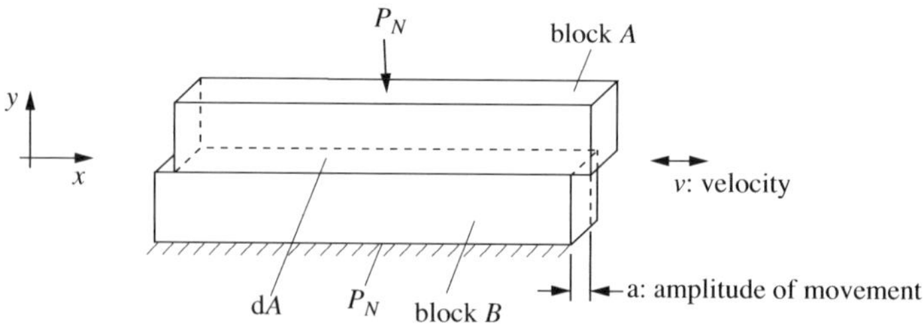

Fig. 7.6 Schematic representation of linear welding of two wood pieces *A* and *B* where *v* indicates the velocity of the oscillatory sliding movement of block *A* on the fixed block *B*.

Figure 7.7 shows that the movement of two plane surfaces during their linear welding can be simulated as a unidimensional movement. Hence, the equation of movement $u(t)$ of a point *M* of the surface s_A in the direction *x* when the block *B* is fixed and block *A* undergoes a linear oscillatory movement with amplitude a and frequency *f* can be written as

$$u(t) = a\sin(\omega t) \tag{7.5}$$

where $\omega = 2\pi f$

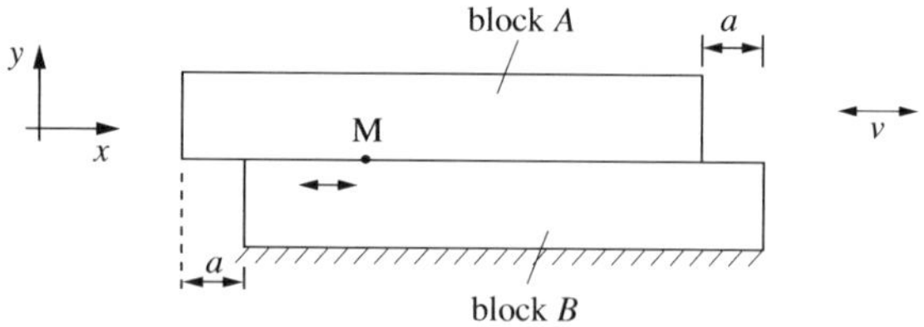

Fig. 7.7 Oscillatory movement of a point M on the surface s_A of block *A* during linear frictional welding.

It is assumed that the movement of all points located on the surface s_A is the same as that of point M without any phase difference among these points. This assumption is valid when blocks *A* and *B* are considered highly rigid in the direction of movement or if the movement of block *A* is considered a solid body motion. The frequency *f* corresponds to the movement of the welding machine. The velocity $v(t)$ of all the points located on the surface s_A is then given by

$$v(t) = \frac{\partial u(t)}{\partial t} \tag{7.6}$$

or

$$v(t) = 2\pi\, a\, f \cos(2\pi f\, t) \tag{7.7}$$

Introduction of equation (7.7) into (7.4) yields the frictional power per unit surface area q and for $t = 1$ second,

$$q = \mu(t)\, P_N\, 2\pi\, a\, f \cos(2\pi f) \qquad [\mathrm{J/(s \cdot m^2)}] \tag{7.8}$$

The frictional power generated by linear friction (which is also valid for any other type of frictional welding) will thus depend on parameters such as the coefficient of friction of the wood during welding, the normal pressure (welding pressure), the welding frequency and the amplitude of the vibratory movement. The welding frequency, the amplitude of the movement and the normal pressure are parameters of the welding machine, whereas the coefficient of friction depends on the type of material and on the condition of the surface during the welding process (e.g., wood species, orientation of annual rings on the welding surface, surface moisture content, surface roughness and temperature). Since the condition of the surface is a function of the parameters of the welding machine, the coefficient of friction during the welding operation is not a constant but varies with time and machine parameters. The formulation for q can be written

$$q = \int_0^1 \mu(\tilde{t}(t), a, f) P_N \cdot 2\pi a f \cos(2\pi f)\mathrm{d}t \;\; [\mathrm{J/(s \cdot m^2)}] \tag{7.9}$$

and the expression for the generated energy E becomes nonlinear and coupled during linear friction welding since μ is a function of $\tilde{t}(t)$, f and a, where $\tilde{t}(t)$ represents the surface condition which is function of time, so that

$$E = \int_0^t [\mu(\tilde{t}(t), a, f) \cdot P_N \cdot 2\pi \cdot a \cdot f \cdot \cos(2\pi \cdot f \cdot t) \cdot \mathrm{d}A] \cdot \mathrm{d}t \tag{7.10}$$

there, the limits of the integral are 0 and t, and dA is the area of the welding surface. As linear frictional welding involves a unidimensional movement, only one $\mu(\tilde{t}(t), a, f)$ corresponding to the coefficient of friction in the direction of movement is included in the expression for E.

Process 2: Circular friction welding

Circular friction welding (Figure 7.5) consists of a bi-dimensional frictional movement of the surface s_A on the surface s_B of the fixed block B. In this movement, all points on s_A undergo an oscillatory circular movement with frequency f and amplitude a. As in linear frictional welding, it is assumed that no phase differences exist among the vibratory movements of the points on the surface s_A. The movement of each point on the surface s_A with reference to its circular trajectory (Figure 7.8), is given by

$$u(t) = 2\pi \cdot a \cdot f \cdot t \tag{7.11}$$

In this case, the velocity v is constant and is equal to

$$v = v(t) = \frac{\partial u(t)}{\partial t} = 2\pi \cdot a \cdot f \tag{7.12}$$

Since the direction of the velocity v at any instance is the direction of the circular movement, the frictional power q per unit surface area and for $t = 1$ second is given by

$$q = 2\pi\, \mu_{eff}(\tilde{t}(t), a, f) \cdot P_N \cdot a \cdot f \qquad [\text{J}/(\text{s} \cdot \text{m}^2)] \tag{7.13}$$

where a is the amplitude (or the radius) of the circular movement (Figure 7.5), f is the frequency of the welding machine in Hz and μ_{eff} is the effective coefficient of friction. μ_{eff} may be calculated as the mean value of μ_x and μ_y, i.e.,

$$\mu_{eff} = 1/2\ (\mu_x + \mu_y) \tag{7.14}$$

The energy generated during the circular welding movement on a surface dA is then given by the integral

$$E = \int_0^t \left[\mu_{eff}(\tilde{t}(t), a, f) \cdot P_N \cdot 2\pi \cdot a \cdot f \cdot \mathrm{d}A \cdot t\right] \mathrm{d}t \tag{7.15}$$

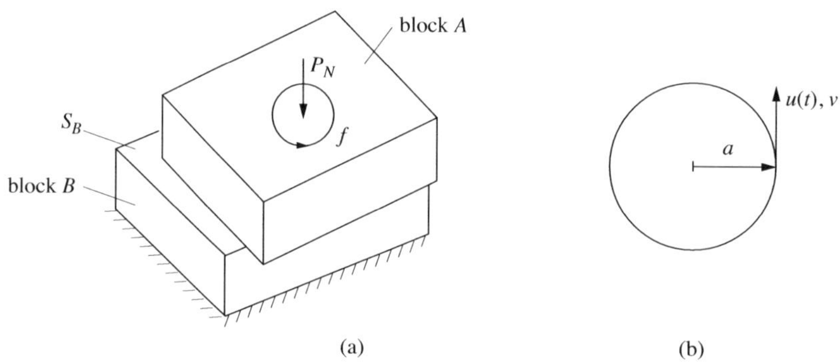

Fig. 7.8 Circular friction welding with (a) the circular oscillatory movement of block A subjected to a compression force P_N during welding, and (b) the direction of $u(t)$ and the velocity v.

Consequently, the power generated depends on the effective coefficient of friction of the wood during welding, the normal pressure (welding pressure), the welding frequency and the amplitude of the circular movement a.

Stamm *et al.* (2005a) have measured the variation in the coefficient of friction and the temperature at the wood interface during the circular welding of Norway spruce and beech. These experiments have shown that the condition of the surface changes considerably during the welding process. Stamm *et al.* (2005a) have separated the coefficient of friction into six phases as shown in Figure 7.9.

Phase I: (From S to point E_1 on the friction coefficient curve). During this phase, two wood pieces are brought into contact under a normal pressure P_N and frictional forces develop. Due to the friction forces, surface asperities are smoothened. The relatively rough surfaces cause the temperature of the surface to increase. At about 370 K, the temperature increases more slowly. Moisture in the surface is evaporated and the polishing of the surfaces leads to a decrease in the coefficient of friction.

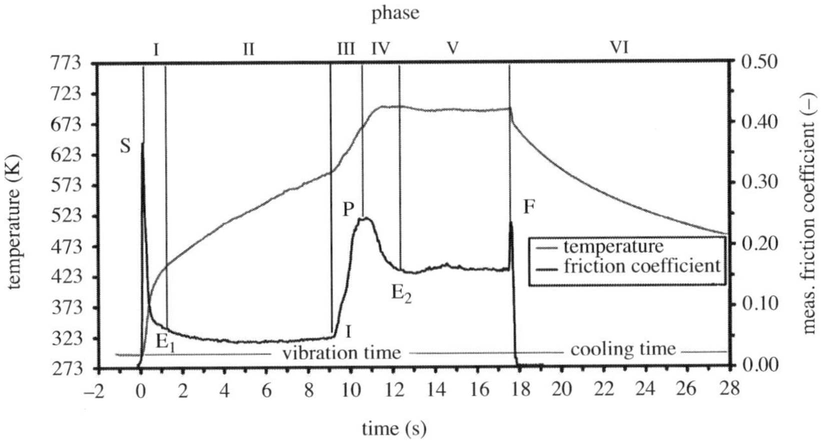

Fig. 7.9 Classification of the welding process into six phases based on the coefficient of friction and the interfacial temperature. Results for Norway spruce during welding with a normal pressure $P_N = 0.78$ MPa and a frequency of circular movement $f = 130$ Hz (Stamm *et al.*, 2005a).

Phase II: (From E_1 to I). This phase shows a constant coefficient of friction (i.e., no change in the state of surface) and a more or less linear increase in temperature.

Phase III: (From I to P). The friction force starts to increase with a high slope and this is accompanied by the generation of smoke. The surface of the wood decomposes at a temperature of about 580 K. The friction coefficient increases continuously up to point P.

Phase IV: (From P to E_2). Both the temperature and friction force reach an equilibrium at point E_2, where the temperature is about 713 K.

Phase V: (From point E_2 to F). Both the temperature and friction force remain constant. The heat generated by frictional energy is consumed by converting the solid wood to a "molten layer" of decomposed wood cells, and hot smoke is expelled from the surface. According to Shafizadeh (1984), the evaporation of levoglucosan and the volatile pyrolysis products takes place in a temperature range between 573 and 773 K. This chemical reaction is highly endothermic. The heat of evaporation leads to a cooling effect.

Phase VI: (From point F until solidification and cooling of the molten surface). The circular movement is stopped at point F. Cooling of the specimen and solidification of the "molten surface" at the interface then occur.

The variations in temperature and coefficient of friction shown in Figure 7.9 are typical for the welding of spruce wood. Stamm (2005) has also studied the welding of several species of wood (beech, spruce, birch and larch), and has found a similar behavior as a function of the welding time. However, the increase in temperature at the beginning of the process is much faster in beech than in the other species, probably due to the different surface properties. Based on the explanation given for each stage of Figure 7.9, it is possible to postulated that phases I + II, phases III + IV and phase V respectively correspond to dry friction, transition friction and viscous friction.

7.5 INFLUENCE OF WELDING PRESSURE AND FREQUENCY ON THE COEFFICIENT OF FRICTION DURING A CIRCULAR FRICTION MOVEMENT

The variation in the coefficient of friction during friction welding is a complex phenomenon that depends not only on the characteristics of the material to be welded but also on the parameters of the welding machine. Stamm (2005) has studied the influence of welding pressure and frequency on the coefficient of friction during the circular friction welding of spruce. Table 7.1 and Table 7.2 show the values of the welding parameters used in the investigation.

Table 7.1 Welding parameters and values of normal pressure (Stamm, 2005).

Parameters	Dimensions	Pressure
Amplitude	(mm)	0.95
Normal pressure	(MPa)	0.31, 0.47, 0.63, 0.79, 0.94
Frequency	(Hz)	130
Thickness specimen	(mm)	10
Welding displacement	(mm)	2
Cooling time	(s)	20
Cooling pressure	(MPa)	1.57

Table 7.2 Welding parameters and values of frequency (Stamm, 2005).

Parameters	Dimensions	Frequency
Amplitude	(mm)	0.95
Normal pressure	(MPa)	0.63
Frequency	(Hz)	80, 95, 110, 125, 150
Thickness specimen	(mm)	10
Welding displacement	(mm)	2
Cooling time	(s)	20
Cooling pressure	(MPa)	1.57

Figure 7.10 presents the influence of normal pressure on the coefficient of friction recorded during welding. Each curve corresponds to a specific normal pressure, given in Table 7.1, while the other parameters were kept constant. Each curve is obtained by averaging five different tests. This figure shows that, although the general form of the variation in the coefficient of friction is similar for various normal pressures, the total vibration time and the length of phases II and V (Figure 7.9) decrease with an increasing normal pressure. During these tests, a preset value (a welding displacement) was considered as a criterion for stopping the frictional movement.

A welding pressure of 0.31 MPa was too low to reach a satisfactory welding, but all other tests showed acceptable results. After stopping the frictional movement, the sample was cooled for about 40 seconds under a normal pressure of 1.57 MPa (cooling pressure). The maximum temperature reached at the welding line during the tests were almost constant with values varying between 693 and 713 K.

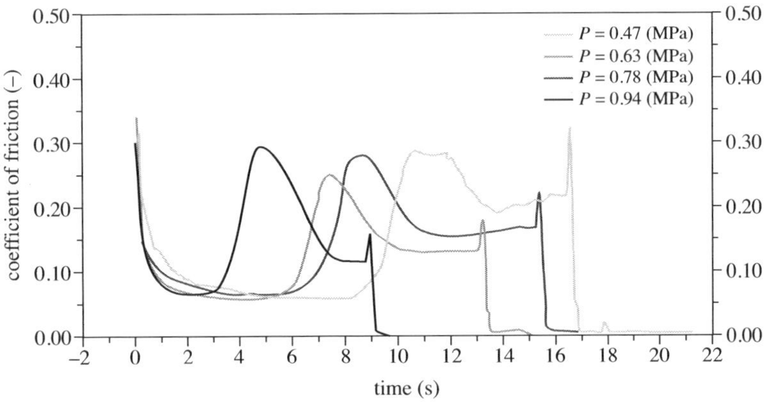

Fig. 7.10 Influence of normal pressure on the coefficient of friction. The curves represent the average of five tests (Stamm, 2005).

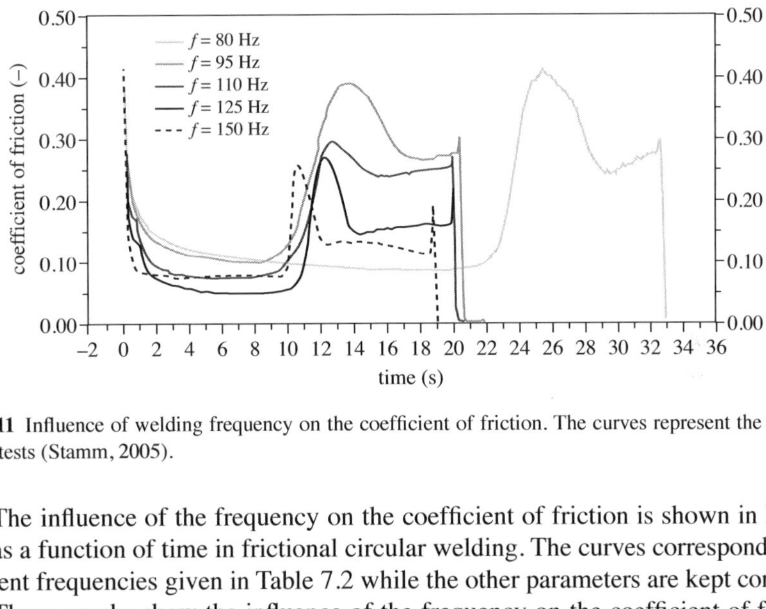

Fig. 7.11 Influence of welding frequency on the coefficient of friction. The curves represent the average of five tests (Stamm, 2005).

The influence of the frequency on the coefficient of friction is shown in Figure 7.11 as a function of time in frictional circular welding. The curves correspond to the different frequencies given in Table 7.2 while the other parameters are kept constant.

These graphs show the influence of the frequency on the coefficient of friction during various phases (Figure 7.9). For instance, the coefficient of friction decreases with an increasing frequency during phases II, III, IV and V. In phase V, also the temperature increases with an increasing frequency, which lowers the viscosity of the "molten" material and, therefore, reduces the coefficient of friction. Contrary to the influence of normal pressure on the vibration time, which is lowered with increasing pressure, the total vibration time seems to be independent of the frequency. This is rather surprising since the energy produced during the circular movement is a function of the frequency f, as given in equation 7.15. This surprising result may be due to the complex relation between the coefficient of friction, the frequency f, the pressure P_N, amplitude of movement and material properties.

Vairis and Frost (2000) have shown that a minimum frictional energy is needed in the welding of metal. It is clear that a certain minimum energy is also necessary to achive wood-to-wood bonding, otherwise no weld can be formed. Moreover, Stamm (2005) has shown that the energy produced at a frequency of 80 HZ and a welding pressure of 0.31 MPa was too low to produce a weld in a specified amount of time.

7.6 MICROSTRUCTURE OF LIGNO-BONDING BY FRICTION WELDING

During friction welding, wood pieces are subjected to a vibratory frictional movement under pressure normal to the welded plane in a welding machine. During the welding process, the interface of the wood heats up, thus inducing a "melted" material. After the frictional movement is stopped and during the cooling phase, the "melted" layer solidifies forming a continuous joint between two wood pieces. Figure 7.12 presents a micrograph of a welded joint of two spruce pieces achieved by circular friction, observed in a confocal microscope. This figure illustrates that the connected interfacial zone consists of two parts: a middle section of a dense amorphous material "fused" with fragments of densified wood fibers and melted material, and a second part of densified fibers surrounding the middle section. The gradient of densification of the second part is gradually decreased until one reaches the un-deformed wood fibers. The total depth of the middle and second zones is less than one millimeter.

To understand the thermal reactions, thermal decomposition and alteration of the wood constituents during the welding process, in the temperature ranging from 140 to 440 °C, Stamm *et al.* (2005c and 2006) studied the formation of the "melted" material by chemical and spectroscopic methods. Their studies demonstrated that the

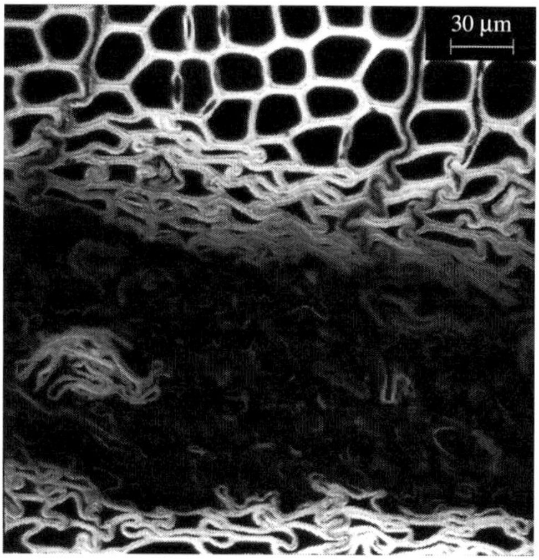

Fig. 7.12 Micrograph of a zone of joined and adjacent areas between two pieces of spruce welded by circular friction (Stamm *et al.*, 2005b).

hemicelluloses were degraded to a great extent while cellulose remained relatively stable. Lignin also underwent distinct changes, as demonstrated by an increase in free phenolic groups and the decrease in the typical bonds between the phenolpropane units. They also detected furan derivatives within the volatiles of the smoke gas, arising mainly from hemicelluloses. The conclusion was reached that reactions between furfural and other furan derivatives with lignin were the main reactions in the friction zone, leading to cross-linking of the "melted" contact material.

This material has also been examined by Gfeller *et al.* (2004), Kanazawa *et al.* (2005) and Ganne-Chédeville *et al.* (2005) and it has been shown that the essential chemical modifications involving the polymeric constituents of wood in friction welding occur in the first 5-6 seconds and that they slow down or even stop afterwards. Analysis by FT-IR and CP-MAS ^{13}C NMR of the welded area of wood has pointed at dehydration and an apparent increase in the crystallinity of cellulose. A certain hemicellulose degradation has been found to occur, accompanied by the generation of furfural. Cellulose degradation is very slight. Both analytical techniques show an increase in the proportion of lignin in the welded interphase. Self-condensation of lignin occurs by internal rearrangement. This progresses throughout the whole welding process. After 6 seconds of welding, the wood begins to carbonize, and an increase in cross-linking of the lignin network is observed. It has been suggested that to produce the frictional energy necessary to make the reaction between furfural and other furan derivatives with lignin and to allow good adhesion between the two wood pieces, the variation in temperature at the interface has to be monitored (Ganne-Chédeville *et al.*, 2006a,b). Thus, the temperature progression at the welding line during linear friction welding has been measured by infrared thermography. Moreover, physical analyses such as fracture and mechanical shear tests (mode II), transmitted light microscopy and X-ray densitometry have been carried out by Ganne-Chédeville *et al.* (2007, 2008a,b).

7.7 SOME NEW RESULTS AND CHALLENGES

Recently, considerable work has been performed to investigate the potentiality of wood joining by various types of frictional techniques: linear friction, circular friction and dowel rotary friction. Their mechanical performances, particularly the resistance to fracture propagation and the influence of water on fracture toughness, have been examined by various tests developed for wood joints by adhesives. Joining by frictional methods still shows certain inconveniences, which should be overcome before their application in timber construction. Experiments have indicated that:

- The shear strength of wood joined by friction is less than the shear strength obtained by adhesives, which is about 10 MPa. According to the European Standard IN 205-D1, the required shear strength for wood joints authorized for use in timber construction is more than 10 MPa. Stamm (2005) has shown that the maximum shear strength of welded wood achieved by circular friction is about 4.2 and 3.3 MPa for beech and spruce, respectively. Gfeller *et al.* (2004) have however reported that a value of more than 7 MPa can be reached for beech by linear friction welding. Pizzi *et al.* (2004) have also shown that the shear resistance of rotary dowel friction welding may be comparable to that of dowels glued by adhesives. Omrani *et al.* (2009) have demonstrated that the shear strength of small pieces of wood joined by linear friction can be up to 10 MPa.

- The joint strength results depend on the wood species, welding machine parameters and wood properties. These results are particularly dependent on the size of the welded pieces. The shear resistance of the welded joint achieved by circular friction decreases with increasing dimensions of the pieces.
- Water absorption by friction welded wood joints lowers the resistance of the joints to shear (mode II) and to tension (mode I) fracture propagation.

Recently, other studies have been carried out on improving the strength of wood joints by optimizing the welding machine parameters (frequency, amplitude, and welding displacement or welding process time), to better understand up-scaling problems and improve the quality of large welded joints for application in timber construction. The state of advances in wood welding processes is shown in the following.

Figure 7.13 shows a multi-layer wooden element consisting of 8 alternating layers of spruce and beech obtained by circular friction welding. The dimensions of this multi-layer element are $6.3 \times 4.5 \times 9.1$ cm^3. This laminated wooden element can be used instead of laminated wooden beams or wooden wall elements made of glued boards. Such timber elements do not require a high shear resistance (Stamm, 2005).

Figure 7.14 presents a photograph of a linear welding machine at BFH-Biel in Switzerland. This welding machine is capable of welding two pieces of wood that are two meters in length.

Figure 7.15 displays a photograph of a wooden surfing board consisting of thin wood elements joined through linear welding.

Fig. 7.13 A wooden multi-layer element made of alternated layers of spruce and beech bonded by circular friction welding (Stamm, 2005).

Fig. 7.14 A linear welding machine at BFH-AHB in Switzerland.

Fig. 7.15 A surfing board made by linear welding, (BFH-AHB in Switzerland).

7.8 REFERENCES

BOONSTRA, M., PIZZI, A., GANNE-CHÉDEVILLE, C., PROPERZI, M., LEBAN, J.-M. & PICHELIN, F. (2006). Vibration welding of heat-treated wood. *Journal of Adhesion Science and Technology, 20*(4):359–369.

COULOMB, C.A. (1785). *Théorie des machines simples, en ayant égard au frottementde lears parties, et à la roideur des cordages*. (Theory of the simple machines, by having regard with the friction of their parts, and the stiffness of the ropes.) Mémoires de Mathématique et de Physique de l'Academie des Sciences, pp. 161-342.

GANNE-CHÉDEVILLE, C., PIZZI, A., THOMAS, A., LEBAN, J.-M., BOCQUET, J.-F., DESPRES, A. & MANSOURI, H. (2005). Parameter interactions, two-block welding and the wood nail concept in wood dowels welding. *Journal of Adhesion Science and Technology, 19*(13/14):1157-1174.

GANNE-CHÉDEVILLE, C., LEBAN, J.-M., PROPERZI, M., PICHELIN, F. & PIZZI, A. (2006a). Temperature and density distribution in mechanical vibration wood welding. *Wood Science and Technology, 40*(1):72–76.

Ganne-Chédeville, C., Properzi, M., Pizzi, A., Leban, J.-M. & Pichelin, F. (2006b). Parameters of wood welding: A study with infrared thermography. *Holzforschung, 60*(4):434–438.

Ganne-Chédeville, C., Properzi, M., Pizzi, A., Leban, J.-M. & Pichelin, F. (2007). Edge and face linear vibration welding of wood panels. *Holz als Roh- und Werkstoff, 65*(1):83–85.

Ganne-Chédeville, C., Properzi, M., Leban, J.-M., Pizzi, A. & Pichelin, F. (2008a). Wood welding: chemical and physical changes according to the welding time. *Journal of Adhesion Science and Technology, 22*(7):761-773.

Ganne-Chédeville, C., Duchanois, G., Pizzi, A., Leban, J.-M. & Pichelin, F. (2008b). Predicting the thermal behaviour of wood during linear welding using the finite element method. *Journal of Adhesion Science and Technology, 22*(12):1209-1221.

Ganne-Chédeville, C., Duchanois, G., Pizzi, A., Pichelin, F., Properzi, M. & Leban, J.-M. (2008c). Wood welded connections: energy release rate measurement. *Journal of Adhesion Science and Technology, 22*(2):169-179.

Gfeller, B., Zanetti, M., Properzi, M., Pizzi, A., Pichelin, F., Lehmann, M. & Delmotte, L. (2003). Wood bonding by vibrational welding. *Journal of Adhesion Science and Technology, 17*(11):1573–1589.

Gfeller, B., Pizzi, A., Zanetti, M., Properzi, M., Pichelin, F., Lehmann, M. & Delmotte, L. (2004). Solid wood joints by in situ welding of structural wood constituents. *Holzforschung, 58*(1):45-52.

Kanazawa, F., Pizzi, A., Properzi, M., Delmotte, L. & Pichelin, F. (2005). Parameters influencing wood-dowel welding by high-speed rotation. *Journal of Adhesion Science and Technology, 19*(12):1025–1038.

Leban, J.-M., Pizzi, A., Wieland, S., Zanetti, M., Properzi, M. & Pichelin, F. (2004). X-ray micro-densitometry analysis of vibration-welded wood. *Journal of Adhesion Science and Technology, 18*(6):673–685.

Leban, J.-M., Pizzi, A., Properzi, M., Pichelin, F., GelHaye P. & Rose C. (2005). Wood welding: A challenging alternative to conventional wood gluing. *Scandinavian Journal of Forest Research, 20*(6):534-538

Omrani, P., Mansouri, H.R. & Pizzi, A. (2009). Linear welding of grooved wood surfaces. *Holz als Roh- und Werkstoff, 67*(4):479-481.

Pizzi, A., Leban, J.-M., Kanazawa, F., Properzi, M. & Pichelin, F. (2004). Wood dowel bonding by high-speed rotation welding. *Journal of Adhesion Science and Technology, 18*(11):1263–1278.

Schlarp, A.K. (1989). *Zum Vibrationsschweißen von Polymerwerkstoffen: Process, Struktur, Eigenschaften.* (Vibration welding of polymer materials: process, structure, properties.) PhD Thesis, Universität-Gesamthochschule Kassel, Faculty of Mechanical Engineering.

Shafizadeh, F. (1984). The chemistry of pyrolysis and combustion. In: *The chemistry of solid wood*, Rowell, R.M. (ed.). Advances in Chemistry Series, Vol. 207, American Chemical Society, Washington DC, pp. 489-529.

Stamm, B. (2005). *Development of friction welding of wood –physical, mechanical and chemical studies.* PhD Thesis No. 3396, Ecole Fédérale de Lausanne, Switzerland.

Stamm, B., Natterer, J. & Navi, P. (2005a). Joining wood by friction welding. *Holz als Roh- und Werkstoff, 63*(5):313-320.

Stamm, B., Natterer, J. & Navi, P. (2005b). Joining of wood layers by friction welding. *Journal of Adhesion Science and Technology, 19*(13/14):1129-1139.

Stamm, B., Windeisen, E., Natterer, J. & Wegener, G. (2005c). Thermal behaviour of polysaccharides in wood during friction welding. *Holz als Roh- und Werkstoff, 63*(5):388-389.

Stamm, B., Windeisen, E., Natterer, J. & Wegener, G. (2006). Chemical investigation on the thermal behaviour of wood during friction welding. *Wood Science and Technology, 40*(7):615-627.

Suthoff, B., Schaaf, A., Hentschel, H. & Franz, U. (1996). *Verfahren zum reibschweißartigen Fügen von Holz.* (Method for joining wood.) Offenlegungungsschrift DE 196 20 273 A1. Deutsches Patent und Markenamt.

Suthoff, B. & Kutzer, H.-J. (1997). *Verfahren zum reibschweißartigen Verbinden von Holz.* (Method for joining wood.) Offenlegungungsschrift DE 19746 782 A1. Deutsches Patent und Markenamt.

Vairis, A. & Frost, M. (2000). Modelling the linear friction welding of titanium blocks. *Material Science and Engineering, A 292*(1):8-17.

CHAPTER 8

SURFACE DENSIFICATION

8.1 INTRODUCTION

There is a considerable interest in developing treatments for densifying wood species that have a low density. Numerous experiments, as well the experimental results given in Tables 6.1, 6.2 and 6.3, indicate that the impact resistance, the longitudinal Young modulus and the strength increase with an increasing wood density. The results presented in Figures 6.18 and 6.19 also indicate that the shear strength at the LR plane and the Brinell hardness of wood are raised by densification. These results as well as those given in Figure 5.29 shows that the mechanical properties of wood are density-dependent.

The goal of the THM wood treatments presented in chapters 5 and 6 is to uniformly densify a block of wood in the transverse direction. This densification process is used during the molding and shaping of wood. A microphotograph of a uniform densification is given in Figure 8.1.

Instead of obtaining a uniform densification, it is sometimes desired to increase the abrasion resistance and hardness of a wood surface by densifying only a thin surface layer of the wood. Figure 8.2 shows a micrograph of a densified surface. For

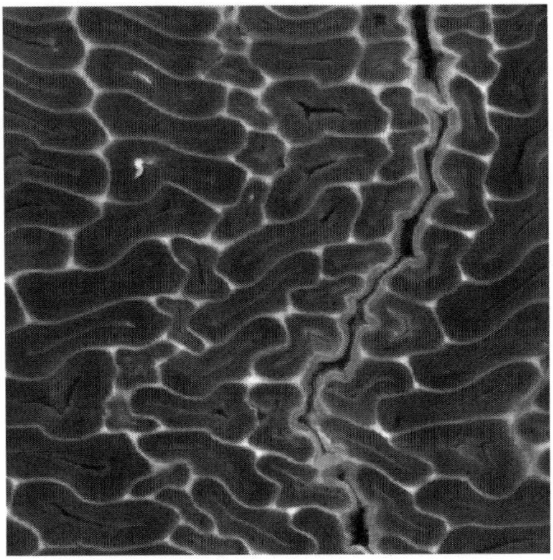

Fig. 8.1 Micrograph of uniformly densified spruce latewood cells (Navi & Heger, 2005).

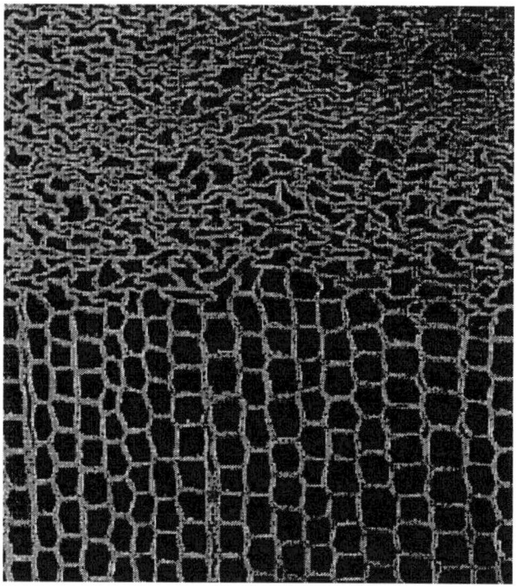

Fig. 8.2 Micrograph of the boundary between the compressed surface and uncompressed regions of wood (Inoue *et al.*, 1990).

decades, numerous attempts have been made to invent novel techniques for producing a form-stable densified wood surface. The main motivation of these studies has been to utilize low-density woods for flooring where the wood surface is exposed to excessive wear. The techniques developed are based mainly on two distinct characteristics of wood, namely the influence of moisture content on the glass transition temperature (T_g), i.e., the thermal softening of wood decreases with an increasing moisture content as discussed in Chapter 5, and the poor thermal conductivity of wood. This chapter briefly presents these techniques and the associated problems.

8.2 THM TECHNIQUES FOR WOOD SURFACE DENSIFICATION

The first attempts to densify the surface of panels go back to the work of Tarkow and Seborg, (1968), who developed a technique to produce two types of compression deformations during densification: a small elastic compression of the whole wood element and a large permanent compressive deformation in a thin layer of the surface of the element. To fix the deformation, the parameters of the densification, i.e., the moisture content of the wood surface layer, the temperature, the duration of processing and the level of the applied load were optimized. The aim was to eliminate the stored elastic strain energy in the densified portion of wood by the combined action of heat and humidity.

The technique consisted in heating a humidified panel to a temperature of 260 °C and then compressing it by up to several hundred kg/cm^2. The panel was moved between a pair of heated plates at a linear speed of about 8 m/min.

Two redwood and maple panels about 20 mm in thickness were densified by this procedure. The maximum density in the surfaces of both the maple and the redwood was measured and both panels showed an increase in density at their surfaces. When the compression-set after prolonged periods of high humidification was studied, it showed a distinct compression recovery that was dependent on the pressing conditions (temperature, humidity, load level and pressing time). The compression-set was stable enough to give the product an industrial application.

Inoue *et al.* (1990) developed a technique to densify the surface layer of coniferous wood. They carved 2-mm wide and 5-mm deep grooves in the surface of coniferous wood specimens at intervals of 150 mm across the grain direction. The grooved surface was filled with water and heated with microwaves. The surface was then compressed by about 45%. They then produced surface densified lumber from sugi, hinoki and western hemlock and found an increase in hardness of 120-150%, however the compressed-set was not fixed during the densification procedure.

Another technique developed to densify the surface of wooden panels was developed by Pizzi *et al.* (2005). Their technique was similar to that used for the friction welding of wood elements but with a small modification in the preparation of the specimens. In surface densification, a wooden panel (specimen) is first cut in the tangential direction to produce two specimens that are then fixed on the friction-welding machine as in wood friction welding. To avoid bonding of the two surfaces during the process, sunflower oil or some other oil type is added to the surfaces prior to the frictional movements. During processing, when the surfaces are "melted", the movement of the machine is stopped but the applied compression force, the welding pressure, is maintained until a complete solidification of the surfaces is achieved. The densified surfaces are then conditioned at 20 °C and 65% RH. The parameters of the welding machine used for this process were adapted from the work of Gfeller *et al.* (2003). To examine the influence of the machine parameters and the wood humidity on the densified surfaces, various experiments have been carried out with four machine vibration times: 3, 4, 6 and 9 seconds. The other parameters used were: contact holding time, i.e. the period in which the specimens were maintained together under pressure after the machine vibration had stopped, 5 seconds; the welding pressure, 1.3 MPa; the holding pressure, 2.0 MPa exerted on the surfaces after the welding vibration had stopped; the amplitude of linear vibration, i.e., the movement of one surface relative to the other surface, 3 mm; and the frequency of the welding machine, 100 Hz.

After conditioning, these samples were used for Brinell hardness measurements and X-ray microdensitometry of sections of the treated surfaces. The experimental results showed that the Brinell hardness for beech wood specimens increased almost linearly with the welding time. The X-ray microdensitometry showed that a thin layer of the surface with a higher density began to appear with a machine vibration time of 4 seconds and that this progressively increased in thickness for 6-second and 9-second welding. The thin high-density surface layer had a thickness between 10 and 400 μm and the density varied between 800 and 1200 kg/m^3. The work indicated that the presence of sunflower oil for friction time of 3 seconds and 6 seconds had formed a good water barrier on the densified layer, but gave no indication of the improvement of surface compression-set recovery.

Rautkari *et al.* (2009) used the friction welding machine (without adding oil to the wood surfaces) to densify the surface of a wooden panel. In their work, a Branson 2700 linear vibrational welding machine was suitably adapted to give a surface densification. The equipment consisted of two platens: a lower fixed plate and an upper plate capable of vibrating linearly in a controlled manner. The lower plate contained electrical heating elements with thermocouples for temperature sensing and to control the heating of the plate. The wood element was fixed to the upper plate, and the surface was heated to 100 °C by the lower platen before the linear vibration of the upper plate was started. The wood welding machine and its adaptation to surface densification is presented in Figure 8.3. The holding pressure exerted on the panel after the welding vibration stopped was 0.58 MPa. The panel was kept in the machine until the surface temperature had decreased to 60 °C. Densification was carried out at a frequency of 100 Hz with an amplitude of about 3 mm, and the process pressure was 2.2 MPa. The machine parameters were able to "melt" the surface of the specimen in 12 seconds.

Fig. 8.3 Welding machine with a wood panel (D) with dimensions $150 \times 20 \times 30$ mm^3 ($L \times T \times W$) fixed to the upper plate, and the lower plate with pre-heating unit (*E*) (Rautkari *et al.*, 2008).

During densification, the surface temperature was measured, and was found to vary between 150 and 200 °C. The overall thickness densification was about 2.7%. The compression in the surface layers could be much larger, but this was not reported by Rautkari *et al.* (2009). Nevertheless, the surface hardness was increased by 137%. These values unfortunately do not show the percentage surface densification or the compression-set recovery under thickness swelling tests in water. Another study by Rautkari *et al.* (2010) using the same densification method shows that, after several high humidity-oven dry cycles, the densification achieved by this process was almost completely reversible, i.e., there occurred full recovery. However, contact angle measurements indicated that the wettability of the densified surfaces was significantly lower than that of the unmodified surfaces.

Tests on thickness swelling in boiling water have demonstrated that the densified surfaces obtained by these techniques were unstable and that the densified cells were widely open as shown in Figure 8.4.

Another wood surface densification method is the implementation of a heated press with a cooling system (Lamason & Gong, 2007). In this technique, the sur-

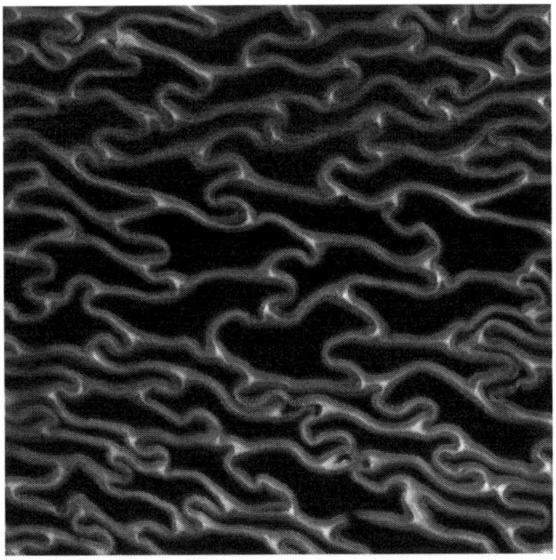

Fig. 8.4 Widely open earlywood cells under very wet conditions (Navi & Heger, 2005).

face of aspen wood was soaked in boiling water for 5 minutes before densification, and different processing parameters (compression level, pressing temperature, press closing time) influencing the mechanical properties, such as hardness, modulus of elasticity (MOE) and nail withdrawal strength, were optimized. The optimum process parameters were found to be: compression 24%, temperature 145 °C, and closing time 7 minutes. Based on these optimum pressing conditions, the hardness, MOE and nail withdrawal resistance of the surface-densified solid wood was improved by on average 140%, 23%, and 132%, respectively.

It has been shown by Gong *et al.* (2010) that, after wood surface densification, the set-recovery in wet or humid conditions can be partially eliminated, using a post-heat-treatment process under normal atmosphere conditions. This was achieved by heat treating surface-densified aspen at temperatures of 190 °C, 200 °C and 210 °C. Unfortunately the treatment times are not mentioned in the report. During the heat-treatment, the wood was found to spring back slightly, probably due to the injection of steam during the process. This phenomenon can be eliminated using a modified heat-treatment process (Rautkari & Hughes, 2009). The heat-treatment typically starts by heating the wood in steam, but in this modified heat-treatment process, steam is injected only when the heat-treatment chamber temperature is raised sufficiently to avoid recovery. Here, the starting temperatures for steam injection were 140 °C, 170 °C and 200 °C and the actual heat-treatment temperature was 200 °C for two hours. A complete or almost complete elimination of the recovery was reported.

It has been shown that wood surface densification (before post-heat-treatment to fix the compression-set recovery) might improve mechanical properties and also decrease wettability. Moreover, it has been studied whether surface densification may enhance fire retardation with or without fire retardation chemicals (Subyakto *et al.*, 1998).

8.3 PROBLEMS IN SURFACE DENSIFICATION

As previously mentioned, almost all the techniques used for wood densification are based on two distinct characteristics of wood, viz: the influence of the moisture content on the glass transition temperature (T_g) of the wood and the poor thermal conductivity of the wood. Experiments have shown that, during wood-surface processing, the large deformation (compression) produced is not permanent, since the densified cells open widely at high humidity and high temperature. As described with respect to uniform densification in Chapter 6, the fixation of the compression-set can be achieved mainly by eliminating the stored elastic strain energy through Thermo-Hydro-Mechanical post-treatments, where the wood is heated in saturated steam to a temperature of 200 °C (in a closed system) for about 4 minutes, or to a temperature of 160 °C or 180 °C for about 60 or 20 minutes, respectively. The consequences of these treatments are a rather limited degradation of the wood and its mechanical properties. The investigation by Rautkari and Hughes (2009) has shown that, to totally fix the compression-set in the densified surface, it was necessary to heat wood in steam to 200 °C for more than 2 hours. Under such conditions, densified wood will obviously degrade significantly and lose a great amount of its mechanical strength. This phenomenon has been clearly shown by Gong *et al.* (2010), who heat-treated undensified wood as well as densified wood at three temperatures: 190 °C, 200 °C and 210 °C. They report that the hardness, MOE, nail withdrawal resistance and wearing resistance of undensified aspen after heat treatment were higher than the values for heat-treated densified aspen. The heat treatment processing time was not reported, but it is evident from the negative impact value on MOE (reducing the values for densified and undensified wood by 24% and 32%, respectively, as compared to that for non-heat-treated densified wood) that this time should be several hours. It is important to note that the level of heating the processing time and moisture content of the element during post-treatment have an important influence on the wood degradation during post-heat-treatment. If the aim of surface densification is to increase the wood surface wear and hardness by densification, then the conditions of post-treatment should be such that it does not greatly reduce the compressed wood properties.

8.4 REFERENCES

GFELLER, B., ZANETTI, M., PROPERZI, M., PICHELIN, F., LEHMANN, M. & DELMOTTE, L. (2003). Wood bonding by vibrational welding. *Journal of Adhesion Science and Technology, 17*(11):1573-1589.

GONG, M., LAMASON, C. & LI, L. (2010). Interactive effect of surface densification and post-heat-treatment on aspen wood. *Journal of Materials Processing Technology, 210*(2):293-296.

INOUE, M., NORIMOTO, M., OTSUKA, Y. & YAMADA. T. (1990). Surface compression on coniferous wood lumber. Part I. A new technique to compress the surface layer. *Mokuzai Gakkaishi, 36*(11):969-975.

LAMASON, C. & GONG, M. (2007). Optimization of pressing parameters for mechanically surface-densified aspen. *Forest Products Journal, 57*(10):64-68.

NAVI, P. & HEGER, F. (2005). *Comportement thermo-hydromécanique du bois*. (Behaviour of thermo-hydro-mechanical processed wood.) Presses polytechniques et universitaires romandes, Lausanne, ISBN 2-88074-620-5.

PIZZI, A., LEBAN, J.-M., ZANETTI, M. & PICHELIN, F. (2005). Surface finishes by mechanically induced wood surface fusion. *Holz als Roh und Werkstoff, 63*(4):251-255.

RAUTKARI, L., PROPERZI, M., PICHELIN, F. & HUGHES, M. (2008). An innovative thermo densification method for wood surfaces. In: *Proceedings of the 10th World Conference on Timber Engineering*, Miyazaki, Japan, ISBN 978-1-61567-088-8.

RAUTKARI, L., PROPERZI, M., PICHELIN, F. & HUGHES, M. (2009). Surface modification of wood using friction. *Wood Science and Technology, 43*(3/4):291-299.

RAUTKARI, L. & HUGHES, M. (2009). Eliminating set-recovery in densified wood using a steam heat-treatment process. In: *Proceedings of the 4th European Conference on Wood Modification*, Englund, F., Hill, C.A.S., Militz, H. & Segerholm, B.K. (eds.). SP Technical Research Institute of Sweden, pp. 393-396, ISBN 978-91-86319-36-6.

RAUTKARI, L., PROPERZI, M., PICHELIN, F. & HUGHES, M. (2010). Properties and set-recovery of surface densified Norway spruce and European beech. *Wood Science and Technology, 44*(4):679–691.

SUBYAKTO, K.T., HATA, T., ISHIHARA, S., KAWAI, S. & GETTO, H. (1998). Improving fire retardancy of fast growing wood by coating with fire retardant and surface densification. *Fire and Materials, 22*(5):207-212.

TARKOW, H.T. & SEBORG, R. (1968). Surface densification of wood. *Forest Products Journal, 18*(199): 104-107.

CHAPTER 9

HEAT TREATMENT

9.1 INTRODUCTION

In resent decades, developments in the area of heat treatment have accelerated considerably. During the 1980s, French and Japanese industries began to heat-treat wood in order to increase the resistance to microbial attack. Since then, the interest in heat treatment has increased all over the world. The underlying reason for applying heat treatment is the growing demand for environmentally friendly high-durability wood, i.e., to increase the service life of wood materials without the use of toxic chemicals. Since heat treatment requires no physical impregnation of the wood, it is a process that can be used on species that are hard to impregnate with preservatives such as Sitka spruce and Norway spruce heartwood.

Heat treatment or thermal modification of wood is wood modification in a strict sense, since the material undergoes chemical changes. The process essentially involves a controlled degradation of the wood, primarily resulting in the destruction of hemicelluloses and includes several methods. These methods are mainly divided into those performed in closed systems and those performed in open ones. In a closed system, all the parameters are controlled, which is not the case in an open system.

Generally, heat treatment involves temperatures between 150 and 260 °C. The high temperature is achieved with superheated steam in a vacuum or with an inert gas such as nitrogen. Pre-heated oil can also be used, in which case the oil acts as a heat-transfer medium and also excludes oxygen from the wood. Heat treatment of wood above 300 °C is of limited practical value due to the degradation of the wood material.

The exact method of treatment can have a significant effect on the properties of the modified wood, and the most important process variables (Hill, 2006) are

- the treatment atmosphere;
- the choice of a closed or open system;
- the choice of wet or dry systems;
- the use of a catalyst;
- the wood species;
- the time and temperature of treatment; and
- the dimensions of the material.

Heat treatments of wood have been investigated for many years and are now commercially available. A simplified picture of the results of the heat treatment is that the increase in stability and durability is accomplished by an increased brittleness and a loss of some strength properties, including impact toughness, fracture modulus and work to failure.

Chapter 2 demonstrated that different heat treatment processes have been used to improve the performance of wood and that the use of these processes dates back many thousands of years. In northern Sweden, the Saami bent the front tips of their wooden skis using heat more than 5200 years ago (Åström, 1993). From hieroglyphic pictures of various operations on different parts of a chariot in Egypt in about 1400 B.C., poles and chariot yokes were made using heat-bent wood members (see Figure 2.2). The purpose of using a THM-process in these cases was to shape the wood, whereas the reason for modern heat treatment processes is to improve physical properties such as durability and shape stability. The use of heat treatment in this manner in ancient times was uncommon. The method employed to increase the durability of wood in the ground, above ground and in water was to burn the outer part of the wood in order to obtain an insulating layer of char. This technique is still in use in the Nordic countries for fence poles, where the part in the ground is burned. However, more than one hundred years ago, Heinzerling (1885) expressed doubts about the advantages of charing wood for increased durability. Nowadays, it is known that burning of the outer surface of wood makes the wood more flame resistant, but it is not known how and if this was applied in ancient times.

Many methods of heat treatment of wood have been reported in the literature from the beginning of the 20th century when it was found that drying wood at a high temperature increased its dimensional stability and reduced its hygroscopicity and strength (Tiemann, 1915; Koehler & Pillow, 1925; Pillow, 1929; Stamm & Hansen, 1937). After the First World War, comprehensive investigations were made on the influence of kiln drying or temperature on the strength of wood for the aviation industry in the United States (Tiemann, 1920; Wilsson, 1920).

Stamm *et al.* (1946) reported on the first systematic attempts to increase the resistance to wood-destroying fungi by heating wood beneath the surface of molten metal. They found that by doing this at temperatures between 140 and 320 °C, swelling was reduced in Sitka spruce by 60% and its resistance to microbial attack was increased. They also found that the raise in stability and durability was accompanied by a greater brittleness and the loss of some strength properties. The heat treatments usually caused a darkening of the wood and the wood had a tendency to crack and split. Thunell and Elken (1948) and Buro (1954, 1955) continued the work by Stamm and studied the heat treatment of wood in various gaseous atmospheres and beneath the surface of molten metal.

Burmester (1973) investigated the effects of temperature, pressure and moisture on wood properties in a closed system and found that the optimum conditions for pine were 160 °C, 20-30% moisture content and 0.6 MPa pressure. The processing time was then about 5 hours. A high dimensional stability, resistance to brown-rot fungi attack as well as minimal loss of strength were reported. Burmester denoted the process FWD (Feuchte-Wärme-Druck).

Several names have been given to the various heat-treated products and treatments for wood including Staypak and Stabwood in the United States, Lignostone and Lignofol in Germany, Jicwood and Jablo in the UK, Thermowood in Finland, Plato in The Netherlands, the OHT –process (Oil Heat Treatment) in Germany, and the retification process and Les Bois Perdure in France.

9.2 THERMAL DEGRADATION OF WOOD AND ITS CONSTITUENTS

Wood exposed to elevated temperatures undergoes a thermal degradation process and the degradation is highly dependent on the temperature, the duration of exposure, the process pressure, the moisture content etc. Shafizadeh and Chin (1977) found that timber heated at 120 °C loses 10% of its strength after one month, whereas at 140 °C the same loss in strength occurs after only one week.

Tiemann (1942) summarized the extensive work carried out between 1905 and 1908 at the Yale Forest School in USA on the influence of drying temperature and humidity on the strength of three species of wood (white ash, loblolly pine and red oak). The effect on the strength was studied after subjecting the wood to different temperatures and various humidities during drying (in dry air at 60-150 °C, in saturated steam at 100-166 °C and in superheated steam at 134-166 °C) for varying lengths of time and then re-soaking or exposing the wood to the atmospheric conditions for several months or a year. Three general effects were noted: all the processes decreased the strength of wood, the hygroscopicity was reduced by most of the treatments, and the color darkened particularly at higher temperatures. In general, it was also noted that white oak was damaged to a larger extent than ash or pine.

The thermal degradation of wood appears to be different depending on whether air is present or not. Generally, the different processes during heating of wood are

- heating without air access (pyrolysis, e.g. producing charcoal in a closed system); and
- heating with air access (combustion),
  - complete combustion (e.g., burning in a fireplace),
  - incomplete combustion (e.g., production of producer gas or charring in a charcoal stack).

In the absence of oxidizing agents or other catalysts, the heat-treatment process of wood is called pyrolysis. This means that the wood does not combust, but thermally degrades into solid, liquid and gaseous components. The production of charcoal is one of the oldest and historically most common pyrolysis processes.

In reporting the exothermic reactions in wood, Kollmann (1960) defined three phase points:

- The flame point, 225 to 260 °C, at which decomposed gases will burn if an ignition source is present.
- The burning point, 260 to 290 °C, at which burning occurs with a steady flame. The decomposition is exothermic during the burning point and causes a self-induced flash.
- The flash point, 330 to 470 °C, where spontaneous ignition occurs.

The result of a heat treatment is highly dependent on the presence or absence of oxygen and water. The presence of oxygen will lead to oxidative reactions that can be prevented by performing the treatment in an inert atmosphere such as oil, nitrogen, water or steam. The use of steam is an efficient and cheap way of creating an inert atmosphere, but it also influences the reactions that take place during the treatment.

9.2.1 Thermal degradation of wood

The changes in wood characteristics take place according to a time-temperature-moisture-oxygen relationship. Evidence of acid hydrolysis has been seen in the interior of living old eucalypts (Stewart *et al.*, 1961) and the acetic acid concentration has been found to increase with time over a two-year period, when green wood is stored at 48 °C (Packman, 1960). The rate of thermal degradation differs for the various constituents of timber: it is the highest for hemicelluloses, much lower for cellulose and the lowest for lignin. The rate of degradation of wood is intermediate between the degradation rates of cellulose and lignin.

The changes that occur in wood at temperatures below 40 °C are mainly attributed to physical changes such as the emission of water and volatile extractives such as terpenes (Manninen *et al.*, 2002). Some minor chemical changes probably start to occur in the interval of 40-90 °C, predominately due to certain extractives. Englund and Nussbaum (2000) found high concentrations of volatile organic compounds (VOC), mainly monoterpenes, when drying Scots pine and Norway spruce using conventional schedules with temperatures up to 60 °C as well as schedules reaching 110 °C. When wood is heated it releases volatile, flammable gases that burn when in contact with a source of ignition. Wood itself does not burn; the wood material decomposes on heating and it is the emitted gases that burn.

The products from completely pyrolyzed wood are reported to be a proportional combination of the pyrolyzates from cellulose, hemicelluloses and lignin, considering each to be pyrolyzed separately (Browne, 1958). However, the chemistry of the thermal degradation of the isolated wood components is undoubtedly different from that taking place within the cell wall, where the various reactions can act synergistically with the wood material. This section deals with wood as whole, while subsequent sections treat the degradation of the constituents of wood.

The degradation or strength loss during heat treatment is due to hydrolysis of hemicelluloses, which is influenced by moisture, temperature, time and oxygen. Wood degrades much faster when it is heated in either steam or water than when it is heated in a dry condition (MacLean, 1951, 1953; Stamm, 1956). Heating wood in the presence of water or steam leads to the formation of organic acids, mainly acetic acid, which catalyze the hydrolysis of the hemicelluloses to soluble sugars, thus speeding up the degradation process (Hillis, 1975; McGinnis *et al.*, 1984).

However, a high moisture content also leads to a decrease in the timber temperature and serves as a solvent for acids formed during hydrolysis. In the case of dry timber, higher temperatures are needed to achieve the same degradation as in moist timber. Differences in degradation have been found between wood treated in a closed system and that treated in an open system (Stamm, 1956, 1964). The heating in a closed system results in a more rapid degradation. This is because there occurs a build-up in a closed system of chemicals such as acetic acid that can interact with the chemical reactions taking place. The degradation products then trigger the degradation process.

The rate of thermal degradation is also dependent on the surrounding atmosphere, especially with regard to the presence or absence of oxygen. Stamm (1956) demonstrated that the thermal degradation of wood heated in the presence of oxygen is more rapid than that of wood heated in an oxygen-free atmosphere. The presence of oxygen during hydrolysis results in a process known as wet oxidation, where the initial reaction is the formation of acids. As these acids increase in concentration, hydrolytic reactions become favorable. The rate of hydrolysis is increased not only for the hemicelluloses, but also for the cellulose component (McGinnis *et al.* 1984). Thus, the combined processes of hydrolysis and oxidation contribute to the thermal degradation of wood. The degradation is greater in oxygen than in vacuo, but this is also related to a lower energy transfer in a vacuum (Stamm, 1964).

During heat-treatment of wood, the chemical wood structure is transformed by autocatalytic reactions of the cell-wall constituents. It is known that during the heat-treatment of wood under moist conditions, carbonic acids, mainly acetic acid, will initially be formed as a result of cleavage of the acetyl groups, particularly of hemicelluloses (Kollmann & Fengel, 1965; Dietrichs *et al.*, 1978; Bourgois & Guyonnet, 1998). Depending on the acid concentration and the temperature, hemicelluloses become hydrolyzed into oligomeric and monomeric structures (Klauditz & Stegmann, 1955; Bobleter & Binder, 1980; Carrasco & Roy, 1992). Subsequently, the monomeric sugar units are dehydrated to aldehydes, and furfural is formed from pentoses and hydroxymethylfurfural from the dehydration of hexose sugar units (Burtscher *et al.*, 1987; Kaar *et al.*, 1991; Ellis & Paszner, 1994).

Lignin is the least reactive wood component, but at high temperatures above 200 °C bonds within the lignin complex are cleaved, resulting in a higher concentration of phenolic groups (Runkel, 1951; Kollmann & Fengel, 1965). This state of increased reactivity of the lignin leads to various condensation reactions of aldehydes and lignin and to the autocondensation of lignin.

In general, the thermal degradation of wood proceeds along one of two competing reaction pathways. At temperatures up to about 200-300 °C, carbon dioxide and traces of organic compounds are formed in addition to the release of water vapor. At this stage, the wood polymers degrade through the breaking of internal chemical bonds that results in the formation of free radicals, carbonyl, carboxyl, and hydroperoxide groups, carbon monoxide, carbon dioxide and the formation of reactive carbonaceous char (Rowell, 2005). The second mechanism, which takes over at temperatures above of 300 °C lead to a much more rapid decomposition and cleavage of secondary bonds. The pyrolysis gases contain 200 or more different components (Brenden, 1967). The degradation is accompanied by a reduction in weight depending on the temperature and duration of heating.

When a piece of wood is heated in the absence of air, zones develop parallel to the heat-absorbing surface, delimited by the temperatures attained. The zones are well marked in wood because of its relatively low thermal conductivity and density and its relatively high specific heat.

Browne (1958) divides the pyrolysis process into four phases, or temperature ranges, all of which can be present simultaneously in wood of an appreciable thickness:

- Phase A, below 200 °C, in which only non-combustible gases, primarily water vapor, with traces of carbon dioxide, formic and acetic acids, and glyoxal are produced. The evaporation of sorbed water is complete.
- Phase B, from 200 to 280 °C, in which the same gases as in phase A are produced, but with much less water vapor and some carbon monoxide. At this

point, the reactions are endothermic, and the products are almost entirely non-flammable.

- Phase C, from 280 to 500 °C, in which active pyrolysis takes place under exothermic conditions leading to secondary reactions among the products. The products are largely combustible (carbon monoxide, methane, etc.), and include highly flammable tars in the form of smoke particles. The charcoal residue catalyzes secondary reactions.
- Phase D, above 500 °C, in which the residue consists primarily of charcoal, which provides an extremely active site for further secondary reactions. Early combustion stages are similar to the pyrolysis stages, modified slightly by oxidation.

The course of events when wood is heated in air can be similarly divided into phases, but these include oxidation reactions and, after ignition, combustion of pyrolysis and oxidation products. Based on the same temperature divisions as in pyrolysis, combustion may be categorized as follows:

- Phase A, which, in addition to being characterized by the evolution of non-combustible gases, is affected by some exothermic oxidation processes.
- Phase B, in which the primary exothermic reaction takes place without ignition. The ignition point may, however, also be defined as the temperature at which the exothermic reactions begin.
- Phase C, in which ignitable combustible gases are produced after secondary pyrolysis. Flaming combustion can then occur if the gases are ignited, but the combustion is restricted to the gas phase. If ignition is not induced, flaming may not occur until near the end of the pyrolysis when the evolved gases cannot insulate the charcoal layer from the oxygen. Spontaneous ignition of the charcoal takes place at a temperature lower than the ignition temperature of any of the products evolved.
- Phase D, above 500 °C, in which the charcoal glows and is consumed. Above 1000 °C, non-luminous flames are supported by the combustion of hydrogen and carbon monoxide.

Thermal analysis comprises a group of techniques in which a physical property of a substance is measured as a function of temperature, while the substance is subjected to a controlled temperature program. Early work in this area has been reviewed by Beall and Eickner (1970). Thermogravimetric analysis (TGA or TG) involves measuring the change in sample mass with changing temperature. In TGA, a mass loss is observed if a thermal event involves the loss of a volatile component. Chemical reactions such as combustion involve mass losses, whereas physical changes such as melting do not. The latter can be studied by differential thermal analysis (DTA), which involves measuring the variation of heat flux in a sample with variation of temperature. The technique is based on the fact that when a substance is heated, it undergoes reactions and phase changes that involve absorption or emission of heat. In DTA, the temperature of the test material is measured relative to that of an adjacent inert material. A thermocouple imbedded in the test piece and another in the inert material are connected so that any difference in temperature generated during the heating cycle is graphically recorded as a series of peaks on a moving chart. The use of thermal analysis techniques has shown that the results obtained are quite variable and strongly dependent upon the method of preparation of the materials as well as on the experimental parameters, particularly the heating rate and atmosphere (Hill, 2006).

Figure 9.1 shows a DTA-curve for solid wood. A minor endotherm reaction can be seen to occur just above 100 °C, corresponding to the removal of water. Generally for wood, the thermographs first show an endothermal maximum in the range of 120-150 °C which is attributed to the evaporation of more strongly absorbed water. The exothermal peaks at 200-250 °C and 280-320 °C and those higher than 400 °C are due to the degradation of the wood components (Fengel & Wegener, 2003).

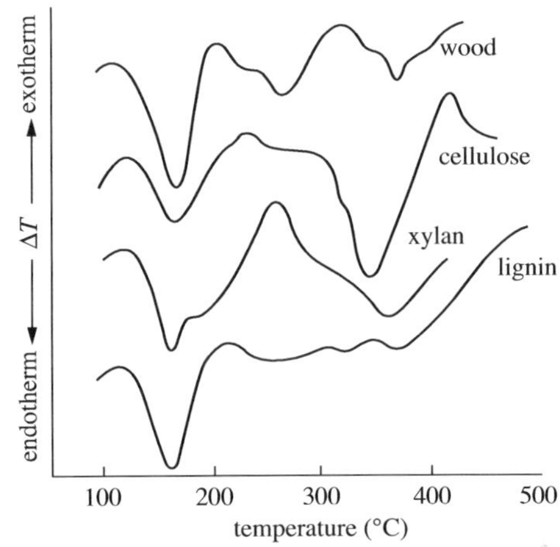

Fig. 9.1 DTA curves for the pyrolysis of beech wood (*Fagus silvatica* L.) *and its components (Košik et al., 1968).*

Figure 9.2 shows the TGA curve for solid pine heated from room temperature to about 500 °C. Below 200 °C, the weight loss is very small and there is no or very little

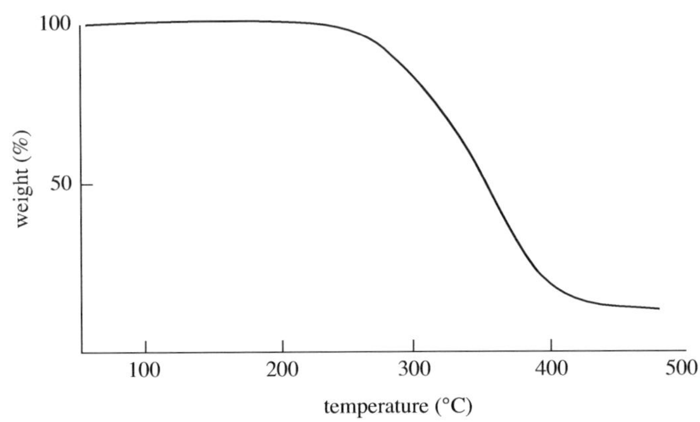

Fig. 9.2 TGA curve for the pyrolysis of solid pine wood (Shafizadeh & Chin, 1977).

degradation of the wood. However, up to about 140 °C a variety of heat-treatment processes have been developed. The results of the process depend on several variables including time and temperature, treatment atmosphere, wood species, moisture content, wood dimensions and the use of a catalyst (Stamm, 1956; Millett & Gerhards, 1972). The temperature and time of treatment are the most critical elements, and treatments in air lead to oxidation reactions that are detrimental to the desired properties of the treated wood. Wood degrades faster when heated in either steam or water as opposed to under dry conditions (MacLean, 1951, 1953; Stamm, 1956; Millett & Gerhards, 1972; Hillis, 1975). Stamm (1956) showed that wood heated in the presence of oxygen degraded more rapidly than wood heated in an oxygen-free atmosphere. With increasing treatment time, the wood became more brittle (Yao & Taylor, 1979, Edlund, 2004). In general, hardwoods are less thermally stable than softwoods, and this is attributable to differences in the hemicellulosic content and composition (Hill, 2006). The weight loss from hardwoods is greater than that from softwoods, probably due to the greater content of acetyl groups in hardwoods which release acetic acid during heat treatment contributing to the acid hydrolysis (MacLean 1951, 1953; Millett & Gerhards, 1972; Hillis 1975).

Fengel (1966a-c) noticed certain anatomical changes in spruce heated at 180 and 200 °C. The S-layer of latewood tissue became fractured, and the encrusting materials in the torus began to flow at these temperatures, which also correspond to the transition temperature between phase A and phase B as defined by Browne (1958).

9.2.2 Thermal degradation of wood components

There are a number of difficulties in describing the changes that take place when wood is heated. A range of chemical reactions which are combinations of both endothermic and exothermic reactions take place simultaneously, making the determination of the onset temperatures for the different reactions nearly impossible. The analysis is complicated further by interactions between reactions in different constituents. This means that an analysis performed on an isolated derivative of one of the components can be very different from what actually takes place inside wood. There is not only an interaction between the components found inside wood, but also interactions between the wood and the treatment atmosphere.

As the wood is heated, the first weight loss is due to the loss of water, followed by the loss of a variety of degradation products and volatile gases (Shafizadeh & Chin, 1977). Mitchell *et al.* (1953) concluded that, in the early stages of degradation, 83% of the weight loss is water. As the temperature increases, cell-wall polymers start to degrade. Pyrolysis of the hemicelluloses takes place at about 270 °C followed closely by that of the cellulose. Lignin is much more stable at high temperatures (Stamm, 1956). Figure 9.3 gives a schematic representation of the changes in wood components due to temperature under humid conditions.

The presence of the thermally labile acetyl groups is a significant factor affecting the thermal degradation of hemicelluloses. The thermal degradation starts with deacetylation and the release of acetic acid as a depolymerization catalyst which further increases the polysaccharide decomposition (Tjeerdsma *et al.*, 1998a). Acid-

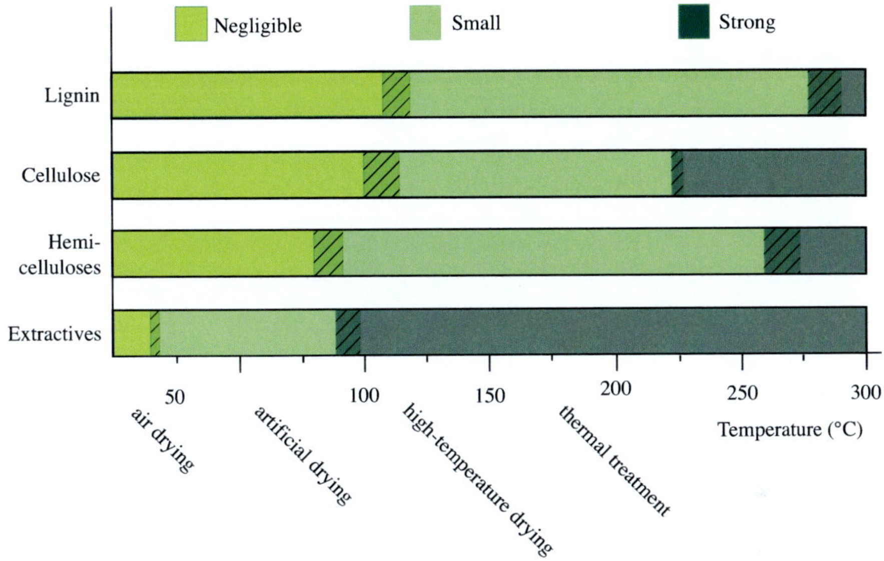

Fig. 9.3 A schematic representation of changes in wood components due to temperature under humid conditions, but without regard to time.

catalyzed degradation leads to the formation of formaldehyde, furfural, and other aldehydes (Tjeerdsma *et al.*, 1998a). At the same time, hemicelluloses undergoe dehydration reactions with the decrease of hydroxyl groups (Weiland & Guyonnet, 2003).

The higher resistance of cellulose than of hemicelluloses has been reported by several authors (Bourgois & Guyonnet, 1988; Yildiz *et al.*, 2006; Esteves *et al.*, 2008c). The amorphous cellulose degrades resulting in the formation of furans such as hydroxymethyl furfural and furfural, Figure 9.4. The degradation of the amorphous cellulose leads to an increase in cellulose crytallinity (Bhuiyan *et al.*, 2000, 2001; Wikberg & Maunu, 2004; Bhuiyan & Hirai, 2005; Boonstra & Tjeerdsma, 2006). The hemicelluloses and cellulose are also degraded to internal ethers and other rearrangement products. The optimum amount of water that can be lost is three molecules of water from two anhydroglucose units, i.e., 16.7% of the carbohydrate polymers or 12% of the wood (Stamm & Baechler, 1960).

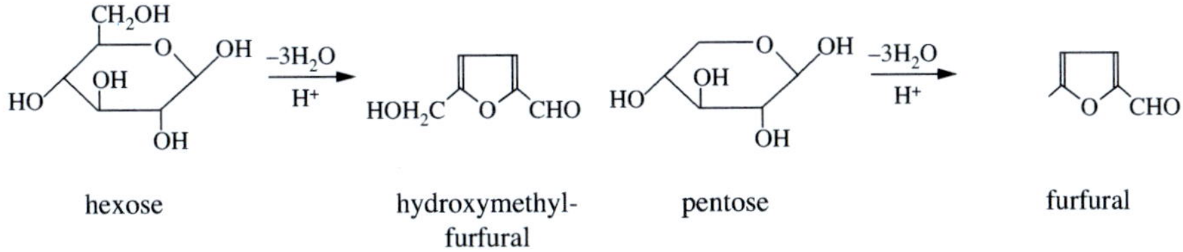

Fig. 9.4 Formation of hydroxymethyl furfural from hexoses and furfural from pentoses as a result of the heat treatment of wood.

Rowell *et al.* (2009) carried out a sugar analysis of aspen before and after treating it with saturated steam at 220 °C for 8 minutes in a closed system. The results show that in the very early stages of weight loss, the hemicellulose polymers broke down while the cellulose remained unchanged. Over half of the arabinan and rhamnans were lost during this initial heating, Table 9.1.

Table 9.1 Sugar analysis of aspen before and after heat-treatment with saturated steam at 220 °C in a closed system for 8 minutes (Rowell *et al.*, 2009).

	Before heat treatment	After heat treatment
Weight loss (%)	0	2.2
Arabinan (%)	0.56	0.20
Galactan (%)	0.68	0.58
Rhamnan (%)	0.30	0.14
Glucan (%)	42.4	42.2
Xylan (%)	16.3	14.6
Mannan (%)	1.6	1.5
Total carbohydrate (%)	61.8	59.2

In lignin, polycondensation reactions with other cell-wall components, resulting in further crosslinking, contribute to an apparent increase in the lignin content (Bourgois & Guyonnet, 1988; Dirol, & Guyonnet, 1993). In lignin, there is a cleavage of ether linkages, especially β-O-4 linkages, and a reduction of the methoxyl content leading to a more condensed structure (Nuopponen *et al.*, 2004; Wikberg & Maunu, 2004).

Most of the extractives disappear or are degraded during the heat treatment, especially the most volatile, but new compounds that can be extracted from wood appear, resulting from the degradation of cell-wall structural components. Bourgois *et al.* (1989) extracted waxes, carbohydrates, tannins, resins, and small amounts of hemicelluloses from *Pinus pinaster* wood treated at temperatures between 240 and 290 °C.

Liquid and gaseous phases are formed in addition to the solid wood during heat treatment. At temperatures between 200 and 300 °C, the liquid phase is almost

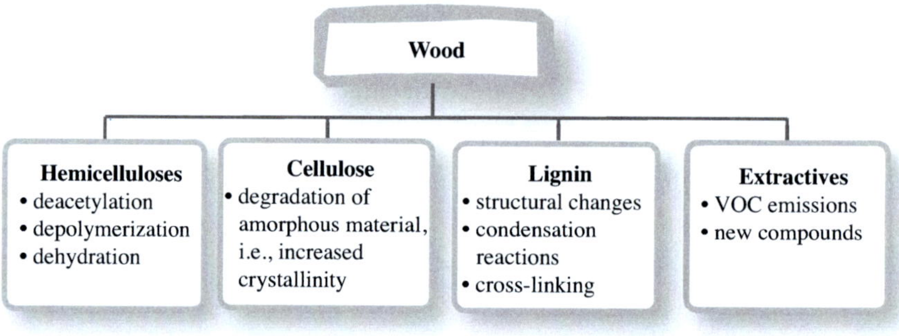

Fig. 9.5 Changes in the main wood constituents during heat treatment.

exclusively composed of water and acetic acid with small amounts of formic acid, furfural, and methanol. The acids catalyze the degradation of polysaccharides and reduce their degree of polymerization (Millitz, 2002).

The chemical changes occurring in the main components of wood due to heat treatment are summarized in Figure 9.5.

9.3 PHYSICAL CHANGES IN WOOD DUE TO HEAT TREATMENT

Heat-treatment, i.e., the heating of green or dried wood to temperatures of 150 °C to 260 °C, significantly influence the properties of the wood, e.g. the hygroscopicity, dimensional stability, strength, decay resistance and durability. On average, the equilibrium moisture content is reduced to about half the value of the untreated wood. Due to the low equilibrium moisture content and partial degradation of the wood components, the susceptibility to biological degradation is reduced. However, the mechanical strength properties are also reduced: certain strength properties are reduced by up to 60% by certain heat-treatment processes. This means that heat-treated wood is not suitable for load-bearing purposes (Jämsä & Viitaniemi, 2001; Millitz & Tjeerdsma, 2001; Rapp & Sailer, 2001; Vernois, 2001). The reduced hygroscopicity, increased dimensional stability and increased decay resistance of heat-treated wood depend on the decomposition of a large portion of the hemicelluloses in the cell wall.

9.3.1 Weight loss

The loss of weight for wood is one of the most important features of heat-treatment and it is commonly referred to as an indicator of quality. Several authors have studied weight loss with heat-treatment and concluded that it depends on wood species,

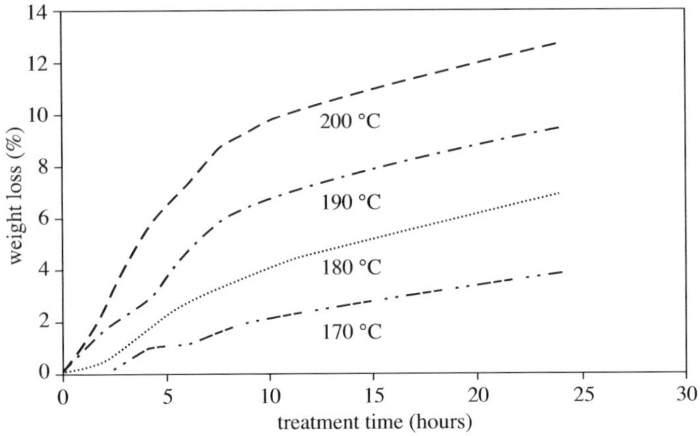

Fig. 9.6 Weight loss vs. heating time at different temperatures during the heat-treatment of maritime pine (*Pinus pinaster* Aiton.) in an open system. The initial moisture content of the wood was 9% (Esteves *et al.*, 2008a).

heating medium, temperature, and treatment time. Figure 9.6 shows the weight loss vs. heating time in an open system at several temperatures, adapted from Esteves *et al.* (2008a), and Figure 9.7 displays similar data from Welzbacher *et al.* (2007).

Several studies have shown that increasing the treatment temperature also leads to a significantly greater loss of weight, which is therefore a valuable and reliable parameter indicating the treatment intensity (Obataya *et al.*, 2002; Tjeerdsma *et al.*, 2002; Del Menezzi & Tomaselli, 2006; Metsä-Kortelainen *et al.*, 2006; Brischke *et al.*, 2007; Welzbacher *et al.*, 2007). The heat-treatment intensity is a product of treatment temperature and treatment time. The influence of the treatment temperature on the treatment intensity has been shown to be significantly stronger than that of the treatment time. A decrease in mass of 12.5% was obtained after 3 hours at 240 °C, but a longer treatment time was necessary to achieve the same weight loss at a lower treatment temperature; viz.: 8 hours at 220 °C, 14 hours at 210 °C, 20 hours at 200 °C, and 72 hours at 180 °C (Welzbacher *et al.*, 2007).

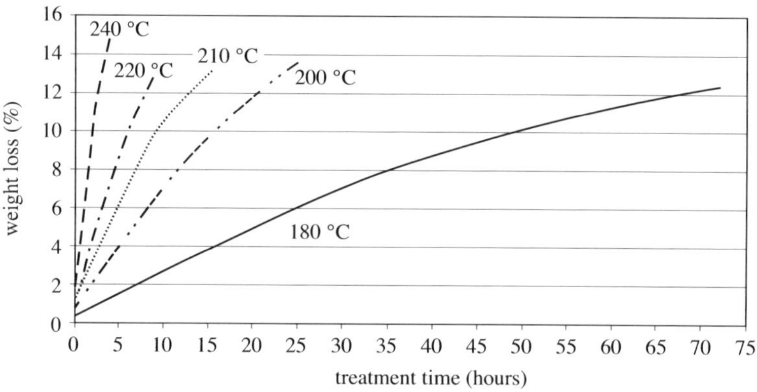

Fig. 9.7 Weight loss vs. heating time for several temperatures during heat-treatment of Norway spruce (*Picea abies* (L.) Karst.) in a open system. The specimens were tightly wrapped in aluminium foil to minimize the oxidation process of wood. (Welzbacher *et al.*, 2007).

The strong influence of the treatment temperature on the treatment intensity has been mentioned by several authors (Burmester, 1970; Mazela *et al.*, 2003; Boonstra *et al.*, 2006; Paul *et al.*, 2006). However, the gradients of the graphs show that the impact of the treatment time on the treatment intensity is greater at higher temperatures.

9.3.2 Hygroscopicity and dimensional stability

Heat treatment has been established as a method by which the hygroscopicity of wood is substantially reduced (Runkel & Witt, 1953; Seborg *et al.*, 1953; Stamm, 1956; Kollmann & Schneider, 1963; Kollmann & Fengel, 1965; D´Jakanov & Konepleva 1967; Nikolov & Encev, 1967; Noack, 1969; Hillis, 1984). Strongly associated with the reduction of the hygroscopicity, the dimensional stability is increased.

For a long time it has been recognized that excessive heating of wood reduces its hygroscopicity. Tiemann (1915) found that drying white ash, loblolly pine and red oak wood in superheated steam at about 150 °C for four hours reduced the subsequent moisture absorption by 10 to 25%. Heating in superheated steam at 200 °C for 3-4 hours reduced the hygroscopicity to almost half of the original value (Tiemann, 1942). Stamm (1964) came to the same result as Tiemann when he passed pieces of wood with a thickness up to 50 mm for a short time beneath molten metal or a salt bath at a temperature of 280-325 °C. Wood treated in this way becomes noticeably darker with reduced strength but with a significantly reduced affinity for water.

In general, most drying treatments, even at mild temperatures, decrease the hygroscopicity of wood, i.e. its capacity for reabsorption of moisture from the air, although this effect may be eliminated upon saturation with water (Tiemann, 1941a-c).

Because of the loss of hygroscopic hemicellulose polymers during heat treatment, the equilibrium moisture content (EMC) is reduced. The dimensional stability of heat-treated wood was first thought to be due to a cross-linking of cellulose chains (Stamm & Hansen, 1937; Burmester, 1975; Tjeerdsma *et al.*, 1998a). The derivates in Figure 9.4 formed during the decomposition of the hemicelluloses have a poor hygroscopicity and have little effect on the EMC. It is also noticed by Seaborg *et al.* (1953) that heat-treated wood shrinks in organic solvents (DMF, DMSO, pyridine, methanol) even more than untreated control samples. Dimensional stabilization can be explained by an important degradation of the hemicelluloses of wood (Weiland & Guyonnet, 2003). This heat-sensitive polymer undergoes dehydration reactions with the loss of numerous hydroxyl groups. The heat treatment principally affects the most hydrophilic polymer of wood (hemicelluloses) and as a result, the treated wood has a low affinity for water and therefore a good dimensional stability (Weiland & Guyonnet, 2003).

In the literature review by Esteves and Pereira (2009), the explanations for the lower equilibrium moisture content of heat-treated wood are summarized as: less water being absorbed by the cell walls as a result of a chemical change with a decrease in the number of hydroxyl groups; the cellulose hydroxyl groups being less accessible to water molecules due to the increase in cellulose crystallinity; and cross-linking occurring in lignin.

Esterification of hydroxyl groups during heat treatment results in a decreased hygroscopicity of the wood, but this is believed to be minor compared to the influence of cross-linking, also known to occur during heat treatment of wood (Tjeerdsma & Militz, 2005).

Table 9.2 shows a summary of the effect of heat-treatment on the EMC according to various authors. In most cases, a 50% reduction in EMC was observed. In most examples given, the treatment was carried out in steam at 190-220 °C with a relative humidity of 35-65%.

Keith and Chang (1978) have shown that the differences in hygroscopicity due to heat treatment are more apparent during absorption compared with the desorption process. Figure 9.8 displays the sorption-desorption behavior of unmodified wood and wood modified in nitrogen gas at 250 °C for two hours (Hill, 2006). The sorption-desorption isotherms of heat-treated wood is more linear compared to the isotherms of unmodified wood and there is also a reduction in the hysteresis for heat-treated wood.

Table 9.2 Reduction in equilibrium moisture content (EMC) as a result of heat treatment of wood.

Wood species	Temperature (°C)	Time (hours)	Reduction in EMC (%)	Reference
Fagus silvatica	180 °C	4	40	Teichgräber, 1966
	190 °C	2.5	60	Giebeler, 1983
Pinus pinaster	190 °C	8-24	50	Esteves *et al.*, 2006
Pinus sylvestris	220 °C	1-3	50	Anon., 2003
Eucalyptus globulus	190 °C	2-24	50	Esteves *et al.*, 2006
Pinus sylvestris	Plato process		50-60	Tjeerdsma, 2006

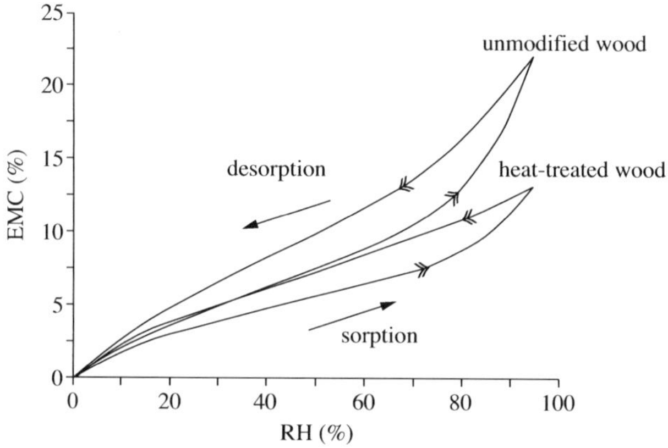

Fig. 9.8 Sorption-desorption isotherms of unmodified and heat-treated wood (after Hill, 2006).

Stamm *et al.* (1946) dimensionally stabilized wood by heating it beneath the surface of molten metal. The product was named Staywood. For this product, Stamm (1964) determined the temperature-time combinations required to give a certain reduction in swelling of the wood. For example, a 40% reduction in swelling can be obtained by heating either at 320 °C for 1 minute or at 150 °C for one week.

Welzbacher *et al.* (2007) studied the impact of the heat-treatment temperature (180, 200, 210, 220 and 240 °C) and various heat treatment times on selected biological, mechanical, optical, and physical properties of Norway spruce (*Picea abies* (L.) Karst.). The anti-swelling efficiency (*ASE*), as a measure of the increased dimensional stability of the heat-treated specimens, showed a strong dependence on the decrease in mass and on the treatment temperature, Figure 9.9. An increase in *ASE* corresponded to a decrease in weight. The average of the tangential and radial swelling (*S*) was used to calculate the anti-swelling efficiency (Eq. (9.1))

$$ASE = \frac{S_{control} - S_{heat\text{-}treated}}{S_{control}} \tag{9.1}$$

The highest *ASE* value of approximately 40% was achieved with a treatment temperature of 240 °C and a decrease in weight greater than 12%. Lower *ASE* values for the same level of decrease in weight were found at other treatment temperatures: 34% ASE at 220 °C, 30% *ASE* at 210 and 200 °C, and 20% *ASE* at 180 °C.

In general, a higher heat-treatment temperature gave a higher *ASE*, and this has been supported by the results of several authors, including Syrjänen and Kangas (2000), Rapp and Sailer (2001), Goroyias and Hale (2002), and Popper *et al.* (2005). In the same way, a prolongation of the heat-treatment time at a given temperature leads to an increase in *ASE* (Dirol & Guyonnet, 1993; Paul *et al.*, 2006). However, a comparsion between the loss of weight and the ASE for all single-treatment temperatures showed that increasing the treatment intensity by extending the treatment time does not significantly increase the *ASE* beyond a certain loss of weight (approx. 8% at 180 °C, 12-13% at 200, 210, 220, and 240 °C). Esteves *et al.* (2006) have studied the dimensional stability of *Eucalyptus globulus* and found an *ASE* value of 77% for heat treatment at 190 °C and a corresponding weight loss of 2% and an *ASE* of 77% for heat treatment at 200 °C and a corresponding weight loss of 10%.

The effects of oil type, temperature and treatment duration on the equilibrium moisture content, water absorption and shrinkage in white pine were studied by Wang and Cooper (2005). Samples with dimensions 25 × 25 × 100 mm^3 were heat-treated at 200 °C or 220 °C in palm oil, soy oil or slack wax for two or four hours. These three oils have different chemical compositions, the main differences being the degree of saturation of the chemical bonds: soy oil is highly unsaturated, and liquid at ambient temperature; palm oil is high in saturated fatty acids and a semi-solid at ambient temperature, while slack wax consists mostly of long straight-chain hydrocarbons and mineral oil, with excellent water barrier properties and with a melting point of 50-70 °C. Results showed that the equilibrium moisture content could be reduced by up to 52%, and that the treatment in wax was the most effective. The higher the temperature and the longer the treatment, the lower was the resulting equilibrium moisture content. The water absorption was also reduced the most by the combination

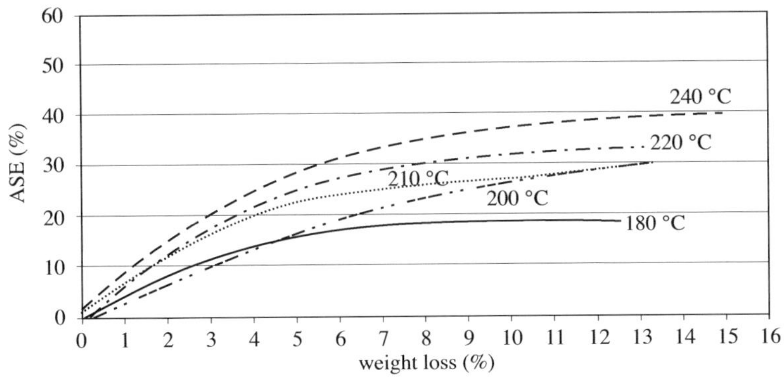

Fig. 9.9 The anti-swelling efficiency (ASE) as a function of the decrease in weight of Norway spruce (*Picea abies* Karst.) specimens (10 ×5 ×20 mm^3) heat-treated in a open system at several temperatures. The specimens were tightly wrapped in aluminum foil to minimize the oxidation of the wood (Welzbacher *et al.*, 2007).

of wax, a high temperature and a long treatment duration. Their results confirmed that chemical reactions in wood resulting from the heat treatment account for the main improvements of wood properties in reduced hygroscopicity and improved dimensional stability, while the oil absorbed by wood reduces the rate of water absorption.

The longitudinal absorption in heat-treated pine, spruce and birch has been studied at temperatures of 170 °C and 200 °C with the help of CT scanning by Johansson *et al.* (2006). The results showed that the longitudinal water absorption in pine sapwood was substantially greater after heat treatment at 170 °C than in untreated pine sapwood. In pine heartwood, the absorption of water was low in both heat-treated and untreated wood. Spruce wood showed a low water absorption in sap- and heartwood in both heat-treated and untreated samples. Birch displayed a decreasing uptake of water when increasing the treatment temperature.

Dvinskikh *et al.* (2011) have evaluated the water uptake kinetics in wood cladding materials by a portable NMR spectrometer. They found that the water uptake in heat-treated spruce samples is in strong contrast to the uptake in samples of non-heat-treated spruce. In the heat-treated samples, a high MC was observed close (< 1 mm) to the wet surface while only a slight increase in MC was detected at depths exceeding 1 mm. For non-heat-treated spruce wood, a increase in moisture content was seen deeper in the wood. This points at the permeability of heat-treated wood being low, in agreement with previous investigations of the properties of heat-treated wood, see for instance Runckel and Witt (1953).

9.3.3 Strength

Heat-treatment has been found to affect the mechanical properties of wood (Stamm *et al.*, 1946; Thunell & Elken, 1948; Seborg *et al.*, 1953; Davis & Thompson, 1964; Noack, 1969; Rusche, 1973; Giebeler, 1983; Weiland & Guyonnet, 1997; Tjeerdsma *et al.*, 1998b; Kubojima *et al.*, 2000; Sundqvist *et al.*, 2006; Scheiding *et al.*, 2005, Boonastra *et al.*, 2007a,b; Gonzalez-Pena & Hale, 2007). The strength and stiffness decrease when wood is heated and increase when it is cooled. For heating during short periods, the temperature effect is immediate and reversible. Permanent (irreversible) reductions in strength and stiffness can result if wood is exposed to elevated temperatures for an extended time. These irreversible changes in the mechanical and physical properties of wood are generally attributed to thermal degradation of the wood substance.

The influence of heat on the strength of wood in various applications has been studied for a long time. We know that in wood, as in any other solid body below the melting temperature or below the temperature of thermal decomposition, the strength and stiffness decrease with increasing temperature due to thermal expansion of the crystal lattice of the cellulose and due to the increased intensity of the thermal molecular oscillations, see Figures 9.10 and 9.11. We also know that the humidity has a strong influence on the properties of wood and that it is always necessary to take the humidity conditions in wood and in the process into consideration when processing wood, e.g., during heat treatment.

The effects of temperature on the strength and elastic properties of wood may be classified as temporary or permanent. Temporary effects are those observed only at the particular temperature involved and are independent of the period of exposure

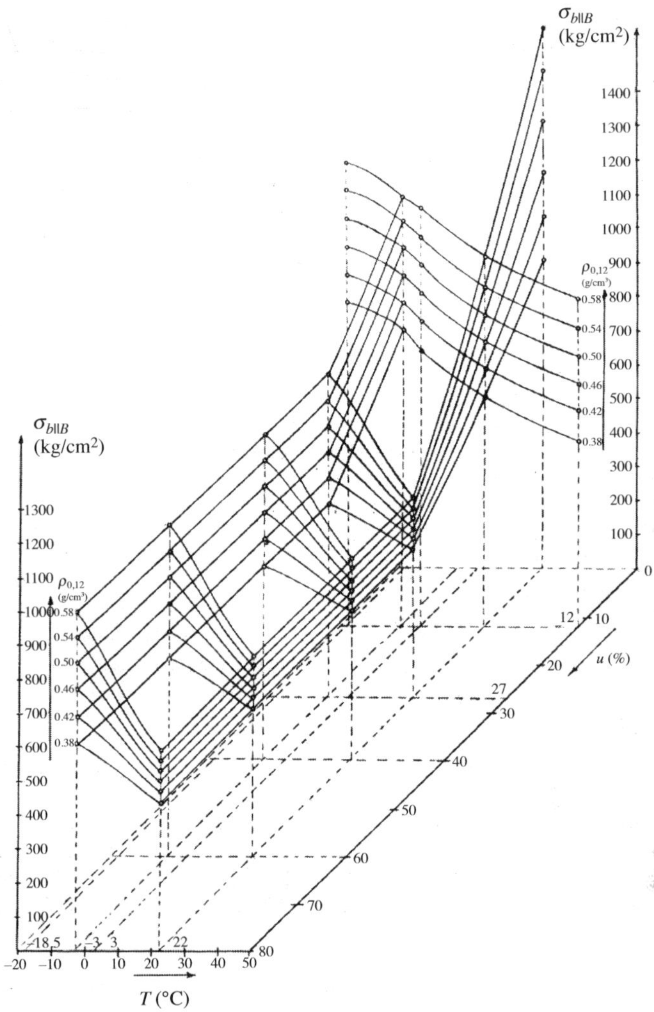

Fig. 9.10 Equivalent fiber stress at maximum load ($\sigma_{b\|B}$) of Scots pine (*Pinus sylvestris* L.) with respect to temperature (T), moisture content (u) and density ($\rho_{0,12}$) (Thunell, 1940).

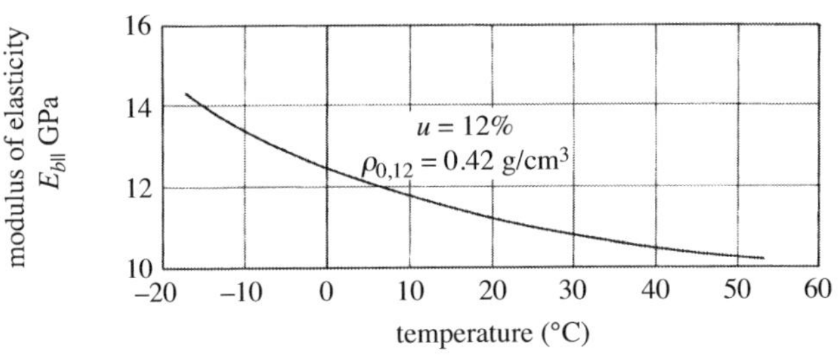

Fig. 9.11 Relationship between modulus of elasticity from a bending test on Scots pine (*Pinus sylvestris* L.) and temperature (Thunell, 1941).

to high or low temperatures. Permanent effects are those that are retained after the wood has been restored to a normal temperature and are a function of the duration at the temperature as well as of the temperature itself. In the case of heat treatment, the increase in temperature is high and the effects on mechanical properties are permanent. It is then of interest to know how the temperature level and how the processing time influence the reduction in mechanical properties. The manner in which the mechanical properties are influenced under thermo-hydro-mechanical actions is also discussed in detail in Chapters 4-6.

Several researchers have investigated the temporary influence of a high or low temperature on wood properties, and the effects of temperatures as low as –160 °C and as high as +200 °C have been observed (Kollmann, 1939). The tests were performed in an autoclave, i.e., in a closed system. In an early study by Kollmann (1940), no appreciable change in crushing strength of dry spruce or beech was found when the wood was subjected to 160 °C for four hours, but at 200 °C the strength of beech dropped rapidly in less than one hour and that of spruce in less than two hours.

During a heat-treatment of wood, the strength properties are reduced and the mode of failure of the heat-treated wood in mechanical tests is always brittle. For this reason, heat-treated wood should never be used in load bearing applications. Strength properties are reduced in heat-treated wood as a result of the degradation of the cell-wall matrix resulting from the hemicellulose polymer degradation.

Stamm *et al.* (1946) report a loss of mechanical properties as a result of the heat treatment of wood. They treated wood at 320 °C for one minute or 150 °C for a week and found a 17 per cent reduction in the modulus of rupture (MOR) when heated beneath a surface of molten metal and 50 per cent when heated in air, i.e., in an open system. Under the same conditions, there was less loss in the modulus of elasticity (MOE).

Mitchell (1988) found that the losses in strength properties were much lower when wood was heated dry rather than wet, with a greater loss in MOR than in MOE. Rusche (1973) found that strength losses were related to the rate and extent of mass loss for the heating beech and pine in the presence and absence of air at temperatures ranging from 100 to 200 °C. Losses in maximum strength and work to maximum load were greater for compressive than for tensile loads. The decrease in MOE was significant only when the mass loss was greater than 8 to 10% and this was the case for both species.

Rapp and Sailer (2001) treated pine and spruce at 180-220 °C for various times in air and in oil and determined the MOR and MOE in a three-point bending test. The highest MOE was 11000 N/mm^2 in oil heating and there was a slight loss in MOE in both air and oil heating. MOR, however, decreased by 30% in oil heating. Moreover, the impact bending strength was reduced and the wood became brittle. Oil-heated wood lost about 50% and air-heated wood lost over 70% of the impact strength compared to controls.

Table 9.3 presents a summary of other data related to strength changes in heat-treated wood. In all cases, there was only a small decrease in MOE but major changes in MOR depending on temperature, time and atmosphere.

The measurement of color is not a good indicator of the strength of heat-treated wood, as the color is very inhomogeneous (Johansson & Morén, 2006). However, Brischke *et al.* (2007) conclude that the color measurement of heat-treated timber could easily be implemented on an industrial scale for quality control.

Table 9.3 Reduction in MOR as a result of the heat-treatment of various woods in relation to untreated wood.

Species	Treatment	Process	Reduction in MOR (%)	Reference
Birch *(Betula pendula)*	Vapour, 200 °C 3 hours	Thermo Wood	43	Johansson & Morén, 2006
Spruce *(Picea orientalis)*	Air, 200 °C 10 hours	Open system	40	Yildiz *et al.*, 2006 (compression strength)
Pine *(Pinus sylvestris)*	Oil, 220 °C 4.5 hours	OHT	30	Rapp & Sailer, 2001
Eucalyptus *(Eucalyptus globulus)*	Vapour, 200 °C 10 hours	Closed system	50	Esteves *et al.*, 2006

9.3.4 Decay resistance

Heat treatment is an alternative to chemical wood preservation systems which aim to interfere with the metabolism of wood-destroying fungi. That heat treatment improves resistance to biological decay has been shown in many studies, e.g., by Stamm *et al.* (1946), Stamm and Baechler (1960), Viitanen *et al.* (1994), Leithoff and Peek (1998), Rapp and Sailer (2000; 2001), Sailer *et al.* (2000), Syrjänen (2001), Vernois (2001), Militz (2002), Brischke and Rapp (2004), Welzbacher and Rapp (2007), and Metsä-Kortelainen & Viitaniemi (2009). Table 9.4 presents the data of Stamm *et al.* (1946). The resistance to decay in heat-treated wood is probably due to the loss of hemicellulose polymers from the cell wall. As was shown earlier, the greater the weight loss in the heating process, the greater is the durability of the heat-treated wood. A reduction in swelling of over 40% did not result in any weight loss in the decay test, Table 9.4. Similar results were obtained when the test fungus was *Lenzites trabea* 517 (Stamm & Baechler, 1960). Lekounougou *et al.* (2009) suggest that chemical modifications caused by heat-treatment disturb the enzymatic systems involved in wood degradation by *Trametes versicolor* and lead to an improvement of durability observed for heat-treated wood. They also conclude that these observations are in good agreement with the fact that an improvement in wood durability can be achieved only if a sufficient level of modification has been reached.

Table 9.4 Relationship between reduction in swelling and decay resistance of heat-treated white pine using *Trametes serialis* fungus (Stamm *et al.*, 1946).

Reduction in swelling (%)	Weight loss due to decay (%)
Untreated control	28.4
30-33	12.5
33-38	< 4.5
40 or more	0

Table 9.5 lists the fungal resistance of heat-treated wood of Scots pine (*Pinus sylvestris* L.) and Norway spruce (*Picea abies* (L.) Karst.) heated in oil or air at three different temperatures and tested for decay during 19 weeks in the EN 113 (1996)

decay test using *Coniophora puteana* fungus (Rapp & Sailer, 2001). The data show that heating in oil is more effective in reducing attack by fungi than heating in air. This may be due to the presence of residual oil in the samples.

Table 9.5 Weight loss (%) according to EN 113 (1996) of Scots pine and Norway spruce heated in oil or air.

Temperature	Oil-heated		Air-heated	
(°C)	Pine	Spruce	Pine	Spruce
Untreated control	40	48		
180	13	15	25.0	31.2
200	1.9	13.1	15.8	26.7
220	2.0	0.0	11.0	5.5

Welzbacher and Rapp (2007) compared several types of industrially heat-treated products using various fungi in laboratory tests and in different field and compost conditions. Table 9.6 shows the weight loss during an EN 113 (1996) test using three types of fungi. The heat treatment in oil was the most effective, but the influence of the oil in the decay test is not known.

Table 9.6 Weight loss (%) of different woods after various heat-treatment processes due to attack by *Postia placenta*, *Coriolus versicolor* and *Coniophora puteana* (Welzbacher & Rapp, 2007).

Material		*Postia placenta*	*Coriolus versicolor*	*Coniophora puteana*
Control species				
Pinus sylvestris L.	1)	31.0	5.1	47.5
Pinus sylvestris L.	2)	26.2	35.7	60.3
Pseudotsuga menziesii F.		14.0	2.6	27.4
Quercus petrea Liebl.		0.8	14.3	3.9
Heat treated wood	Process			
Pices abies (L.) Karst.	Plato	10.0	6.8	3.7
Pinus sylvestris L.	Thermowood	16.0	9.0	1.9
Pinus maritima Mill.	Retification	13.3	7.8	12.2
Pinus sylvestris L.	OHT	7.4	5.6	3.4

1) including both sap- and heartwood
2) including only sapwood

9.3.5 Durability on weathering

The equilibrium moisture content (EMC) of heat-treated wood was reduced by up to 50% (Table 9.2) which signifies that the heat-treated wood had a much lower moisture content than untreated controls. Consequently, the swelling and shrinkage of heat-treated wood were drastically reduced (Burmester, 1981; Dirol & Guyonnet, 1993; Goroyias & Hale, 2002; Popper *et al.*, 2005; Repellin & Guyonnet, 2005). Thus, heat-treated wood may be suitable for outdoor applications due to its greater durability and dimensional stability. The reduction in the moisture content in heat-treated wood in outdoor applications will also have a positive effect on the resistance to degradation of the wood due to weathering. However, this is not the most important factor for the weathering behavior of heat-treated wood.

The short-term color stability of heat-treated wood exposed to artificial UV radiation was better than that of untreated wood (Ayadi *et al.*, 2003). Heat-treatment also reduced the erosion of modified spruce (*Picea abies* (L.) Karst.) and beech (*Fagus sylvatica* L.) wood during artificial accelerated weathering (Feist & Sell, 1987). Lignin at the surface of heat-treated wood has been reported to be slightly less susceptible to photodegradation than lignin in unmodified wood, possibly due to an increased condensation of the lignin induced by the heat treatment (Nuopponen *et al.*, 2004). Feist and Sell (1987) suggested that the low equilibrium moisture content of heat-treated wood might reduce its susceptibility to photodegradation due to the moisture content strongly influencing the photodegradation. The increased weathering resistance of heat-treated wood could also be due in part to its increased water-repellency, which would restrict the leaching of photodegraded lignin and hemicelluloses from exposed surfaces. Nevertheless, the color of heat-treated wood fades quite quickly when the wood is outdoors and it eventually becomes grey (Jämsä *et al.*, 2000), indicating photodegradation and loss of lignin at the exposed wood surfaces.

Heat-treated wood was less susceptible to colonization by *Aureobasidium pullulans* and other staining fungi than unmodified wood (Feist & Sell, 1987). There are discrepancies in the literature as to whether or not heat treatment reduces the susceptibility of wood to surface cracking when it is exposed outdoors. Vernois (2001) stated that "cracking due to dimensional movement is reduced in comparison with natural wood". In contrast, Jämsä *et al.* (2000) found that heat-treated pine and spruce surfaces cracked to the same extent as unmodified wood when exposed outdoors, and the application of unpigmented stains and oils did not reduce the severity of cracking of the modified wood.

However, Sandberg (1999) and Sandberg and Söderström (2006) have shown that the tendency of wood to crack during weathering is strongly related to the annual ring orientation in its cross section. To avoid cracks occurring on wood used outdoors, wood with an annual ring orientation perpendicular to the exposed wood surface should be selected. A variation in the annual ring orientation in the wood used for different studies is probably the main reason for the discrepancies in the literature as to whether or not heat-treatment reduces the susceptibility of wood to crack when exposed outdoors. Feist and Sell (1987) found that there was considerably more grain-raising and cracking on heat-treated spruce exposed for 14 months to natural weathering than on unmodified spruce samples. Furthermore, the surfaces of heat-treated spruce samples after weathering were much rougher than those of unmodified samples. In contrast, the surfaces of heat-treated beech were smoother than those of unmodified wood after the beech samples had been exposed outdoors for 14 months, and there was little difference in the cracking of the samples. Feist and Sell (1987) also found that semi-transparent and film-forming stains performed slightly better on heat-treated beech, whereas their performance on heat-treated spruce was slightly worse on unmodified spruce. Jämsä *et al.* (2000) concluded that heat-treatment does not adversely affect the performance of coating on wood and that no alterations in coating protocols are necessary when finishing heat-treated wood.

Heat-treated wood contains residual acetic acid produced by the breakdown of the hemicelluloses and for this reason it is important that only stainless steel fixings are used with heat-treated wood. Jermer and Andersson (2005) have studied the corrosion of fasteners in wood heat-treated according to the Thermowood process. The results show that the corrosion of fasteners in heat-treated wood was more severe than in its untreated counterpart. Mild steel and zinc-coated steel are the most susceptible. Stainless steel is hardly attacked at all.

9.3.6 Durability in contact with the ground

The applicability of heat-treated wood for use in contact with the ground is a matter of controversial discussion. Some investigations have reported that heat-treated wood has a limited applicability for use in ground contact owing to insufficient durability (Kamdem *et al.*, 1999; Westin *et al.*, 2004). Jämsä & Viitaniemi (1998), Wienhaus (1999) and Rapp *et al.* (2000) have recommended the application of heat-treated wood entirely above ground. However, Boonastra and Doelman (1999) and Tjeerdsma *et al.* (2000) reported a smaller mass loss of heat-treated wood after fungal attack, and have classified heat-treated wood as very durable to durable, i.e., according to durability class 1-2 in EN-350-2 (1994). This would in theory allow its application in soil contact according to EN-460 (1994). In addition, various industrial suppliers promote their heat-treated wood as having unlimited suitability for use in ground contact (Welzbacher & Rapp, 2007).

In soil block tests involving brown-rot fungi, it has been shown that heat-treatment can modify the durability from non-resistant to moderately resistant, depending on the wood species and on the fungus (Kamdem *et al.*, 2002). Even so, it has been demonstrated that heat-treated wood in ground contact is only marginally better than untreated wood.

9.3.7 Emissions from heat-treated wood

Volatile compounds are associated with heat-treated wood, but it has been found that the levels present are lower than those found in native wood.

Heat-treated woods are characterized by considerable changes in their chemical properties, resulting in a significant odor (Kamdem *et al.*, 2000). The degradation products of a range of wood components cause this typical odor (Boonstra & Tjeerdsma, 2006; Windeisen *et al.*, 2007; Peters *et al.*, 2008). The formation of toxic polyaromatic compounds during heat treatment has been discussed in the literature. It has been suggested that the formation of these products leads to the resistance of heat-treated wood to fungal decomposition and microorganisms (Kamdem *et al.*, 2000). Increasing indoor applications of heat-treated wood has increased an interest in elucidating the chemical composition of these emissions and in developing ways to reduce them. Kamdem *et al.* (2000) studied emissions from maritime pine and poplar samples heat-treated at 200 to 250 °C for 6 hours. The amounts of toxic and non-toxic compounds in the heat-treated wood were not quantified, but these authors stated that their proportions were small. They also suggested further investigations before claiming that heat-treated wood is completely clean and deserving of a health safety label.

Peters *et al.* (2008) found that furfural and 5-methylfurfural are the main emission products from heat-treated wood. The characterization and removal of odorous substances caused by the heat treatment of beech wood (*Fagus sylvatica* L.) were investigated by Peters *et al.* (2009) using chromatographic and spectroscopic methods. They found that pentanal and hexanal, probably formed by the oxidation of unsaturated fatty acids, were the prevailing aldehydes in untreated wood samples. Their amounts decreased during heat treatment, while furfural and 5-methylfurfural, degradation products of hemicelluloses, accounted for almost 100% of the aldehyde fraction. It can be assumed that odorous emissions from heat-treated wood can be strongly reduced by solvent extraction with an ethanol/cyclohexane mixture that leads to wood samples without significant odor emissions. The VOCs emissions from heat-treated wood are less than those from air dried wood, since the emission of terpenes such as pinene, camphene, and limonene during drying are higher than from treated pine. Manninen *et al.* (2002) found that VOCs emissions from air-dried pine (*Pinus sylvestris* L.) were about eight times higher than those from heat-treated wood.

9.3.8 Color changes in heat-treated wood

During heat treatment, the color of wood is modified and becomes darker, often because of the formation of colored degradation products from hemicelluloses (Sehlstedt-Persson, 2003; Sundqvist, 2004; Esteves *et al.*, 2008b) and from the extractives that seem to participate in the development of color in heat-treated wood (McDonald *et al.*, 1997; Sundqvist & Morén, 2002). The formation of oxidation products such as quinones is also referred to as a reason for color change (Tjeerdsma *et al.*, 1998a; Mitsui *et al.*, 2001; Bekhta & Niemz, 2003). Moreover, Sehlstedt-Persson (2003) suggested that the change in color resulting from hemicellulose degradation might be due to hydrolysis by a reaction similar to the Maillard reaction, which is a well-known process in the food industry. Figure 9.12 shows examples of color changes in different wood species having been heat-treated at two different temperature levels in the Thermowood process.

9.4 COMMERCIAL HEAT-TREATMENT PROCESSES

Industrial heat-treatment processes typically aim at improving the biological durability of less durable wood species and at enhancing the dimensional stability of wood or wood-based products, e.g., particle boards. The properties of industrially produced heat-treated wood in general have recently been intensively investigated (Leithoff & Peek, 1998; Sailer *et al.*, 2000; Jämsä & Viitaniemi, 2001; Rapp & Sailer, 2001; Boonstra & Tjeerdsma, 2006; Del Menezzi & Tomaselli, 2006) and for OSB-strands particularly by Paul *et al.* (2006, 2007).

Fig. 9.12 Color of heat-treated wood species at 185 °C (left) and 200 °C (right); (a) spruce, (b) birch, (c) aspen, (d) oak, (e) ash, (f) beech. All samples are oiled after the final surface preparation.

On the European market, several industrial heat-treatment processes have been introduced during recent decades. The most common processes are

- the Thermowood process;
- the Plato wood process;
- the retification process;
- le Bois Perdure; and
- the OHT-process (oil heat-treatment).

The basic difference between these processes resides in their choice of oxygen-excluding and heat-transporting media: the Thermowood process uses steam, the Plato process uses liquid water in a first step followed by a conditioning phase, the Retification process uses nitrogen gas and the OHT-process uses rapeseed oil. However, the treatment conditions are very similar in that these processes operate in the temperature range between 150 and 240 °C to modify the chemical composition of the cell wall.

Modified heat treatment processes are also emerging in Denmark, e.g., the WTT process working at 160-180 °C, and in Austria, e.g., the Huber holz process working at 170-230 °C. Both processes employ heated steam as the medium, but the WTT process can also add oil in the process. The different processes have been published in several patents (e.g., Bluhm *et al.*, 1979; Ruyter, 1989; Viitaniemi *et al.*, 1993, 1994; Rem *et al.*, 1994; Boonstra *et al.*, 2005)

9.4.1 Thermowood

Since the 1990s, comprehensive research at the Finnish Research Center (VTT) together with the Finnish industry on heat-treatment has led to a commercial process under the trade name of Thermowood. Today, the process is licensed to the members of the Finnish Thermowood Association. The process is probably the most successful in Europe and provided about 90% of the heat-treated wood in Europe in 2008.

In the Thermowood process, the wood is heated in the presence of steam, and the steam is used to replace air by water vapor. The treatment is made with steam, with less than 3 to 5% oxygen, without pressure, and with an air speed of at least 10 m/s (Syrjänen & Kangas, 2000). The content of oxygen in the air is typically below 3.5% (Syrjänen, 2001) and such a low oxygen content limits the oxidative degradation of wood.

The Thermowood process is divided into three main stages: high-temperature drying, heat treatment, and a cooling phase with conditioning, as seen in Figure 9.13. The treatment can be undertaken with either green or dried wood, and drying is the longest phase in the process. The method is suitable for both softwoods and hard-

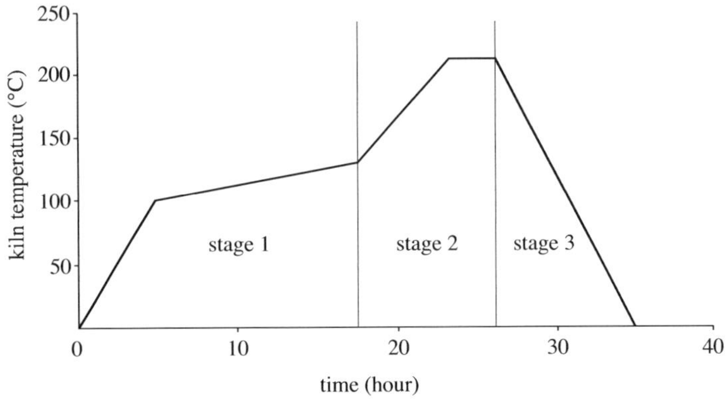

Fig. 9.13 The stages of the Thermowood process. Stage 1: high-temperature drying, stage 2: heat treatment, stage 3: cooling and conditioning.

woods, and needs to be optimized for each wood species. The wood's starting moisture content has no significance for the success of the heat-treatment. In any case, the wood is dried until it is absolutely dry in the first phase of the treatment, during which the kiln temperature is rapidly raised using heat and steam to a level of around 100 °C. Thereafter, the temperature is increased steadily to 130° C during which time the high-temperature drying takes place and the moisture content in the wood is reduced to approximately 0%. The duration of this drying phase depends on the initial moisture content of the wood, on the wood species, and on the wood thickness. If the process starts from green wood, the wood can be dried in a very fast steam-drying process. Fast drying is possible since it is not necessary to consider the color changes and because resins will flow from the wood in the heat-treatment process anyway. Successful drying is important in order to avoid internal cracks.

The second stage starts when the high-temperature drying phase is completed. The temperature is then increased to 185-215 °C and, once the target level has been reached, the temperature is kept constant for 2-3 hours depending on the end-use application.

In the final stage, the temperature is lowered to 80-90 °C using water spray systems and the wood is then conditioned to a moisture content of 4-7% or other moisture content appropriate for its end use.

When the temperature is being raised or lowered, a special adjustment system is used in order to prevent surface and internal cracking. The wood's internal temperature regulates the temperature rise in the kiln. The difference between the kiln temperature and that of the wood is dependent on the dimensions of the wood specimens.

The quality of the raw material has a significant effect on the quality of the final heat-treated wood product, but the parameters used for the process must be optimized separately for each wood species. In Finland, the species used for heat-treatment are pine (*Pinus sylvestris* L.), spruce (*Picea abies* (L.) Karst.), birch (*Betula pendula* L.), and aspen (*Populus tremula* L.). In addition, some experience has been gained in the treatment of Radiata pine (*Pinus radiata* D. Don), ash (*Fraxinus excelsior* L.), larch (*Larix sibirica* Ledeb.), alder (*Alnus glutinosa* (L.) Gaertn.), beech (*Fagus silvativa* L.) and eucalyptus.

The process conditions are corrosive, as are the constituent compounds evaporating from the wood. Heat-treatment equipment must be made of stainless steel. In addition, the high-temperature treatment requires non-standard blower and radiator solutions and safety devices. To generate the heat required in the Thermowood process, hot-oil heating systems can be used and fueled by biofuel, fuel oil, or gas. Other heating systems, such as direct electric heating, are also utilized. The equipment employed must also include a steam generator for generating the steam required by the process. Gases evaporating from the wood during the process are processed by methods such as burning. The primary purpose of this processing is to prevent an odor nuisance from being imposed on the environment due to compounds evaporating from the wood.

Thermowood has two standard treatment classes: Thermo-S and Thermo-D where "S" stands for stability and "D" for durability. These two processes work at different temperature levels, Table 9.7.

Table 9.7 Standard process temperatures for Thermo-S and Thermo-D heat treatment classes of Thermowood (Anon., 2003).

Class	Softwoods	Hardwoods
Thermo-S (°C)	190	185
Thermo-D (°C)	212	200

In addition to the appearance, the stability is a key property in the end use applications of products in the Thermo-S class. The average tangential swelling and shrinkage due to moisture for the treated wood is 6-8%, i.e., at the same level as that of untreated wood of Scots pine and Norway spruce. Thermo-S products are classified as being relatively durable according to the standard EN 113, i.e., its natural resistance to decay meets class-3 requirements. Recommended end-use applications for Thermo-S heat-treated softwoods are building components, furnishings in dry conditions, fixtures in dry conditions, furniture, garden furniture, sauna benches, door and window components, and recommended uses for Thermo-S hardwoods include furnishings, fixtures, furniture, flooring, sauna structures, and garden furniture.

The biological durability is in addition to the appearance a key property in the end-use applications of products in the Thermo-D class. The average tangential swelling and shrinkage due to moisture for Thermo-D wood is 5-6%. Thermo-D products are classified as durable according to the EN 113 standard, i.e., its natural resistance to decay meets class-2 requirements. Recommended end-use applications for Thermo-D heat-treated softwood include claddings, outer doors, shutters, environmental constructions, sauna and bathroom furnishings, flooring and garden furniture. Thermo-D hardwood end-use applications are the same as for Thermo-S but, if a darker color is desired, Thermo-D should be used.

Starting with non-durable (class 5) softwood species, the durability classes 1-3 can be obtained depending on the process. In general, the bending strength is reduced by up to 30% by the higher temperatures. Because of the material's limited resistance against decaying soil organisms, it is not recommended for use with ground contact (Jämsä & Viitaniemi, 2001).

9.4.2 Plato wood

The Plato process was invented in the 1980's by Shell (Ruyter, 1989) and is now used by the Plato Company in the Netherlands. The Plato process uses dried or green wood and has five stages. When green wood is to be processed, the procedure starts with a pre-drying stage. This stage can be excluded if one intends to employ already dried wood. The time of treatment is about two weeks, depending on the species, thickness, and shape of the wood pieces, as well as on the final use. The heating medium can be steam or hot air (Militz & Tjeerdsma, 2001).

In general, the process involves five steps.

1. A pre-drying stage in a conventional industrial wood kiln to a moisture content of 14-18%.
2. A hydrothermolysis stage in a stainless steel reactor for 4-5 hours, where the timber is heated to 150-180 °C in an aqueous environment at superatmospheric pressure (including saturated steam as the heating medium).

3. A drying stage in a conventional industrial wood kiln for 3-5 days using common procedures to a moisture content of 8-9%.
4. A curing stage in a special stainless steel curing kiln, where the timber is again heated to 150-190 °C for 14-16 hours, but under dry and atmospheric conditions.
5. A conditioning stage, where the moisture content of the timber is elevated for 2-3 days to the level which is required for manufacturing (4-6%). Conditioning is done in the same conventional industrial wood kiln as the drying stage, including the use of saturated steam to increase the moisture content of the treated timber.

The hydrothermolysis stage in the plato-treatment leads to various chemical transformations. One aim of this two-stage process is to ensure abundant moisture in the woody cell wall during the hydrothermolysis. This promotes an increased reactivity of the cell-wall components at a relatively low temperature. In order to reach a selective degree of depolymerization of the hemicelluloses during the hydrothermolysis, relatively mild conditions can be applied to limit undesired side reactions which can influence the mechanical properties negatively (Tjeerdsma *et al.*, 1998b).

A relatively mild heat treatment of wood according to a two-step process leading to improved dimensional stability and improved timber performance were investigated by solid phase CP-MAS 13C-NMR, in order to identify at the molecular level the reasons for the reported improvements (Tjeerdsma *et al.*, 1998a). All the changes occurring in the heat-treated wood are consequences of reactions known in wood chemistry and described earlier in this chapter.

9.4.3 The retification process

The retification process is the result of research carried out in France at the Ecole des Mines de Saint-Etienne, in collaboration with the license owner NOW S.A. (New Option Wood) and the kiln manufacturer Fours & Brûleurs Rey. The process has been industrialized since 1997 under the brand Retiwood.

In the retification process, wood with a moisture content of about 12% is treated in an oven at temperatures of 200 to 240 °C. Once the maximum temperature is reached, the wood is cooled by sprinkling. Retification is a mild pyrolysis process (T < 280 °C) conducted under an inert atmosphere (nitrogen gas) containing less than 2% oxygen (Vernois, 2001). The mild pyrolysis of wood mainly cracks hemicelluloses and begins to modify lignin. By-products of the hemicellulose pyrolysis condense and polymerize on lignin chains, hence the notion of reticulation (the creation of chemical bonds between polymeric chains) which led to the name “retification”. These reactions create a new “pseudo-lignin” which is more hydrophobic and rigid than the initial material. An infrared spectroscopy study has indeed revealed that the chemical bonds are modified in the treated wood, the number of oxygen-containing groups (mainly hydroxyl groups) decreases while the number of carbon double bonds increases. The cellulose crystallinity does not seem to be affected (Avat, 1993).

Bending strength losses of up to 40% have been reported and the process is very sensitive to slight temperature changes, i.e., the process temperature must be precisely controlled. The adhesion to paint of this wood is drastically reduced. This is believed be due to the exudation of resin from resinous species. The durability is very dependent on species, process temperature, process time, and the accuracy of the process. Swelling and shrinkage are reduced by a factor 2. The equilibrium moisture content is typically 4-5%, instead of 10-12% for untreated wood.

9.4.4 Le Bois Perdure

A second French process is named Le Bois Perdure. This technique starts with the drying of green wood in an oven. The wood is then heated to 230 °C in a steam atmosphere (low O_2), where the steam has been generated from the water from the green wood. Wood processed by this method have about the same properties as wood from the the retification process.

9.4.5 Oil heat treatment

Heat-treatment usually takes place in an inert gas atmosphere at a temperature between 180 and 240 °C. Since the boiling points of many natural oils and resins are higher than the temperature required for the heat treatment of wood, the heat treatment of wood in a hot oil bath is a feasible option. Improvements in various wood characteristics can be expected using an oil-bath heat-treatment rather than a heat-treatment in a gaseous atmosphere, due to the behavior of oils in conjunction with the effect of heat. The Menz Holz Company in Germany started to use a hot-oil treatment in 2000 (Vernois, 2004).

Oil enables both a fast and equal transfer of heat to the wood, providing the same heat conditions over the whole vessel and perfect isolation of the wood from oxygen. The oil medium can be obtained from plants, e.g., rape seed oil, sunflower oil, soybean oil. Also tall oils or tall oil derivatives in addition to drying oils such as linseed oil are conceivable. Linseed oil has proved to be unproblematic, although the smell that develops during the heat treatment may be a drawback. The smoke point and the tendency to polymerization are also important for the drying of the oil in the wood and for the stability of the oil batch. The ability of the oil to withstand heating to a temperature of at least 230 °C is a prerequisite. The consistency and color of the oil change during heat treatment. The oil becomes thicker due to volatile components evaporating, and products from the decomposition of the wood accumulate in the oil and change its composition. This obviously leads to an improved setting of the oils.

The durability of the product depends on wood species and temperature. For spruce, temperatures up to 220 °C are needed to reach durability class 2, whereas 200 °C is sufficient for pine. With the Menz process, no reduction in stiffness has been observed, although the strength was reduced by 30% after heating at 220 °C. Paintability with acrylic waterborne systems has been reported to be good and even better than with gas-heat-treated material. The typical improvement in dimensional stability was 40%. It proved to be necessary to maintain the desired temperature (e.g. 220 °C) in the wood core for 2-4 hours (Rapp & Sailer, 2001). The principle design of this type of plant is shown in Figure 9.14.

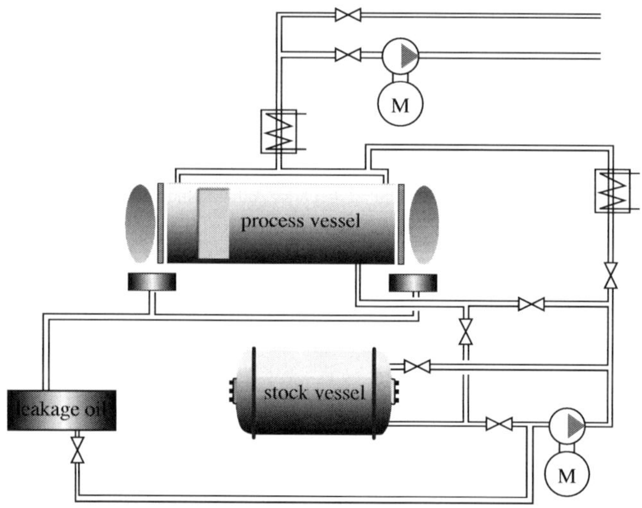

Fig. 9.14 Principle of a heat-treatment plant based on oil as the heating medium.

The process is performed in a closed process vessel. After loading the process vessel with wood, hot oil is pumped from the stock vessel into the process vessel where the hot oil is kept at a high temperature, circulating around the wood. Before unloading the process vessel, the hot oil is pumped back into the stock vessel.

For different degrees of heat treatment, various temperatures are used. To obtain maximum durability and minimum oil consumption, the process is operated at 220 °C. To obtain a maximum durability and maximum strength, temperatures between 180 and 200 °C are used in addition to a controlled oil uptake.

It was found necessary to maintain the desired process temperature (e.g. 220 °C) for 2-4 hours in the middle of the wood pieces being treated. Additional time for heating and cooling is required depending on the dimensions of the wood. The typical time for a whole treatment cycle (including heating and cooling) for logs with a cross section of 100×100 mm^2 and a length of 4 meters is 18 hours.

9.5 REFERENCES

ANON. (2003). *Thermowood handbook*. Finnish thermowood association, Helsinki, Finland.

AYADI, N., LEJEUNE, F., CHARRIER, F., CHARRIER, B. & MERLIN, A. (2003). Colour stability of heat-treated wood during artificial weathering. *Holz als Roh- und Werkstoff, 61*(3):221-226.

AVAT, F. (1993). *Contribution à l'étude des traitements thermiques du bois jusqu'à 300 °C: transformations chimiques et caractérisations physico-chimiques*. (Contribution to the study of the heat treatments of wood up to 300 °C: chemical transformation and physicochemical characterizations.) PhD. Thesis, No. 93 ENMP 0439, Ecole nationale supérieure des mines de Paris, France.

BEALL, F.C. & EICKNER, H.W. (1970). *Thermal degradation of wood components: a review of the literature*. U.S. Department of Agriculture, Forest Service Research Paper FPL 130.

BEKHTA, P. & NIEMZ, P. (2003). Effect of high temperature on the change in color, dimensional stability and mechanical properties of spruce wood. *Holzforschung, 57*(5):539-546.

BHUIYAN, T., HIRAI, N. & SOBUE, N. (2000). Changes of crystallinity in wood cellulose by heat treatment under dried and moist conditions. *Journal of Wood Science, 46*(6):431-436.

BHUIYAN, T., HIRAI, N. & SOBUE, N. (2001). Effect of intermittent heat treatment on crystallinity in wood cellulose. *Journal of Wood Science, 47*(5):336-341.

BHUIYAN, T. & HIRAI, N. (2005). Study of crystalline behaviour of heat-treated wood cellulose during treatment in water. *Journal of Wood Science, 51*(1):42-47.

BLUHM, B., ALSCHER, A., MORAW, K., COLLIN, G. & NILLES, H. (1979). *Verfahren zur Vergütung von Holz.* (Process for improving wood.) Patent No. EP18446, European Patent Office.

BOBLETER, O. & BINDER, H. (1980). Dynamischer hydrothermaler Abbau von Holz. (Dynamic hydrothermal degradation of wood.) *Holzforschung, 34*(2):48–51.

BOONSTRA, M. & DOELMAN, P. (1999). Die Eigenschaften von Plato veredeltem Holz. (Properties of Plato refined wood.) PLATO HOUT BV – Plato veredeltes Holz.

BOONSTRA, M.J. & TJEERDSMA, B.F. (2006). Chemical analysis of heat treated softwoods. *Holz als Roh- und Werkstoff, 64*(3):204-211.

BOONSTRA, M.J., KEGEL, E.V. & RIJSDIJK, J.F. (2005). *Process for upgrading wood parts.* Patent No. EP1597039, European Patent Office.

BOONSTRA, M.J., PIZZI, A., ZOMERS, F., OHLMEYER, M. & PAUL, W. (2006). The effects of a two stage heat treatment process on the properties of particleboard. *Holz als Roh- und Werkstoff, 64*(2):157–164.

BOONASTRA, M.J., VAN ACKER, J., TJEERDSMA, B.F. & KEGEL, E.V. (2007a). Strength properties of thermally modified softwoods and its relations to polymeric structural wood constituent. *Annals of Forest Science, 64*(7):679-690.

BOONASTRA, M.J., VAN ACKER, J. & KEgel, E.V. (2007b). Effect of a two-stage heat treatment process on the mechanical properties of full construction timber. *Wood Material Science and Engineering,* 2(3/4):138-146.

BOURGOIS, J. & GUYONNET, R. (1988). Characterisation and analysis of torrefied wood. *Wood Science and Technology, 22*(2):143-155.

BOURGOIS, J., BARTHOLIN, M. & GUYONNET, R. (1989). Thermal treatment of wood: analysis of the obtained products. *Wood Science and Technology, 23*(4):303-310.

BOURGOIS, J. & GUYONNET, R. (1998). Characterization and analysis of torrified wood. *Wood Science and Technology, 22*(2):143-155.

BRENDEN, J.J. (1967). *Effect of fire-retardant and other inorganic salts on pyrolysis products of Ponderosa pine at 250 °C and 350 °C.* U.S. Department of Agriculture, Forest Service Research Paper FPL 80.

BRISCHKE, C. & RAPP, A. (2004). *Investigation of the suitability of silver fir (Abies alba Mill.) for thermal modification.* International Research Group on Wood Preservation, Doc. No. IRG/WP 04-40275.

BRISCHKE, C., WELZBACHER, C.R., BRANDT, K. & RAPP, A.O. (2007). Quality control of thermally modified timber: Interrelationship between heat treatment intensities and CIE L*a*b* color data on homogenized wood samples. *Holzforschung, 61*(1):19-22.

BROWNE, F.L. (1958). *Theories of the combustion of wood and its control.* U.S. Department of Agriculture, Forest Service Technological Report No. 2136.

BURMESTER, A. (1970). *Formbeständigkeit von Holz gegenüber Feuchtigkeit – Grundlagen und Vergütungsverfahren.* (Influnce of moisture on dimension stability of wood – theory and improvemnts.) Bundesanstalt für Materialforschung, BAM Berichte No. 4.

BURMESTER, A. (1973). Effect of heat-pressure treatments of semi-dry wood on its dimensional stability. *Holz als Roh- und Werkstoff, 31*(6):237-243.

BURMESTER, A. (1975). Zur Dimensionsstabilisierung von Holz. (The dimensional stabilization of wood.). *Holz als Roh- und Werkstoff, 33*(9):333-335.

BURMESTER, A. (1981). *Dimensational stabilisation of wood.* International Research Group on Wood Protection, Document No. IRG/WP/3171.

BURO, A. (1954). Die Wirkung von Hitzebehandlung auf die Pilzresistenz von Kiefern- und Buchenholz. (The influence of heat treatment on decay resistance of pine and beech wood.) *Holz als Roh- und Werkstoff, 12*(8):297-304.

BURO, A. (1955). Untersuchungen über die Veränderung der Pilzresistenz von Hölzern durch Hitzebehandlung in Metallschmelzen. (Investigation of the change in decay resistance of wood during heat treatment in a metal melt.) *Holzforschung, 9*(6):177-181.

BURTSCHER, E., BOBLETER, O., SCHWALD, W., CONCIN, R. & BINDER, H. (1987). Chromatographic analysis of biomass reaction products produced by hydrothermolysis of Poplar wood. *Journal of Chromatography, 390*(2):401–412.

CARRASCO, F. & ROY, C. (1992). Kinetic study of dilute-acid prehydrolysis of xylan-containing biomass. *Wood Science and Technology, 26*(3):189–208.

DAVID, W.H. & THOMPSON, W.S. (1964). Influence of thermal treatments of short duration on the toughness and chemical composition of wood. *Forest Products Journal, 14*(8):350-356.

DEL MENEZZI, C.H.S. & TOMASELLI, I. (2006). Contact thermal post-treatment of oriented strandboard to improve dimensional stability: A preliminary study. *Holz als Roh- und Werkstoff, 64*(3):212–217.

DIETRICHS, H.H., SINNER, H. & PULS, J. (1978). Potential of steaming hardwoods and straw for feed and food production. *Holzforschung, 32*(6):193-199.

DIROL, D. & GUYONNET, R. (1993). *The improvement of wood durability by retification process*. International Research Group on Wood Protection, Document No. IRG/WP/93-40015.

D´JAKANOV, K. & KONEPLEVA, T. (1967). Moisture absorption by Scots pine wood after heat treatment. (in russian) *Izvestija vyssih ucebnyh zavedenij, 10*(1):112-114.

DVINSKIKH, S.V., FURÓ, I, SANDBERG, D. & SÖDERSTRÖM, O. (2011). Moisture content profiles and uptake kinetics in wood cladding materials evaluated by a portable nuclear magnetic resonance spectrometer. *Wood Material Science and Engineering, 6*(3):29-37.

EDLUND, M.-L. (2004). *Durability of some alternatives to preservative treated wood*. International Research Group on Wood Preservation, Document No. IRG/WP 04-30353.

ELLIS, S. & PASZNER, L. (1994). Activated self-bonding of wood and agricultural residues. *Holzforschung, 48*(1):82–90.

ENGLUND, F. & NUSSBAUM, R. (2000). Monoterpenes in Scots pine and Norway spruce and their emission during kiln drying. *Holzforschung, 54*(5):449-456.

ESTEVES, B., DOMINGOS, I., VELEZ MARQUES, A., NUNES, L. & PEREIRA, H. (2006). Variation of dimensional stability and durability of eucalypt wood by heat treatment. In: *Proceedings of ECOWOOD 2006, 2nd International Conference on Environmentally-Compatible Forest Products*. Fernado, Pessoa University, Oporto, Portugal, pp. 185-194.

ESTEVES, B., DOMINGOS, I. & PEREIRA, H. (2008a). Pine wood modification by heat treatment in air. *BioResources, 3*(1):142-154.

ESTEVES, B., VELEZ MARQUES, A., DOMINGOS, I. & PEREIRA, H. (2008b). Heat-induced colour changes of pine (Pinus pinaster) and eucalypt (Eucalyptus globulus) wood. *Wood Science and Technology, 42*(5):369-384.

ESTEVES, B., GRACA, J. & PEREIRA, H. (2008c). Extractive composition and summative chemical analysis of thermally treated eucalypt wood. *Holzforschung, 62*(3):344-351.

ESTEVES, B. & PEREIRA, H. (2009). Wood modification by heat treatment: A review. *BioResources, 4*(1):370-404.

EN 350-2, European Committee for Standardization (1994). *EN 350-2: Durability of wood and wood-based products. Part 2: Guide to natural durability and treatability of selected wood species of importance in Europe*.

EN 460, European Committee for Standardization (1994). *EN 460: Durability of wood and wood-based products – Natural durability of solid wood – Guide to the durability requirements for wood to be used in hazard classes*.

EN-113, European Community for Standardization (1996). *EN-113: Wood preservatives – Determination of toxic values of wood preservatives against wood destroying Basidiomycetes cultures on agar medium*.

FEIST, W.C. & SELL, J. (1987). Weathering behavior of dimensionally stabilized wood treated by heating under pressure of nitrogen gas. *Wood and Fiber Science, 19*(2):183-195.

FENGEL, D. (1966a). Über die Veränderungen des Holzes und seiner Komponenten im Temperaturbereich bis 200 °C. Erste Mitteilungen: Heiβ und Kaltwasserextrakte von termisch behandeltem Fichtenholz. (On the changes of the wood and its components within the temperature range up to 200 °C. Part I: Hot and cold water extracts of thermally treated sprucewood.) *Holz als Roh- und Werkstoff, 24*(1):9-14.

FENGEL, D. (1966b). Über die Veränderungen des Holzes und seiner Komponenten im Temperaturbereich bis 200 °C. Zweite Mitteilungen: Die Hemicellulosen in unbehandeltem und in termisch behandeltem Fichtenholz. (On the changes of the wood and its components within the temperature range up to 200 °C. Part II: The hemicelluloses in untreated and thermally treated sprucewood.) *Holz als Roh- und Werkstoff, 24*(3):98-109.

FENGEL, D. (1966c). Über die Veränderungen des Holzes und seiner Komponenten im Temperaturbereich bis 200 °C. Dritte Mitteilungen: Thermisch und mechanisch bedingte Strukturänderungen bei Fichtenholz. (On the changes of the wood and its components within the temperature range up to 200 °C. Part III: Thermally and mechanically caused structural changes in sprucewood.) *Holz als Roh- und Werkstoff, 24*(11):529-536

FENGEL, D. & WEGENER, G. (2003). *Wood: chemistry, ultrastructure, reactions*. Verlag Kessel, Remagen, Germany, ISBN 3935638-39-6.

GIEBELER, E. (1983). Dimensionsstabilisierung von Holz durch eine Feuchte/Wärme/Drück-Behandlung. (Dimensional stabilization of wood by moisture-heat-treatment.) *Holz als Roh- und Werkstoff, 41*(1):87-94.

GONZALEZ-PENA, M.M. & HALE, D.C. (2007). The relationship between mechanical performance and chemical change in thermally modified wood. In. *Proceeding of the 3rd European Conference on Wood Modification,* Hill, C.A.S., Jones, D., Militz, H. & Ormondroyd, G.A. (eds.). Cardiff, UK, pp.169-172.

GOROYIAS, G.J. & HALE, M.D. (2002). *Heat treatment of wood strands for OSB production: Effect on the mechanical properties, water absorption and dimensional stability.* International Research Group on Wood Protection, Document No. IRG/WP/02-40238.

HEINZERLING, C. (1885). *Konservirung des Holzes*. (Conservation of wood.) Halle a. S., Wilhelm Knapps Verlag, Germany.

HILL, C.A.S. (2006). *Wood modification: chemical, thermal and other processes*. John Wiley & Sons, Chichester, England. ISBN978-0-470-02172-9.

HILLIS, W.E. (1975). The role of wood characteristics in high temperature drying. *Journal of the Institute of Wood Science, 7*(2):60-67.

HILLIS, W.E. (1984). High temperature and chemical effects on wood durability. *Wood Science and Technology, 18*(4):281-293.

JERMER, J. & ANDERSSON, B-L. (2005). *Corrosion of fasteners in heat-treated wood – progress report after two years exposure outdoors*. International Research Group on Wood Protection, Document No. IRG/WP/05-40296.

JOHANSSON, D. & MORÉN, T. (2006). The potential of colour measurement for strength prediction of thermally treated wood. *Holz als Roh- und Werkstoff, 64*(2):104-110.

JOHANSSON, D., SEHLSTEDT-PERSSON, M. & MORÉN, T. (2006). Effect of heat treatment on capillary water absorption of heat-treated pine, spruce and birch. In: *Proceedings of the 5th IUFRO Symposium on Wood Structure and Properties ´06,* Kurjatko, S., Kúdela, J. & Lagana, R. (eds.). Sielnica, Slovakia, pp. 251-255.

JÄMSÄ, S. & VIITANIEMI, P. (1998). Heat treatment of wood. Better durability without chemicals. In: *Proceedings from Nordiske Trebeskyttelsesdager,* pp. 47-51.

JÄMSÄ, S., AHOLA, P. & VIITANIEMI, P. (2000). Long-term natural weathering of coated ThermoWood. *Pigment and Resin Technology, 29*(2):68-74.

JÄMSÄ, S. & VIITANIEMI, P. (2001). Heat treatment of wood: better durability without chemicals. In: *Review on heat treatments of wood, Proceedings of the special seminar of COST Action E22,* Rapp, A.O. (ed.). Antibes, France, pp. 21-26.

KAAR, W.E., COOL, L.G., MERRIMAN, M.M. & BRINK, D.L. (1991). The complete analysis of wood polysaccharides using HPLC. *Journal of Wood Chemistry and Technology, 11*(4):447–463.

KAMDEM, D.P., PIZZI, A., GUYONNET, R. & JERMANNAUD, A. (1999). *Durability of heat-treated wood*. International Research Group on Wood Protection, Document No. IRG/WP/99-40145.

KAMDEM, D.P., PIZZI A. & TRIBOULOT M.C. (2000). Heat-treated timber: potentially toxic byproducts presence and extent of wood cell degradation. *Holz als Roh- und Werkstoff, 58*(4):253-257.

KAMDEM, D.P., PIZZI A. & JERMANNAUD, A. (2002). Durability of heat-treated wood. *Holz als Roh- und Werkstoff, 60*(1):1-6.

KEITH, C.T. & CHANG, C.I. (1978). *Properties of heat-darkened wood. I. Hygroscopic properties*. Eastern Forest Products Laboratory, Canada, Report No. OPX213E.

KLAUDITZ, W. & STEGMANN, G. (1955). Beiträge zur Kenntnis des Ablaufes und der Wirkung thermischer Reaktionen bei der Bildung von Holzwerkstoffen. (Contributions to the knowledge of the sequence and the effect of thermal reactions in the formation of wood materials.) *Holz als Roh- und Werkstoff, 13*(11):434–440.

KOEHLER, A. & PILLOW, M.Y. (1925). Effect of high temperatures on the mode of fracture of a softwood. *Southern Lumberman, 121*(1576):219-221.

KOLLMANN, F. (1939). Vorgänge und Änderungen von Holzeigenschaften beim Dämpfen. (Operations and changes in wood properties during steaming.) *Holz als Roh- und Werkstoff, 2*(1):1-11.

KOLLMANN, F. (1940). *Die Mechanischen Eigenschaften Verschieden Feuchter Holzer in Temperatur-bereich von -100 bis +200 °C.* (The mechanical properties of different wet woods in the temperature region from 100 to +200 °C.) VDI-Forschungsheft 403, VDI-Verlag, Berlin.

KOLLMANN, F. (1960). Zur Frage des Auftretens exothermer Reaktionen bei Holz. (Occurence of exotermic reaction in wood.) *Holz als Roh- und Werkstoff, 18*(6):193-200.

KOLLMANN, F. & SCHNEIDER, A. (1963). Über das Sorptionsverhalten wärmebehandelter Hölzer. (On the sorption behaviour of heat stabilized wood.) *Holz als Roh- und Werkstoff, 21*(3):77-85.

KOLLMANN, F. & FENGEL, D. (1965). Änderungen der chemischen Zusammensetzung von Holz duerch thermische Behandlung. (Changes in the chemical composition of wood by thermal treatment.) *Holz als Roh- und Werkstoff, 23*(12):461-468.

KOŠIK, M., GÉRATOVÁ, L., RENDOŠ, F & DOMANSKÝ, R. (1968). Pyrolysis of beech wood and its components at low temperatures. II. Thermography of the beech wood ans its components. *Holzforschung und Holzverwertung, 20*(1):15-19.

KUBOJIMA, Y., OKANO, T. & OHTA, M. (2000). Bending strength and toughness of heat treated wood. *Journal of Wood Science, 46*(1):8-15.

LEITHOFF, H. & PEEK, R.-D. (1998). Hitzebehandlung – eine Alternative zum chemischen Holzschutz. (Heat treatment – an alternative to chemical treatment.) In: *Tagungsband zur 21. Holzschutz-Tagung der DGfH,* Rosenheim, pp. 97-108.

LEKOUNOUGOU, S., PÉTRISSANS, M., JACQUOT, J.P., GELHAYE, E. & GÉRARDIN, P. (2009). Effect of heat treatment on extracellular enzymatic activities involved in beech wood degradation by Trametes versicolor. *Wood Science and Technology, 43*(3/4):331-341.

MACLEAN, J.D. (1951). *Rate of disintegration of wood under different heating conditions.* American Wood Preservers' Association Proceedings, 47:155-168.

MACLEAN, J.D. (1953). *Effect of steaming on the strength of wood.* American Wood Preservers' Association Proceedings, 49:88-112.

MANNINEN, A.-M., PASANEN, P. & HOLOPAINEN, J.K. (2002). Comparing the VOC emissions between air-dried and heat-treated Scots pine wood. *Atmospheric Environment, 36*(11):1763-1768.

MAZELA, B., ZAKRZEWSKI, R., GRZESKOWIAK, W., COFTA, G. & BARTKOWIAK, M. (2003). Preliminary research on the biological resistance of thermally modified wood. In: *Proceedings of the first European Conference on Wood Modification,* v. Acker, J. & Hill, C. (eds.). Ghent University (RUG), pp. 113-120.

MCDONALD, A., FERNANDEZ, M. & KREBER, B. (1997). Chemical and UV-VIS spectroscopic study on kiln brown stain formation in Radiata pine. In: *Proceedings of the 9th International Symposium of Wood and Pulping Chemistry,* Montreal, Vol. 70, pp.1-5.

MCGINNIS, G.D., WILSON, W.W. & BIERMAN, C.J. (1984). Biomass conversion into chemicals using wet oxidation. In: *Bioconversion systems.* Wise, D.L. (ed.). CRC Press, Boca Raton, FL, pp. 89-109.

METSÄ-KORTELAINEN, S., ANTIKAINEN, T. & VIITANIEMI, P. (2006). The water absorption of sapwood and heartwood of Scots pine and Norway spruce heat-treated at 170 °C, 190 °C, 210 °C and 230 °C. *Holz als Roh- und Werkstoff, 64*(3):192–197.

METSÄ-KORTELAINEN, S. & VIITANIEMI, P. (2009). Decay resistance of sapwood and heartwood of untreated and thermally modified Scots pine and Norway spruce compared with some other wood species. *Wood Material Science and Engineering, 4*(3/4):105-114.

MILITZ, H. (2002). *Thermal treatment of wood: European processes and their background.* International Research Group on Wood Preservation, Doc. No. IRG/WP 02-40241.

MILITZ, H. & TJEERDSMA, B. (2001). Heat treatment of wood by the Plato process. In: *Review on heat treatments of wood, Proceedings of the special seminar of COST Action E22,* Rapp, A.O. (ed.). Antibes, France, pp. 25-35.

MILLETT, M.A. & GERHARDS, C.C. (1972). Accelerated aging: residual weight and flexural properties of wood heated in air at 115 °C to 175 °C. *Wood Science, 4*(4):193-201.

MITCHELL, P.H. (1988). Irreversible property changes of small loblolly pine specimens heated in air, nitrogen, or oxygen. *Wood and Fiber Science, 20*(3):320-355.

MITCHELL, R.L., SEBORG, R.M. & MILLETT, M.A. (1953). Effect of heat on the properties and chemical composition of Douglas-fir wood and its major components. *Journal of the Forest Products Research Society, 3*(4):38-42.

MITSUI, K., TAKADA, H., SUGIYAMA, M. & HASEGAWA, R. (2001). Changes in the properties of light-irradiated wood with heat treatment: Part 1. Effect of treatment conditions on the change in colour. *Holzforschung, 55*(6):601-605.

NIKOLOV, S. & ENCEV, E. (1967). Effect of heat treatment on the sorption dynamics of beech wood. (in russian) *Naucni trudove, 14*(3):71-73.

NOACK, D. (1969). Über die Heisswasserbehandlung von Rotbuchenholz im Temperaturbereich von 100 bis 180°C. (Hot water treatment of beech in the temperatures of 100 to 180°C.) *Holzforschung und Holzverwertung, 21*(5):118-124.

NUOPPONEN, M., VUORINEN, T., JAMSÄ, S. & VIITANIEMI, P. (2004). Thermal modification in softwood studied by FT-IR and UV resonance Raman spectroscopies. *Journal of Wood Chemistry and Technology, 24*(1):13-26.

OBATAYA, E., HIGASHIHARA, T. & TOMITA, B. (2002). Hygroscopicity of heat treated wood III. Effect of steaming on the hygroscopicity of wood. *Mokuzai Gakkaishi, 48*(5):358–355.

PACKMAN, D.F. (1960). The acidity of wood. *Holzforschung, 14*(6):178-183.

PAUL, W., OHLMEYER, M., LEITHOFF, H., BOONSTRA, M.J. & PIZZI, A. (2006). Optimising the properties of OSB by a one-step heat pre-treatment process. *Holz als Roh- und Werkstoff, 64*(3):227–234.

PAUL, W., OHLMEYER, M. & LEITHOFF, H. (2007). Thermal modification of OSB-strands by a one-step heat pre-treatment – Influence of temperature on weight loss, hygroscopicity and improved fungal resistance. *Holz als Roh- und Werkstoff, 65*(1):57-63

PETERS, J., FISCHER, K. & FISCHER, S. (2008). Characterization of emissions from thermally modified wood and their reduction by chemical treatment. *BioResources, 3*(2):491-502.

PETERS, J., PFRIEM, A., HORBENS, M., FISCHER, S. & WAGENFÜHR, A. (2009). Emissions from thermally modified beech, their reduction by solvent extraction and fungicidal effect of the organic solvent extracts. *Wood Material Science and Engineering, 4*(1/2):61-66.

PILLOW, M.Y. (1929). Effect of high temperatures on the mode of fracture and other properties of a hardwood. *Southern Lumberman, 137*(1766):58-60.

POPPER, R., NIEMZ, P. & EBERLE, G. (2005). Untersuchungen zum Sorptions- und Quellungsverhalten von thermisch behandeltem Holz. (Investigations on sorption and swelling properties of thermally treated wood.) *Holz als Roh- und Werkstoff, 63*(2):135–148.

RAPP, A.O., SAILER, M. & WESTIN, M. (2000). Innovative Holzvergütung – neue Einsatzbereiche für Holz. ("Innovative improvement of wood") In: *Tagungsband zur 17. Dreiländer-Holztagung 2000. Schweizerische Arbeitsgemeinschaft für Holzforschung,* Holz ART, Zürich, pp. 153-160.

RAPP, A.O. & SAILER, M. (2000). Heat treatment of wood in Germany: state of the art. In: *Proceedings of the seminar of production of heat treated wood in Europe.* Tekes, Helsinki.

RAPP, A.O. & SAILER, M. (2001). Oil heat treatment of wood in Germany: state of the art. In: *Review on heat treatment of wood. Proceedings of special seminar of Cost Action E22: Environmental Optimization of Wood Protection,* Rapp, A. (ed.). Antibes, France, pp. 45-62.

REM, P.C., VAN DER POEL, H. & RUYTER, H.P. (1994). *Process for upgrading low-quality wood.* Patent No. EP0612595, European Patent Office.

REPELLIN, V. & GUYONNET, R. (2005). Evaluation of heat-treated wood swelling by differential scanning calorimetry in relation to chemical composition. *Holzforschung, 59*(1):28–34.

ROWELL, R.M. (2005). *Handbook of Wood Chemistry and Wood Composites.* Taylor and Francis, Boca Raton, USA, ISBN 0-8493-1588-3.

ROWELL, R.M., IBACH, R.E., MCSWEENY, J. & NILSSON, T. (2009). Understanding decay resistance, dimensional stability and strength changes in heat treated and acetylated wood. *Wood Material Science and Engineering, 4*(1/2):14-22.

RUNKEL, R.O.H. (1951). Zur Kenntnis des thermoplastischen Verhaltens von Holz. (Thermoplastic behaviour of wood.) *Holz als Roh- und Werkstoff, 9*(2):41–53.

RUNKEL, R.O.H. & WITT, H. (1953). Zur Kenntnis des thermoplastischen Verhaltens von Holz. III. Mitteilung: Über die wasser- und alkohollöslichen Anteile in hitzeplastiziertem Holz. Ergebnis einer papierchromatographischen Untersuchung. (To the attention of the thermoplastic behavior of wood. III: About the water-and alcohol-soluble components in heat-plastizied wood. Result of a paper chromatography investigation.) *Holz als Roh- und Werkstoff, 11*(12):457-461.

RUSCHE, V.H. (1973). Die thermische Zersetzung von Holz bei Temperaturen bis 200 °C. Erste Mitteilung: Festigkeitseigenschaften von trockenem Holz nach thermischer Behandlung. (Thermal degradation of wood at temperatures up to 200 °C – Part I: Strength properties of dried wood after heat treatment.) *Holz als Roh- und Werkstoff, 31*(7):273-281.

RUYTER, H.P. (1989). *Thermally modified wood.* Patent No. EP89-2031709, European Patent Office.

SANDBERG, D. (1999). Weathering of radial and tangential wood surfaces of pine and spruce. *Holzforschung, 53*(4):355-364.

SANDBERG, D. & SÖDERSTRÖM, O. (2006). Crack formation due to weathering of radial and tangential sections of pine and spruce. *Wood Material Science and Engineering, 1*(1):12-20.

SAILER, M., RAPP, A.O., LEITHOFF, H. & PEEK, R.-D. (2000). Vergütung von Holz durch Anwendung einer Öl-Hitze-Behandlung. (Upgrading of wood by application of an oil-heat treatment.) *Holz als Roh- und Werkstoff, 58*(1/2):15-22.

SCHEIDING, W., KRUSE, K., PLASCHKIES, K. & WEISS, B. (2005). Thermally modified timber (TMT) for playground toys: Investigations on 13 industrially manufatured products. In: *Proceeding from the 2nd European Conference on Wood Modification,* Militz, H. & Hill, C. (eds.). Göttingen, Germany.

SEBORG, R.M., TARKOW, H. & STAMM, A.J. (1953). Effect of heat upon the dimensional stabilisation of wood. *Journal of Forest Products Research Society, 3*(9):59-67.

SEHLSTEDT-PERSSON, M. (2003). Colour responses to heat-treatment of extractives and sap from pine and spruce. In: *Proceeding of the 8th International IUFRO Wood Drying Conference, Ispas,* M., Mihaela Campean, M., Cismaru, I., Marinescu, I. & Budau, G. (eds.). Brasov, Romania, pp. 459–464.

SHAFIZADEH, F. & CHIN, P.P.S. (1977). Thermal deterioration of wood. In: *Wood Technology: Chemical Aspects.* Goldstein, I.S. (ed.). ACS Symposium Series 43, American Chemical Society, Washington D.C., pp. 57-81.

STAMM, A.J. (1956). Thermal degradation of wood and cellulose. *Industrial and Engineering Chemistry, 48*(3):413-417.

STAMM, A.J. (1964). *Wood and cellulose science.* Ronald Press, New York.

STAMM, A.J. & HANSEN, L.A. (1937). Minimizing wood shrinkage and swelling. Effect of heating in various gasses. *Industrial and Engineering Chemistry, 29*(7):831-833.

STAMM, A.J., BURR, H.K. & KLINE, A.A. (1946). Staybwood. Heat stabilized wood. *Industrial and Engineering Chemistry, 38*(6):630-634.

STAMM, A.J. & BAECHLER, R.H. (1960). Decay resistance and dimensional stability of five modified woods. *Forest Products Journal, 10*(1):22-26.

STEWART, C.M., KOTTEK, J.F., DADSWELL, H.E. & WATSON, A.J. (1961). The process of fibre separation. Part III. Hydrolytic degradation within living trees and its effects on the mechanical pulping and other properties of wood. *Tappi, 44*(11):798-813.

SUNDQVIST, B. (2004). *Colour changes and acid formation in wood during heating.* PhD-Thesis No. 2004:10, Luleå University of Technology, Sweden.

SUNDQVIST, B. & MORÉN, T. (2002). The influence of wood polymers and extractives on wood colour induced by hydrothermal treatment. *Holz als Roh- und Werkstoff, 60*(5):375-376.

SUNDQVIST, B., KARLSSON, O. & WESTERMARK, U. (2006). Determination of formic-acid and acetic acid concentrations formed during hydrothermal treatment of birch wood and its relation to colour, strength and hardness. *Wood Science and Technology, 40*(7):549-561.

SYRJÄNEN, T. (2001). Production and classification of heat treated wood in Finland. In: *Review on heat treatment of wood. Proceedings of special seminar of Cost Action E22: Environmental Optimization of Wood Protection,* Rapp, A. (ed.). Antibes, France, pp. 7-15.

SYRJÄNEN, T. & KANGAS, E. (2000). *Heat treated timber in Finland.* International Research Group on Wood Protection, Document No. IRG/WP/00-40158.

TEICHGRÄBER, R. (1966). Beitrag zur Kenntnis der Eigenschaftsänderungen des Holzes beim Dämpfen. (On the alteration of the properties of wood during steaming). *Holz als Roh- und Werkstoff, 24*(11):548-551.

THUNELL, B. (1940). *Temperaturens inverkan på böjhållfastheten hos svenskt furuvirke*. (The effect of temperature on the bending strength of Swedish pine-wood.) Statens Provningsanstalt, meddelande 80.

THUNELL, B. (1941). *Hållfasthetsegenskaper hos svenskt furuvirke utan kvistar och defekter.* (Strength of Swedish pine wood without knots and defects.) The Royal Swedish Academy of Engineering Sciences (Ingenjörsvetenskapsakademien, IVA), Report No. 161.

THUNELL, B. & ELKEN, E. (1948). *Värmebehandling av trä för minskning av svällning och krympning*. (Heat treatment of wood for decreased swelling and shrinkage.) The Swedish Wood Technology Research Institute, Report No. 18.

TIEMANN, H.D. (1915). The effect of different methods of drying on the strength of wood. *Lumber World Review, 28*(7):19-20.

TIEMANN, H.D. (1920). *The kiln drying of woods for airplanes*. National Advisory Committee for Aeronautics, Washington DC, Report No. 65.

TIEMANN, H.D. (1941a). Does temperature injure wood? *Southern Lumberman, 163*(2047):49-50.

TIEMANN, H.D. (1941b). Does temperature injure wood? *Southern Lumberman, 163*(2049):49-51.

TIEMANN, H.D. (1941c). Does temperature injure wood? Reduction in hygroscopicity. *Southern Lumberman, 163*(2051):55-57.

TIEMANN, H.D. (1942). *Wood technology: constitution, properties and uses*. Pitman Publishing Corporation, New York, Chicago.

TJEERDSMA, B.F., BOONSTRA, M., PIZZI, A., TEKELY, P. & MILITZ, H. (1998a). Characterisation of thermal modified wood: molecular reasons for wood performance improvement. *Holz als Roh- und Werkstoff, 56*(3):149-153.

TJEERDSMA, B.F., BOONSTRA, M. & MILITZ, H. (1998b). *Thermal modification of non-durable wood species. 2. Improved wood properties of thermally treated wood.* International Research Group on Wood Preservation, Document No. IRG/WP 98-40124.

TJEERDSMA, B.F. STEVENS, M. & MILITZ, H. (2000). *Durability aspects of hydrothermal treated wood*. International Research Group on Wood Preservation, Document No. IRG/WP 00-4.

TJEERDSMA, B.F., STEVENS, M., MILITZ, H. & v. ACKER, J. (2002). Effect of process conditions on moisture content and decay-resistance of hydro-thermally treated wood. *Holzforschung und Holzverwertung, 54*(5):94-99.

TJEERDSMA, B.F. & MILITZ, H. (2005). Chemical changes in hydrothermal treated wood: FTIR analysis of combined hydrothermal and dry heat-treated wood. *Holz als Roh- und Werkstoff, 63*(2):102-111.

TJEERDSMA, B.F. (2006). *Heat treatment of wood – thermal modification*. SHR Timber Research, Wageningen, The Netherlands.

VERNOIS, M. (2001) Heat treatment of wood in France – State of the art. In: *Review on heat treatment of wood. Proceedings of special seminar of Cost Action E22: Environmental Optimization of Wood Protection,* Rapp, A. (ed.). Antibes, France, pp. 37-44.

VERNOIS, M. (2004) Menz Holz: une première unité industrielle de traitement oléothermique. (Menz Holz: a first industrial processing unit for heat treatment in oil.) *CTBA Info, 104*:25-2

VIITANEN, H., JÄMSÄ, S., PAAJANEN, L., NURMI, A. & VIITANIEMI, P. (1994). *The effect of heat treatment on the properties of spruce*. International Research Group on Wood Preservation, Document No. IRG/WP 94-40032.

VIITANIEMI, P., JÄMSÄ, S., Ek, P & VIITANEN, H. (1993). *Method for increasing the resistance of cellulosic products against mould and decay.* Patent No. EP06954087, European Patent Office.

VIITANIEMI, P., RANTA-MAUNUS, A., JÄMSÄÄ, S. & Ek, P. (1994). *Method for processing of wood at elevated temperatures*. Patent No. EP0759137, European Patent Office.

WANG, J.Y. & COOPER, P.A. (2005). Effect of oil type, temperature and time on moisture properties of hot oil-treated wood. *Holz als Roh- und Werkstoff, 63*(6):417-422.

WEILAND, J.J. & GUYONNET, R. (1997). Retifiziertes Holz. (Retification process for wood) In: *Verdichter Holzbau in Europa. Motivation, Erfahrung, Entwicklung. Dreilander Holztagung. 10. Joanneum Research Fachtage, 2-5 November,* Graz, Austria, pp. 109-117.

WEILAND, J.J. & GUYONNET, R. (2003). Study of chemical modifications and fungi degradation of thermally modified wood using DRIFT spectroscopy. *Holz als Roh- und Werkstoff, 61*(3):216-220.

WELZBACHER, C.R. & RAPP, A. (2007). Durability of thermally modified timber from industrial-scale processes in different use classes: Results from laboratory and field tests. *Wood Material Science and Engineering, 2*(1):4-14.

WELZBACHER, C.R., BRISCHKE, C. & RAPP, A. (2007). Influence of treatment temperature and duration on selected biological, mechanical, physical, and optical properties of thermally modified timber. *Wood Material Science and Engineering, 2*(2):66-76.

WESTIN, M., RAPP, A.O. & NILSSON, T. (2004). *Durability of pine modified by 9 different methods*. International Research Group on Wood Protection, Document No. IRG/WP/04-40288.

WIKBERG, H. & MAUNU, S. (2004). Characterisation of thermally modified hard- and softwoods by 13C CPMAS NMR. *Carbohydrate Polymers, 58*(4):461-466.

WINDEISEN, E., STROBEL, C. & WEGENER, G. (2007). Chemical changes during the production of thermo treated beech wood. *Wood Science Technology, 41*(6):523-536.

WIENHAUS, O. (1999). Modifizierung des Holzes durch eine milde Pyrolyse – abgeleitet aus den allgemeinen Prinzipien der Thermolyse des Holzes. (Modification of wood by mild pyrolysis – derived from the general principles of the thermolysis of wood.) *Wissenschaftliche Zeitschrift der Technischen Universität Dresden, 48*(2):17-22.

WILSON, T.R.C. (1920). *The effect of kiln drying on the strength of airplane woods*. National Advisory Committee for Aeronautics, Washington DC, Report No. 68.

YAO, J. & TAYLOR, F. (1979). Effect of high-temperature drying on the strength of southern pine dimension lumber. *Forest Products Journal, 29*(8):49-51.

YILDIZ, S., GEZER, E.D. & YILDIZ, U.C. (2006). Mechanical and chemical behavior of spruce wood modified by heat. *Building and Environment, 41*(12):1762-1766.

ÅSTRÖM, K. (1993). Skidan från Kalvträsk. (The ski from Kalvträsk.) *Västerbotten, 74*(3):129-131, ISSN 0346-4938.

CHAPTER 10

WOOD BENDING

10.1 INTRODUCTION

Wood bending is a chip-less manufacturing method for shaping wood. In the case of solid wood, the bending is generally done in the fiber direction, whereas for laminated wood products, it is performed both in the fiber direction and in the transverse direction of the wood.

Wood bending is one of the techniques used in the THM wood process from a long time ago. Even though is has special features, it is actually not a difficult technique. Some classic examples include boats, barrels and containers. The native American Indian traditionally bent wood to make snowshoes, sledges, luges and canoes. In Europe, this technique has long been used for the manufacture of chairs such as the slat back chair and the ladder back chair that already existed in the middle ages (1100-1450 A.C.) and can often be seen in wood-block prints and copperplate prints from that period. Here, we can only imagine how old the technique is. In northeast Japan, it was used by the old community in the making of their traditional snowshoes called "kanjiki". Their method of bending the wood involved placing the material directly on the pot, putting a cover on it, and after steaming it for a while just pressing it to the desired shape.

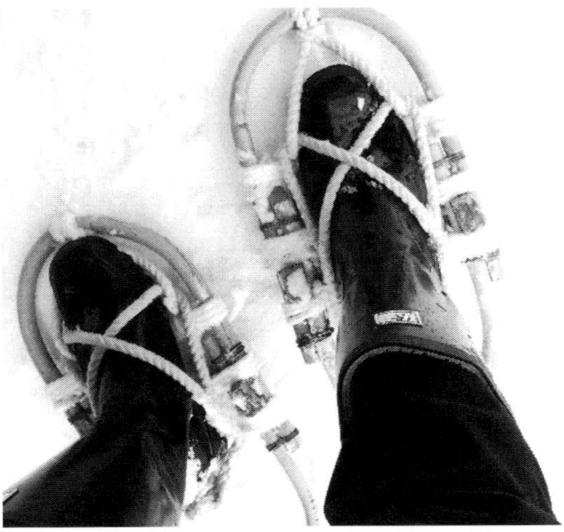

Fig. 10.1 Kanjiki, traditional snowshoes from Japan.

The principles of bending solid wood for industrial applications were described early in literature (see for instance Karmarsch, 1839), and less than fifty years ago wood bending was of great importance in many industries, especially for the manufacture of furniture, boats and ships, agricultural implements, tool handles, and sporting goods. Of the several methods commonly used to produce curved pieces of wood, bending was the most economical with regard to material, the most productive for members of high strength, and perhaps the cheapest. Today, the CNC technique and new materials, especially plastics, have replaced the wood-bending technique.

Now, wood bending is a method necessary in furniture making and in its design strategy, enabling us to manufacture and design furniture accompanied with a new value-added element. Laminated wood is also used for construction purposes in, for example, beams.

The different wood bending processes can be classified into solid-wood bending, laminated bending and other methods, as shown in Figure 10.2. Solid-wood bending can be planar or bi-planar and in most cases the wood has to be plasticized by a chemical or by a thermo-hydro-mechanical process before it can be bent.

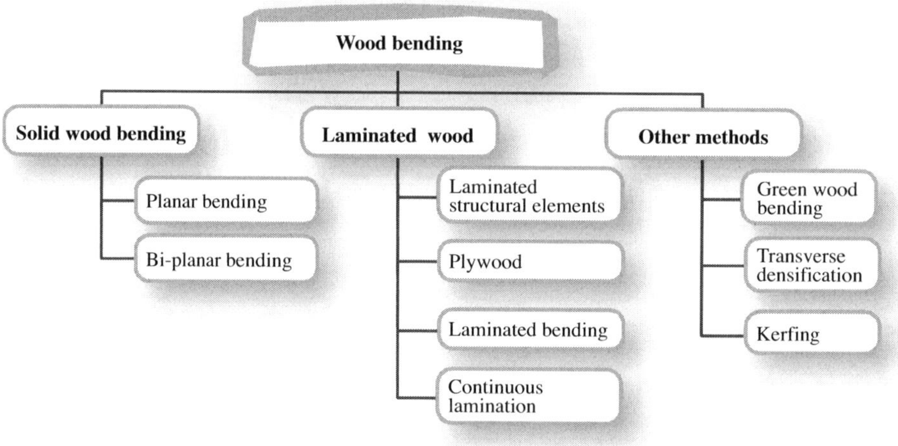

Fig. 10.2 Classification of different methods of wood-bending.

Plane or shaped members of laminated wood can be manufactured from boards, sheets or strips of veneer that are glued together to restrict moisture movement or to prevent spring-back. There are several methods for producing laminated wood from dried wood, and they lead to in four product categories as shown in Figure 10.2. The methods are: lamination of veneers or boards for structural purposes, plane and crosswise lamination of veneers (plywood), forming and lamination of veneers essentially for furniture and interior purposes (laminated bending), and manufacture of continuous laminated shapes.

The terms used to describe products and production processes of laminated wood are often vague and interchangeable and, referring alternately to materials and methods of manufacture. In this chapter, we use the above definitions related to laminated wood.

Laminated wood can be an assembly of layers of wood laid either with the fibers in the same direction and glued together, see Figure 10.3(a), or with the fiber direction of at least some of the plies, not necessarily every other veneer, perpendicular to that of the veneers above and below, Figure 10.3(b), the veneers being glued together under pressure. For a desired structural effect, the fiber direction of some veneers can be oriented in other directions than 0 or 90 degrees.

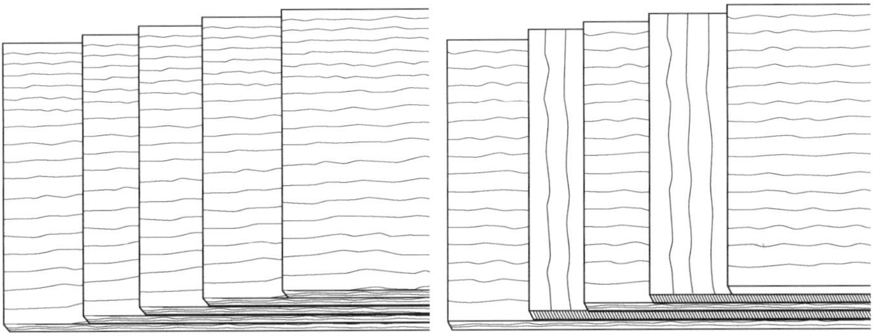

Fig. 10.3 Fiber orientations of individual layers in laminated wood: (a) same direction of grain and (b) crosswise lamination of plies.

In laminated structural elements, e.g., glulam, laminated timber or laminated veneer lumber (LVL), the layers consist of boards or veneer and the products are used not only as structural members in architecture but also in furniture design such as for legs or frames. Other types of laminated structural elements are presented in Chapter 11.

Curved laminated members for structural purposes can be produced by bending the lamella before gluing. Straight laminated members can also be steamed after they are glued by operations similar to those used for solid wood (McKean *et al.*, 1952). However, this type of procedure requires an adhesive that will not be affected by the steaming treatment and it complicates the conditioning of the finished product. It is more common to bend and glue the boards or veneers simultaneously. The cross-section dimensions of a glulam beam are in principle unlimited, but for practical reasons the maximum height of a beam is about 2 meters and this makes glulam an ideal material for large spans. Glulam is also very competitive in modern construction. Laminated boards are stronger than solid wood, and it is also possible to produce beams that are longer than the lengths of the individual boards.

Plywood is a panel consisting of an assembly of veneers bonded together with the directions of the grain in alternate veneers, usually at right angles. There are an odd number of veneers, as symmetry makes the panel less prone to distortion. Flat plywood can be shaped, but once the veneers have been glued to each other no further movement relative to each another can take place without a breakdown of the glue bond and destruction of the plywood as such. If a curved shape is formed from this sort of material, it is clear that the difference in length between the convex and concave faces of the bend cannot be brought about by the veneers sliding over one another, but must result from tensile and compressive strains induced in the material by the bending operation. Flat plywood is often bent by methods that are somewhat similar to those used in bending solid wood. The curvature that can be introduced into a flat piece of plywood depends on numerous variables, such as moisture content, fiber direction, thickness and the number of veneers, species and quality of the veneer, as well as the technique applied in producing the bend.

Laminated bending is produced by forming and gluing the veneers simultaneously without end pressure against a mold. This method is described in detail in section 10.5. The terms *bent* and *molded* are often used interchangeably in reference to laminated bending, but technically bent refers to a the bending of veneers in a single direction, whereas molded usually represents a bending of the veneer in two directions, creating not just an undulation but a surface that is three-dimensional. Figures 10.35 and 10.39 show a molded and a bent laminated bending, respectively.

Continuous lamination processes are used for making continuous wooden forms such as seat rings, hoops, and steering wheels (Stevens & Turner, 1970). Continuous bends are made by splicing end-to-end lengths of lamellae (veneer) to form a continuous strip, which are then glued and wound around the form under pressure. Stevens and Turner (1970) described various techniques that are available for making continuous laminated shapes, such as split female form, revolving form and tension strap, and the continuous strip method. In Chapter 11, a method is described for producing cylinders by gluing strips of veneers together to a length up to 12 meters. The product is called cylindrical LUL.

The class of "other methods", includes bending that cannot be related to solid wood bending or laminated wood, but are interesting because no chemical or thermo-hydro mechanical modification of the wood is required before bending. The methods that are briefly described here are bending wood in the green state, wood densification in the transverse direction and kerfing.

It is difficult to bend unmodified wood to a small curvature without fracture occuring in the wood material, it is common knowledge that a very thin strip of wood can easily be bent by hand to quite a small curvature. Green wood has traditionally been used to form curved members for wooden baskets among other things. Thin strips of wood are bent by hand and held in place by weaving them together or attaching them to other parts. Such bending is achieved without treating the wood. However, the stability of the final product may not be as permanent as that of products made by treatments in which plasticizing methods are used.

If a panel of solid wood is compressed in the transverse direction, it will be more easily bent across the fibers as opposed the uncompressed panel. This is especially noticeable if the compressed panel has vertical annual rings and if the compression occurs in the radial direction, as shown in Figure 10.4. In the same way, compression in the axial direction of wood gives a material that is easier to bend along the fibers. The greater flexibility of the compressed wood is a result of a buckling of the fiber cell wall during compression.

To increase the flexibility of dried wood, kerfing can be adopted, i.e., grooves are sawn in the wood so that the wood member can be bent more easily. Figure 10.5 shows a method of kerfing or laminated inserts invented by Alvar Aalto at the end of the 1920s and patented in 1933, used for bending stool legs. The technique known

Fig. 10.4 Compressed wood is more flexible than its uncompressed counterpart. The photograph shows two samples of wood of the same dimensions under the same load. The deflexion of the sample of wood with vertical annual rings and compressed in the radial direction (left) is greater than that of the uncompressed wood (right).

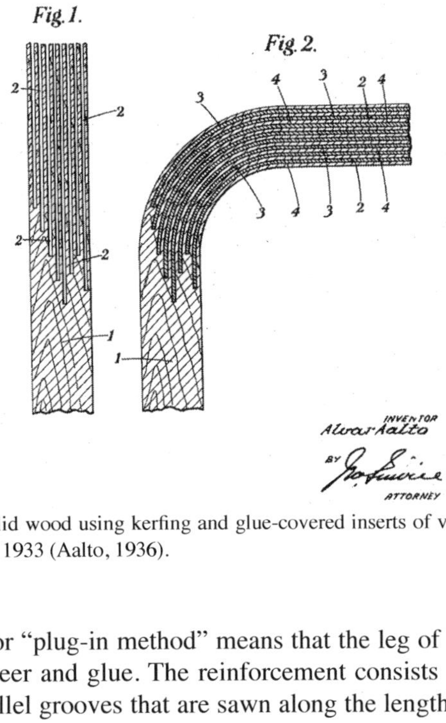

Fig. 10.5 Bending of solid wood using kerfing and glue-covered inserts of veneer. Picture from a patent granted to Alvar Aalto in 1933 (Aalto, 1936).

as the "bent knee" or "plug-in method" means that the leg of solid wood is partially reinforced with veneer and glue. The reinforcement consists of thin veneers placed in a number of parallel grooves that are sawn along the length of the part to be bent, spaced so that the tongues left between the slots are not likely to fracture in the bending operation. In order to avoid excessive weakening of the sections, the slots should be of varying length so that the ends of the inserts are not along a cross-section. This kerfing method of producing bent materials is often adopted when the finished wooden member is mostly straight, but a curved portion at one or both ends.

10.2 PRINCIPLES OF PLASTICIZATION

Wood is a material that one can render plastic or semi-plastic, which means that wood that can be soften and bent so that it keeps its curved shape after it cools. Wood contains areas of a crystalline molecular structure that is hard to plasticize and is therefore also called semi-plastic. The word "plastic" comes from the Greek word *plastikos* meaning "to form".

In the bending of thick pieces of solid wood, however, plasticization of the wood is essential and is an extremely important part of the process. The purpose of all plasticization treatments is to soften the wood sufficiently so that it can withstand, without fracture, the compressive deformation necessary to make the curve. Hot wood is more plastic than cold wood, and wet wood is more plastic than dry wood. A combination of heat and moisture (a TH method) is therefore an effective way of softening wood, but other techniques based on chemicals can also be used with excellent results, cf. Figure 10.6. In Chapter 5, the theory of wood plasticisation was described in more detail and here plasticization is presented in the context of practical wood bending.

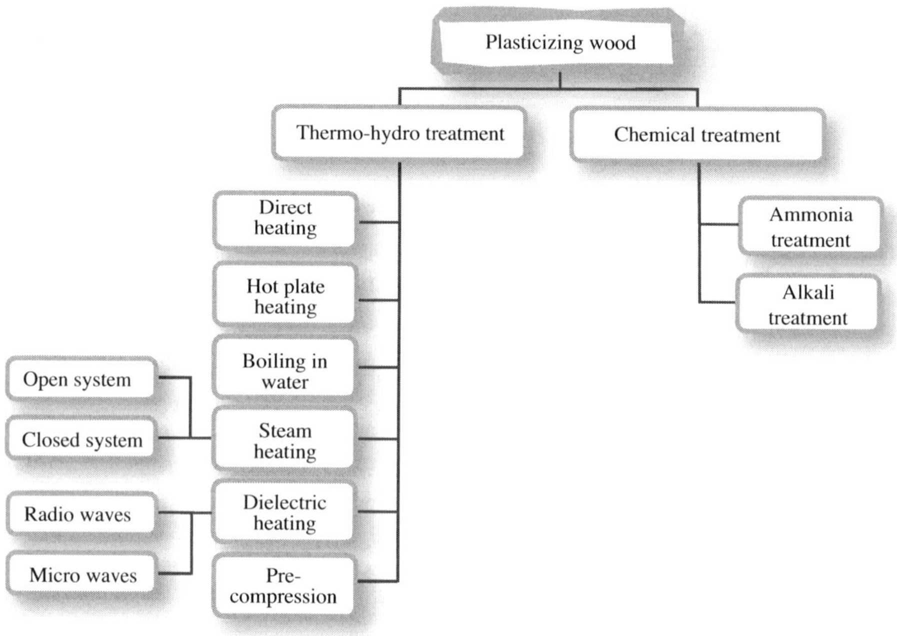

Fig. 10.6 Classification of methods for making solid wood more flexible for bending.

10.2.1 Thermo-hydro softening of wood

Softening by steaming is the most usual means of making wood plastic for bending, but any treatment that heats the wood to a temperature that makes the wood components semi-plastic without detriment to its structure and without causing it to dry appreciably can be used.

Softening by direct heating is when wood is softened by heating over an open fire, in a naked gas flame, or in hot and wet sand. These are traditional softening methods that are seldom used in industrial applications. There is nevertheless an example of an industrial application in the cooperage industry at one time, where the staves for casks were softened by placing the partly shaped piece over a fire of wood shavings, yet keeping the inside wet by sponging it with water (see Chapter 2).

Softening wood by boiling it in water is a simple method, but a special problem arises when extractive substances become dissolved in the water and can subsequently discolor the wood (Stevens & Turner, 1948, 1970).

In practice, plasticization is generally achieved with saturated steam at a temperature close to 100 °C. Dielectric heating technologies, often referred to as radio frequency heating or micro-wave heating, are nowadays used to prepare wood for bending. Examples of such methods are described in section 10.4.3.

Wood lacks plasticity, which means that it cannot be softened sufficiently for molding and cannot be, melted or dissolved (Hon & Shiraishi, 1991). However, in the case of wood welding, we have seen that the thermal alteration of the wood cell leads to the formation of a viscous layer or "molten wood" and this zone provides adhesion between the pieces after cooling. Chow & Pickles (1971) have studied thermal softening of wood and found that the glass transition temperature (T_g) for oven-dry wood occurs in the vicinity of 180 to 200 °C. An increase in moisture content decreases the glass transition temperature.

The thermo-hydro plasticization of wet solid wood is governed mainly by the lignin response. In the wet state, lignin softens independently of the crystalline cellulose. Olsson and Salmén (1993) determined the glass transition temperature (T_g) of wet wood and found that only lignin exceeded its T_g in the range of 10 °C to 100 °C. The main components of the wood cell wall, i.e. lignin, hemicelluloses and cellulose, are together responsible for the fundamental properties of wood, such as its thermoplasticity. These properties are not a simple summation of the characteristics of the individual components but an integration of these properties as a result of mutual interactions between the components.

What actually happens in the wood during plasticization has been explained in more detail in recent years. Briefly, the various amorphous components of the wood (e.g., lignin, hemicelluloses and parts of the cellulose) each have their own glass transition temperature. The values of the T_g of the wood components found in the literature vary considerably, chiefly because of differences in the methods of isolating the components. Above the glass transition temperature, the modulus of elasticity of the material is much lower, i.e., the components soften upon heating. The glass transition temperatures of the wood constituents in completely dry conditions are 230 °C for cellulose, 205 °C for lignin and ca. 170 °C for hemicelluloses (Back & Salmén, 1982). This means that the T_g of the unmodified matrix is so high that some fiber decomposition occurs if elevated temperatures are maintained for a lengthy period. The T_g's of lignin, hemicelluloses and cellulose can be decreased by the addition of moisture or through the use of plasticizing chemicals.

The high hydrogen-bonding capacity of the carbohydrates makes the wood polymers highly sensitive to the moisture content. This is due to moisture functioning as a plasticizer, i.e., a low molecular weight component lowering the glass transition temperature (see Figure 5.3). In the case of hemicelluloses, the transition temperature is lowered to even below zero at high moisture contents approaching 100% RH, whereas in the case of lignin, the branched structure limits the transition temperature to about 70 °C under fully wet conditions (Salmén, 1984). In the disordered cellulose regions, the softening is not particularly pronounced and the transition is shifted only at high moisture contents, as shown in Figure 5.3. This is presumably due to the restriction of the crystalline regions. The influence of moisture on the T_g of the crystalline parts of the cellulose is small. The decrease in T_g of cellulose is only about 6-9 °C in the wet state (Hon & Shiraishi, 1991).

Above the matrix T_g, the lignin may undergo thermoplastic flow and, upon cooling, reset in the same or a modified configuration. This is the principal behind the bending of wood. Wood at a moisture content of 20-25% has a transition temperature close to 100 °C.

Heating of the wood to a temperature close to 100 °C with a moisture content of 25% is sufficient, but steam is used to avoid desiccation. The time must be sufficient to completely heat the material. A good rule-off-thumb is to allow two minutes per millimeter thickness of the wood. Prolonged steaming has not been shown to lead to any improvement, but for wood with a lower moisture content the time usually has to be increased. Steaming at a high pressure causes wood to become plastic in a shorter time, but wood treated with high-pressure steam does not generally bend as successfully as wood treated at atmospheric or low pressure (Peck, 1957).

10.2.2 Softening by chemical agents

Methods for plasticizing by chemical agents have been developed by several researchers, see for instance Schuerch (1963), Davidson (1969), Davidson and Baumgardt (1970), Bach (1974), but these methods are used to only a limited extent. Common chemicals that plasticize wood include water, urea, dimethylol urea, low molecular weight phenol-formaldehyde resin, dimethyl sulfoxide, and liquid ammonia. Urea and dimethylol urea have received limited commercial attention, and a bending process using liquid ammonia has been patented, cf. Schuerch (1966). Plasticization with urea was studied by Thunell and Hongslo (1948) among others.

The most promising method so far is plasticization with anhydrous (i.e., without water) ammonia, which gives a very rapid and effective plasticization. Ammonia can be used in three forms: liquid (anhydrous), at a temperature of –33 °C (needs a refrigerating installation), in solution with a very high concentration of 25% at a temperature of approximately of 20 °C, and gas, whose performance depends on the pressure.

Liquid ammonia has solvent properties that suggest that it should penetrate the lignin fraction of wood. It is also known to swell cellulose and to penetrate its crystal lattice. The most remarkable feature of this technique is the fact that wood can be distorted into complex shapes without clamping. For exact control of the shape, restraint is needed for a short period, but the wood shows no tendency, once significantly bent, to spring back to its original shape. The method can probably become of significance for highly qualified objects and it may be particularly applicable for wood species that are difficult to treat in steam and are thus unsuitable for use with present-day technologies. When using liquid ammonia, the bending process is somewhat different from the bending procedure for softening wood with for instance heat and moisture, in that of the wood is clamped during bending and drying is not commonly necessary. The different process stages during bending with the liquid ammonia process are the following:

- cutting and sizing the article to be worked;
- impregnating the article with liquid ammonia;
- bending the article to the desired shape or embossing a pattern in the surface of the wood; and
- evaporating the ammonia from the shaped or embossed article. The article must be fixed into its final shape during the evaporation period.

The following explanation has been given for the mechanism of the liquid ammonia process. Anhydrous ammonia is a strong hydrogen-bonding, low molecular weight solvent. Ammonia is able to penetrate not only into the amorphous regions of the cell walls but also into the crystal structure of the cellulose and into the phenolic lignin bonding substance (Barry *et al.*, 1936). Ammonia breaks up some of the hydrogen bonds mainly in the lignin, and the resultant softening of the lignin structure enables the chain structures to slide past each other if an external force is applied. After evaporation of the ammonia, hydrogen linkages are once more formed between the polymer chains, and the wood structure becomes recross-linked to its former rigidity.

There is no tendency for the formed article to return to its original shape, but the ammonia treatment process is accompanied by some transverse and longitudinal shrinkage. Add to this changes in the moisture and the article may regain characteristics of wood. Qualitatively, though, the mechanical properties of the treated wood are similar to those of the original untreated wood and the changes in such properties are minimized by making the effective treatment time as short as possible.

A great drawback of the ammonia method is the low boiling point of the impregnating fluid, which makes the use of refrigerating and liquefying equipment essential. Moreover, ammonia is poisonous and thus requires special precautions for this reason.

The exact temperature of the ammonia employed is not critical, but the wood becomes somewhat brittle if working is attempted at too low a temperature (Schuerch, 1966). The impregnation is preferably carried out under conditions of temperature and pressure such that the ammonia is kept in the liquid state. The forming operations may be performed under sufficient pressure to maintain the ammonia in the liquid state at the temperature of the operation but this is not necessary. However, the forming operation should be completed before all of the ammonia escapes from the impregnated wood and the wood resumes its rigidity.

Sufficient time must be allowed for the article being worked to become thoroughly impregnated with ammonia. No precise time can be specified for the impregnation step, as it will naturally be dependent on the nature and thickness of the wood used. After forming, the ammonia is evaporated from the body of the wood, either by heating or by releasing the pressure or both. The ammonia, with very little loss, can be readily recovered for re-use in the process.

Treating wood with liquid ammonia is significantly different from treating wood with gaseous ammonia at atmospheric pressure or with aqueous ammonia (Schuerch *et al.*, 1966). When aqueous ammonia is employed as opposed to liquid ammonia, the maximum amount of bending that can be achieved is much less and the bent wood has to be clamped in its new shape during drying. The aqueous ammonia process results in little or no change in the permanent set or fixing of the bent shape, i.e., the bent wood will regain its original form if it is remoistened. Gaseous ammonia at atmospheric pressure does not penetrate into the wood substance to any great extent, especially if the wood is dry. Schuerch (1966) exposed various strips of wood to a stream of gaseous ammonia for one to five hours, and he noted that the amount of plasticity induced in the wood was substantially greater for shorter rather than longer treatment periods. An explanation for this seemingly reverse effect is that the wood contains moisture at the beginning of the treatment. The small amount of aqueous ammonia thus produced partially plasticizes the wood, with results, at the very best, comparable to those obtained when treating the wood with aqueous ammonia. Compared to treatment in liquid ammonia, the use of aqueous or gaseous ammonia at atmospheric pressure will result in far less plasticity, a greater tendency for the wood to spring back to its original shape, and a less permanent set.

Wood treated with anhydrous ammonia followed by drying is more flexible and denser than dried untreated wood (Pentoney, 1966). The tensile strength is increased but the compressive strength is decreased in proportion to the increase in density. The treated wood is more hygroscopic and swells more than untreated wood.

Davidson (1969) suggested the following advantages of the ammonia process in comparison with steam forming:

- Wood may be formed into somewhat sharper curves with ammonia than with steam.
- The breakage rate during forming should be somewhat reduced.
- Forming with ammonia results in a new permanent shape to which the wood will return even after soaking in water.
- The force required to produce various curves is reduced.
- In some instances, the time required for the bent piece to maintain the form is materially reduced.
- The use of the ammonia process may permit some forming of wood species that are presently considered to be poor bending woods.

A method using hydrazine (N_2H_4) for softening wood has been patented by Huttunen (1973). Hydrazine is a colorless liquid with an ammonia-like odor and is derived from the same industrial chemistry processes used for the manufacture ammonia. However, hydrazine has physical properties that are closer to those of water. It is highly toxic and dangerously unstable, and is usually handled in solution for safety reasons.

The wood is impregnated with a 3-15% aqueous solution of hydrazine and is thereafter shaped in the usual manner by heat and pressure, after which the excess hydrazine is removed. The patent claims that wood treated by this method may be used successfully in for example furniture, skis, packing cases and the boat industry and on the whole in any application where the bending of wood is required. It may also be employed when the natural hardness or strength of wood does not suffice.

10.3 THEORY OF SOLID WOOD BENDING

The reason for the difficulties in bending solid wood is the low extension of wood in tension failure, see for example Prodehl (1931a). Wood in its natural state exhibits elastic properties over a limited stress range, Figure 10.7. When the stresses are removed from the wood within this limited elastic range the wood will return to its original shape. If the deformation in tension exceeds the limit stress in the longitudinal direction of wood, the wood will remain bent. If the deformation strain exceeds the strength of the wood, it will break.

However, when wood is plasticized it becomes plastic. Its compressibility in the longitudinal direction is then greatly increased, as much as 30-40%, although its ability to elongate under tension is not appreciably affected, Figure 10.7. After plasticization, a combination of bending and compression in the longitudinal direction of the wood can be used to limit its extension in tension and it then becomes possible to bend the wood through a relatively sharp curvature. In practice, this means that the manufacturer has to control the length of the pieces during bending. Some type of end stops (strap-and-stop) must be used on the tensile side of the pieces being bent to prevent them from elongating by more than 1-2%.

The theory of solid wood bending is relatively complicated, for the following reasons:

- Wood is an elasto-viscoplastic material with no distinct yield point.
- The yield points in tension and in compression are numerically very different, as shown in Figure 10.7, and the difference becomes greater with increasing moisture content as well as with increasing temperature.
- The strain to failure is much greater in compression than in tension.

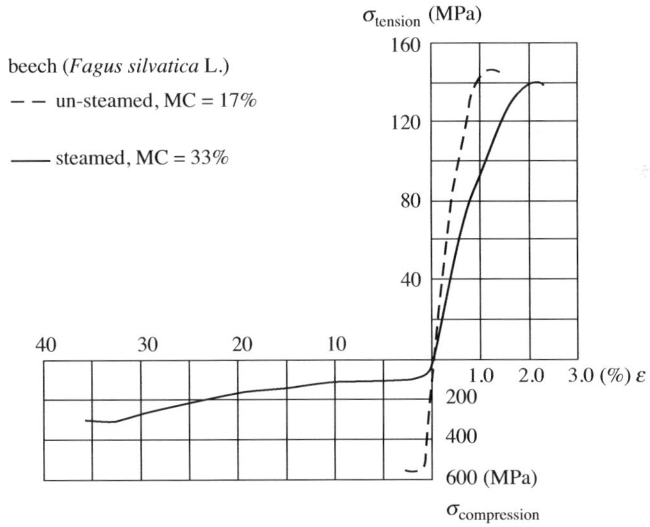

Fig. 10.7 Stress-strain (σ-ε) diagrams for air-dry non-plasticized beech and plasticized beech. Note that the stress and strain axes have different scales to the right and to the left of the origin (after Prodehl, 1931a,b). MC: moisture content.

10.3.1 The macro-mechanics of wood bending

The micro-mechanics of wood bending were discussed in detail in Chapter 6. Let us unstead consider what happens with respect to the deformation and stress if a bending moment is applied to a wooden beam, as illustrated in Figure 10.8. As long as the tensile and compressive stresses in the bent beam are below the yield point, the neutral plane (or elastic line) lies in the middle of the beam. If the bending increases, however, the yield point will be exceeded on the compressed side and the neutral plane will be displaced towards the side under tension. A stress-bending situation is then reached as shown in Figure 10.8a, where the wood begins to flow fairly rapidly while the stress increases on the side under tension. Since the strain to failure is greater on the compressed side than on the side under tension, the material fractures on the latter side long before the material plasticity has been fully utilized on the compressed side, see Figure 10.7.

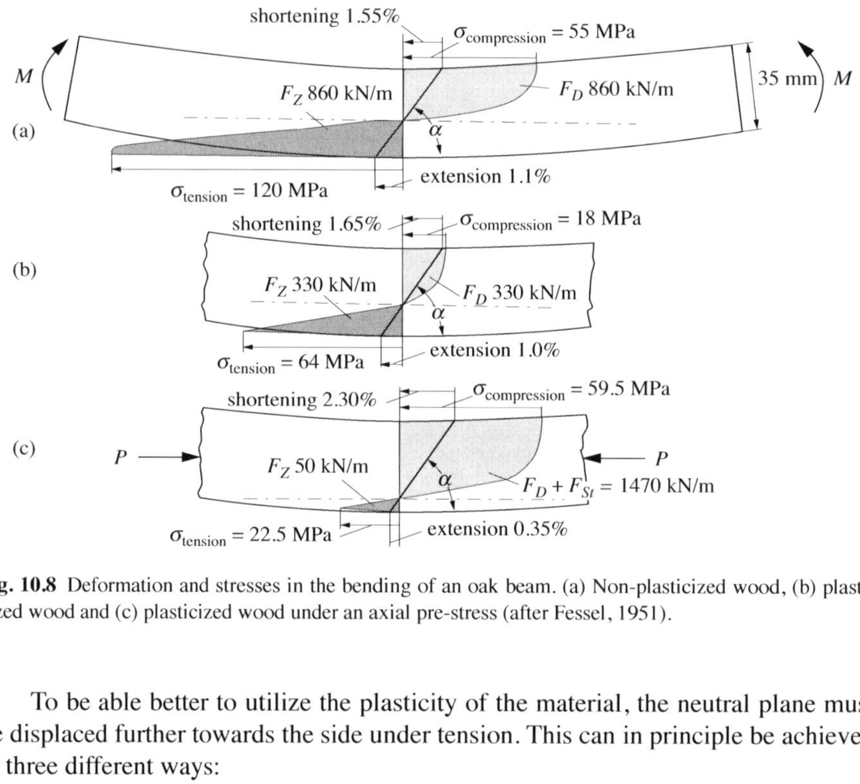

Fig. 10.8 Deformation and stresses in the bending of an oak beam. (a) Non-plasticized wood, (b) plasticized wood and (c) plasticized wood under an axial pre-stress (after Fessel, 1951).

To be able better to utilize the plasticity of the material, the neutral plane must be displaced further towards the side under tension. This can in principle be achieved in three different ways:

1. By changing the cross section of the beam so that a greater amount of material is on the side under tension.
2. By plasticization of the wood, particularly on the compressed side, by treating the wood with for example steam. When wood is plasticized in this way, the compressive strain-to-failure increases drastically but rise in the tensile strain-to-failure is very moderate, see Figure 10.7. In the bending of a plasticized wood beam, the stress-strain diagram resembles that shown in Figure 10.8b.
3. By subjecting the wood to an axial pre-stress. It has been shown, both theoretically and in practice, that, even when the wood is plasticized, fracture

> occurs on the side under tension before the plasticity of the material on the compressed side has been fully utilized. However, this can be counteracted by an axial pre-stress, achieved by placing a flexible band of metal (a strap-and-stop) along the wood piece on the convex side under tension and anchoring the wood with the band at the ends, as shown in Figure 10.9. By varying the degree of pre-stressing, the neutral plane can in theory be placed in any desired position.

The stress-strain diagram for an axially compressive pre-stressed wooden beam is in principle that shown in Figure 10.8c. In this case, the strain can be fully utilized both in the side under tension and in the side under compression.

One important parameter in wood bending is the smallest radius to which a piece of wood can be bent. This limiting radius is reached when the wood is pre-stressed to such an extent that the strain-to-failure is attained simultaneously both on the side under tension and on the side under compression.

When a piece of wood is bent to the smallest radius, as in Figure 10.9, the length of the side under tension is $L\,(1 + \varepsilon_+)$ and the length of the side under compression is $L\,(1 + \varepsilon_-)$, where L is the original length of the wood piece between A and B, ε_+ is the tensile strain-to-failure, and ε_- is the compressive strain-to-failure. Note that ε_- has a negative value.

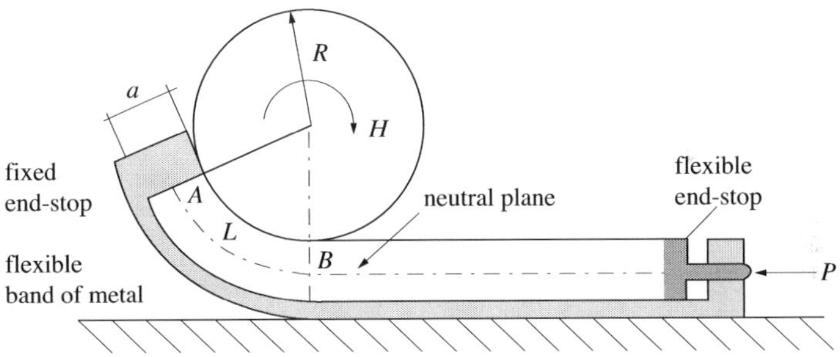

Fig. 10.9 Sketch of wood bent around a circular drum. L is the original length of the wood piece between A and B, a is the thickness of the wood piece and R is the radius of the drum, i.e., the inner radius of the bent piece (after Prodehl, 1931a, b).

It is evident in Figure 10.9 that

$$\frac{L(1+\varepsilon_+)}{L(1+\varepsilon_-)} = \frac{R+a}{R} \tag{10.1}$$

Which means that

$$\frac{a}{R} = \frac{\varepsilon_+ - \varepsilon_-}{1+\varepsilon_-} \tag{10.2}$$

where a is the thickness of the wood, and R is the internal radius of curvature.

This relationship applies for a pre-stressed beam when the plasticity of the material is fully utilized both on the side under tension and on the side under compression, by displacing the neutral axis towards the side under tension. It is also evident that, for a given species, the limiting radius of curvature is directly proportional to the thickness of the board.

In the bending of a non-pre-stressed beam, the tensile strain-to-failure ε_+ is the limiting factor. If, in this case, it is assumed that $\varepsilon_+ \approx \varepsilon_-$ and ε_- is considered negligible in the denominator in equation 10.2, the following approximate relationship is obtained for the bending of a non-pre-stressed wooden beam:

$$\frac{a}{R} = 2\varepsilon_+ \tag{10.3}$$

If the values for ε_+ and ε_- for beech (taken from Figure 10.7) are inserted into equation 10.2, we obtain

$$\frac{a}{R} = \frac{0.012 + 0.36}{1 - 0.36} \quad \Updownarrow \quad \frac{a}{R} = 0.58 \tag{10.4}$$

This means that, for a 25-mm thick board of beech, the smallest bending radius is 43 mm. This value can be compared with the smallest bending radii for beech measured by Stevens and Turner (1970) of 43 mm, Table 10.1. In bending of a non-pre-stressed beam, insertion of the values for the tensile strain-at-failure for plasticized wood from Figure 10.7 in equation 10.3 leads to the following relationship:

$$\frac{a}{R} = 2 \times 0.023 \tag{10.5}$$

which gives a smallest bending radius of 543 mm for a 25-mm thick board.

For non-plasticized beech, the corresponding value is obtained by inserting the tensile strain-at-failure for unplasticized wood in equation 10.3, which leads to

$$\frac{a}{R} = 2 \times 0.012 \tag{10.6}$$

and this gives a smallest bending radius of 1042 mm for a 25 mm thick board.

In the bending of plasticized beech without any pre-stress, Stevens and Turner (1970) found a smallest bending radius of 370 mm, as shown in Table 10.1. This analysis is of course a simplification. Amongst other things, the model presupposes that the cross-section of the wood piece remains unchanged during bending, which is not the case. Nevertheless, the equations give some guidance into the properties of various species of wood in bending. It should, however, be pointed out that in many situations, it can be just as easy to find published values of the smallest bending radius for plasticized wood as it is to find values for ε_+ and ε_-.

A very general rule-of-thumb for the estimation of the critical bending radius of a piece of wood is the following: untreated wood can be bent to a radius equal to 50 times the wood thickness, plasticized wood can be bent to a radius equal to 30 times the thickness, and plasticized wood where the strain on the tensile side is limited by a strap-and-stop can be bent to a radius equal to the thickness of the piece to be bent. In reality, of course, the species has a significant influence on the curvature radius.

If both the moment and the pre-stress on the wood, as shown in Figure 10.8, are released immediately after the bending, the wood piece springs back strongly. To be able to retain the bent shape given to the wood piece, it must be allowed to stay in the clamped position while the plasticization ceases and the material is dried and conditioned.

Table 10.1 The smallest bending radius for different species of wood in the bending of a 25-mm thick board with and without support by a strap-and-stop. The bending has been carried out with fully plasticized wood steamed at atmospheric pressure. The smallest bending radius is that at which breakage occurs in no more than 5% of the wood samples (Stevens & Turner, 1970).

Species	Latin name	Smallest bending radius (mm)	
		Supported	Unsupported
Alder	*Alnus glutinosa*	360	460
Ash, (American)	*Fraxinus sp.*	110	330
Ash (European)	*Fraxinus excelsior*	64	300
Beech (Danish)	*Fagus sylvatica*	43	370
Birch (Canadian yellow birch)	*Betula alleghaniensis*	76	430
Douglas fir	*Pseudotsuga menziesii*	460	840
Elm	*Ulmus hollandica*	13	240
Elm	*Ulmus glabra*	43	320
Hickory	*Carya sp.*	46	380
Iroko	*Chlorophora excelsa*	380	460
Jarrah	*Eucalyptus marginata*	440	990
Larch	*Larix decidua*	330	460
Lime	*Tilia vulgaris*	360	410
Mahogany (African)	*Khaya grandifoliola*	840	910
Oak (American white oak)	*Quercus spp.*	13	330
Oak (European)	*Quercus robur*	51	330
Pine (Carribean pitch)	*Pinus caribaea*	360	710
Robinia	*Robinia pseudoacacia*	38	280
Spruce	*Picea abies*	760(1)	740
Sycamore	*Acer pseudoplatanus*	38	370
Teak (Burma)	*Tectona grandis*	460	890
Walnut (European)	*Juglans regia*	25	280

(1) Data from Kollmann and Coté (1968).

Throughout this process, a certain relaxation takes place in the material, and the stresses are reduced. The neutral plane of bending is again displaced towards the middle, and the spring back should then be small. In addition, strongly compressed wood has been created primarily on the concave side of the bend. Compressed wood has a higher density and a greater movement, i.e., shrinkage and swelling, when the moisture content changes. This also applies to axial shrinkage. When the wood is dried, shrinkage on its concave side strives to further increase the bending. This means that if the bending radius is small in relation to the board thickness, there is no spring-back when the board is taken out of the mold. Instead there are sometimes problems due to the radius tending to decrease further. After withdrawal from the mold, the wood strives to bend more up on drying but to straighten out when moistened.

The curvature of a bent member is not permanent after it has been dried and fixed, because changes in moisture content set up stresses that change the curvature (Peck, 1957). The various zones through the thickness of the piece differ in the extent of deformation and in the amount of longitudinal shrinkage and swelling they will undergo with changing moisture content. In addition, shrinking and swelling across the thickness tend to alter the curvature of the piece. The wrinkled and folded fibers on the concave side, which have been considerably compressed in bending, shrink and swell appreciably in the lengthwise direction, whereas the convex side undergoes negligible lengthwise shrinking and swelling.

There are however large variations from board to board, although no great internal stresses seem to remain in the wood after the bending. In practice, material can be cut away fairly freely without the bending radius changing to any great extent. Another important observation is that the compression of the wood in the bend leads to a greater impact work, although the tensile and bending strengths are lower than in the original wood. This is illustrated in Figure 10.10 which shows how the bending

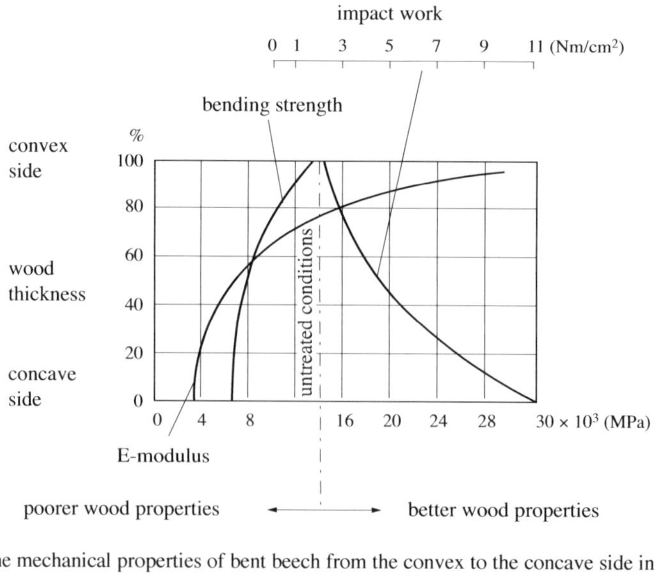

Fig. 10.10 The mechanical properties of bent beech from the convex to the concave side in relation to the properties of the unbent material (Teichgräber, 1953).

process changes the mechanical properties of a sample of beech. The reduction in strength brought about by bending is however seldom serious enough to affect the utility value of the bent member.

The results in Figure 10.10, for example display that the crushing strength is practically always lessened wheras in comparsion to untreated solid wood the modulus of elasticity and the shock resistance become higher. This is especially true if the thickness of the bent parts is smaller and the measurements approach the outer zones.

10.4 SOLID WOOD BENDING IN PRACTICE

For the industrial production of bent parts of solid wood, the hydro-thermal softening method is most common. Here, three methods for bending solid wood are described; the steam bending method, the dielectric field method, and the pre-compressed wood method. The raw-material input to these three processes is the same and the methods all use heat and moisture to soften the wood. In the pre-compressed method, in contrast to the other two methods, the wood can be bent in a cold state.

10.4.1 The selection and preparation of bending material

Ideally, straight-grained, knot-free wood materials should be used for bending work, particularly if the bend is to be of a small radius. Cracks or all other defects in the wood that lead to disturbances in the fiber structure, such as knots, ingrown bark, pitch pockets etc. will cause losses and failure in the wood when comparatively little strain has been applied. Knots should not be present in a wood material to be bent, although small knots can be tolerated to some extent on the compressed side, unless the bending radius is extremely small. Knots on the side under tension have a disastrous effect on the strength.

In selecting a species suitable for bending, several factors need to be taken into consideration, such as the availability of the species in question, its bending properties, the strength properties of the material after bending, and its appearance etc. It is clear that the risk that the wood will break during bending is strongly dependent on the ability of the wood to support a change in shape, i.e., its plasticity. The choice of wood species and the pre-treatment of the boards are therefore extremely important.

In general, hardwood is more suitable for plastic shaping than softwood, and ring-porous hardwood types are in general better than their diffuse-porous conterparts. The most important reason why softwood has poorer bending properties than hardwood is probably the great difference in mechanical properties between the earlywood and latewood in softwood. Hardwood with a high density often has superior bending properties than hardwood with a low density.

After the wood species has been selected, the next question is whether the growth conditions of the tree influence the bending properties and also from where in the tree stem the boards should be taken. Among the experts, there are many often differing opinions about the influence of factors such as the age of the tree, the annual ring width, the soil character etc. Stevens and Turner (1970) say that these factors certainly have an influence, but tests have shown that they are of secondary importance. Most important, as already mentioned, is to have straight-grained wood, free from defects.

An idea of the bending properties of a numbers of species may be obtained from an examination of the list given in Table 10.1. In practice, a few species of wood dominate and beech is the most common, followed by oak and birch. For special purposes, ash and elm are used. Ash, although undoubtedly an excellent bending material, is unusually susceptible to the presence of pin knots near the compression face, and these are very prone to buckling or compression failure. Moreover, if the grain runs out on the tension face, there is a tendency for it to lift when the strap is finally removed. Elm, on the other hand, has exceptionally good bending qualities and generally appears to be remarkably tolerant of defects, although, as in all woods, these naturally have an adverse effect on the bending properties.

The selection of the boards is a critical feature of the whole production chain. In the case of products for which the material is to be bent to a small radius, straight tree stems with a circular cross section must be chosen. A non-circular cross section indicates the presence of reaction wood (tension or compression wood), which cannot be used. Spiral grained trees and tree stems that taper strongly give unsuitable wood.

Within any particular tree stem, straight-grained parts with a long fiber length should be chosen. The juvenile wood closest to the pith is directly unsuitable. The best pieces are obtained at the external regions of the butt log, with the exception of the lowermost part which contains buttress or similar irregularities in the fibers. Wood exhibiting decay is completely useless for bending purposes.

It is often easy to bend a completely moist piece of wood. Less force is necessary than when dry wood is bent. There are, however, four reasons why wood with a moisture content above the fiber saturation point is not used for bending:

- There is a risk that the wood may burst due to the hydraulic liquid pressure in liquid-filled lumens.
- It is difficult to store moist wood (risk of fungus attack etc.).
- It is difficult to process (e.g., planing and sandning) very moist wood with a good result.
- Drying after the bending takes an unnecessarily long time.

The disadvantages of using air-dried wood material are that

- the occurrence of surface checks from seasoning may result in compression failures;
- the steaming of the dried wood may cause distortion of the pieces before bending; and
- the bending force is greater than that required to bend green wood.

As a compromise, wood with a moisture content of 20-25% is usually employed. At a moisture content below 20%, bending in two planes is almost impossible.

A smooth surface is very desirable on the wood pieces prepared for the softening treatment since ridges tend to induce unnecessary buckling. For this reason, planed and smooth-finished material is to be preferred to material straight from the saw. Experience has shown that slash-sawn material, cut and bent with the annual rings parallel to the face of the form, gives results slightly better than quarter-sawn material bent with the rings normal to the face of the form.

The bending must compensate for the shrinkage of the wood after bending. During bending to a small radius of curvature, it happens that the inner part of the edges of a piece swell and the outer part of the edges shrink and further that the inner and outer parts of the edges become convex and concave, respectively. This effect is minimized when the piece is oriented so that the annual rings are concave relative to the strap.

10.4.2 Industrial steam bending of solid wood

In steam bending, to obtain the required conditions of plasticization of the wood before bending, the wood is subjected to saturated steam at atmospheric pressure in a steam chest such as the one illustrated in Figure 10.11. The essentials of such steaming equipment are that sufficient steam is made available to maintain a temperature of 100 °C and that means are provided for readily introducing and removing the wood to be bent. The size of the steam chest should be adapted to the type and size of the production. As previously mentioned, there is no great advantage in using steam at a pressure higher than atmospheric pressure, as the bending properties of wood are not markedly increased at higher pressures.

According to Thunell (1943) steaming should be done with steam at temperatures between ca. 70 °C to 120 °C. Higher temperatures are not appropriate due to the wood becoming brittle. The duration of the steaming varies partly according to the wood's dimensions and species. Table 10.2 shows target times for 100-mm thickness or diameter of the wood piece. For 50 mm thickness, the time is about 70% and for 150 mm, approximately 200% of the specified times in the table.

Table 10.2 Times for steaming at 100 °C for different species of 50 mm thickness.

Species	Time (hour)
Birch, ash	5-6
Scots pine, Norway spruce	6-7
Beech	7-12
Oak	15-18

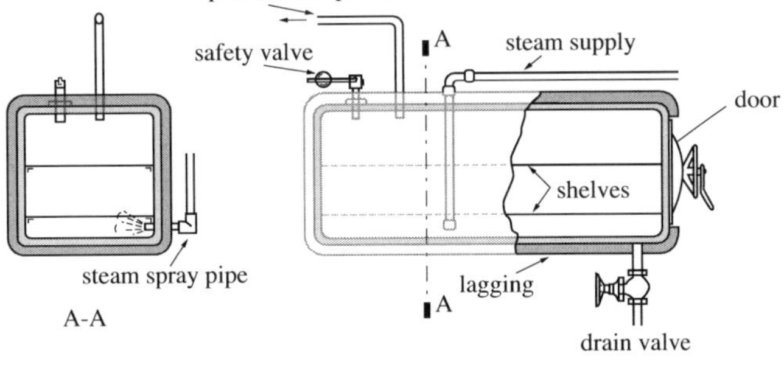

Fig. 10.11 Steam chest for softening the wood before bending.

Methods of shaping

The actual shaping of the steamed materials can be carried out in many different ways. If the dimensions are small and the number of pieces is limited, the shaping can be carried out with simple hand tools. However, in the case of thicker wood dimensions and long series, machines are employed.

Regardless of whether the bending is carried out by hand or with the help of a machine, the principles are the same. It should however always be borne in mind that, since the wood is still visco-elastic after steaming, the risk of fracture increases significantly with a higher rate of bending. It is therefore important to maintain a slow such rate also in the case of machine bending.

The technology, tools and specially adapted machines used for the bending of solid wood can be of various kinds. In general, it can be said that the methods and techniques employed for the relatively uncomplicated uni-planar bending, i.e., bending in a single plane, have undergone considerable development, whereas the more advanced bi-planar bending is still carried out with the same craftsman methods as in Thonét's time (see Chapter 2).

In the conventional bending of solid wood, four different techniques are used:

- bending without a restraining band (strap-and-stop);
- lateral bending with a restraining band in one plane;
- planar bending with a restraining band in one plane; and
- bi-planar bending with a restraining band.

In all these bending techniques, the wood is plasticized before the bending is started.

Bending without a restraining band, i.e. without the support of a metal strap with fixed end-stops, is the simplest method and is used during bending to a large radius (radii ca. 30 times the thickness of the material). The wood piece is usually fastened at one end and then bent over a mold by hand or with some assisting device, as shown in Figure 10.12. The wood piece is then fixed and dried to the desired final moisture content. For thin pieces, the shaping can also be carried out between male and female molds, Figure 10.13. Examples of materials that are bent using this technique are front

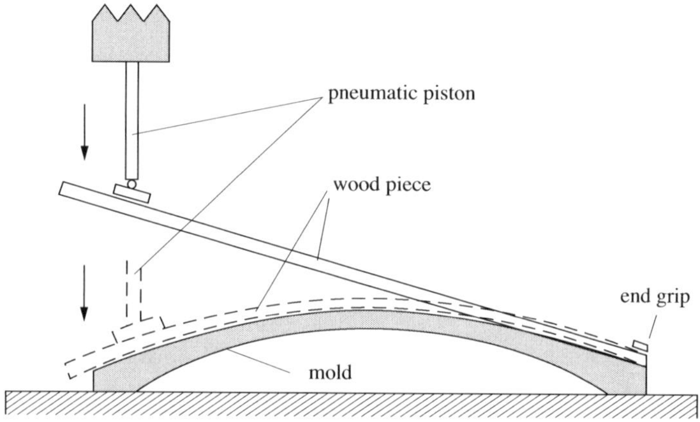

Fig. 10.12 Shaping over a convex mold without restraining bands.

legs, back legs and back rests for chairs, Figure 10.14. The ease of handling of the molds determines how many pieces can be bent in a single bending operation. In the case of turned back legs with a diameter of 35 mm, 10 legs can be bent per bending operation and the production capacity is then on the order of 150-200 legs per hour. The drying time from a moisture content of 25-30% to 7% is approximately two days.

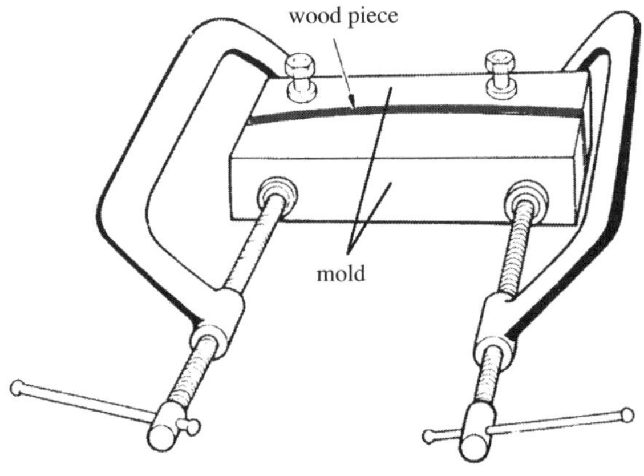

Fig. 10.13 Shaping between male and female molds (Stevens & Turner, 1970).

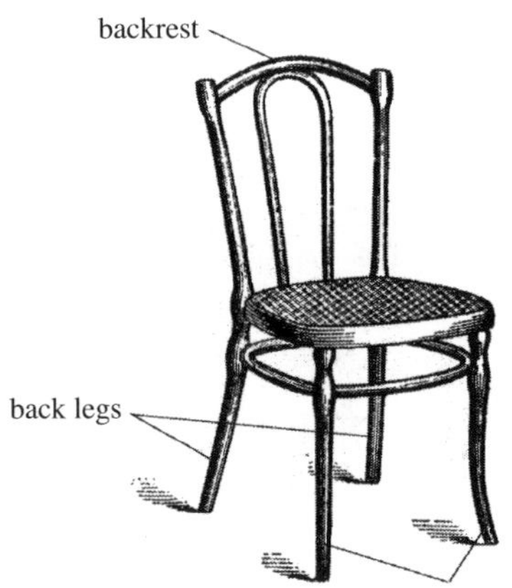

Fig. 10.14 Examples of items that can be bent without a restraining band.

Lateral bending with a restraining band in one plane is used in the bending of U-shaped pieces, and the principle of the equipment commonly used for this purpose is shown in Figure 10.15. In the bending procedure, the wood piece is fixed between two end stops providing a compressive force along the fibers during the actual bending. The bending is carried out around a template with the desired radius. If the material is thin and the bending radius is large, the bending can be carried out between

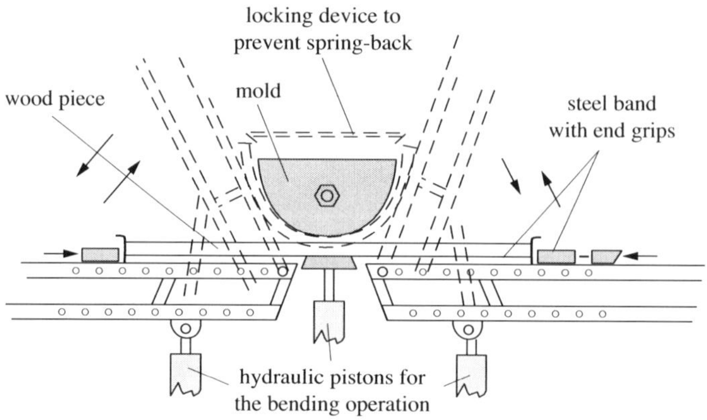

Fig. 10.15 Diagram showing the principle parts of a machine for lateral bending with a restraining band in one plane (lever arm bending machine).

male and female molds (see Figures 10.13 and 10.22). In the case of thick materials (i.e., a thickness of 40-70 mm), the material must remain in the mold for 1.5 to 2 minutes until it has "solidified". Thereafter, the material is loosened with the restraining band fixed to prevent spring-back. The capacity for thick materials is 2-30 items per hour, whereas for thin dimensions, 50-100 items can be made per hour. The drying time varies from 1 day to 15-18 days, depending on the dimensions of the wooden item. The method is used to produce for instance arms, back pieces, back rests, seat frames, arm supports etc. for chairs, Figure 10.16.

Fig. 10.16 U-shaped pieces can be suitably bent according to the lateral bending technique with a restraining band in one plane.

Plane bending with a restraining band in one plane is usually carried out in some type of planar bending machine, Figure 10.17. The method is used for planar O-, S- and U-shaped pieces. The mold is set up and fixed in adjustable holders on the bending table. The restraining band (strap) is stretched and fastened in a strap stretcher,

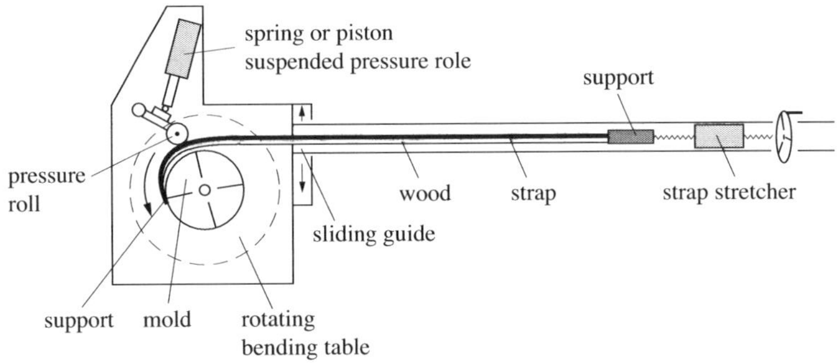

Fig. 10.17 Diagram of a planar bending machine(revolving table machine) for making O-shaped items.

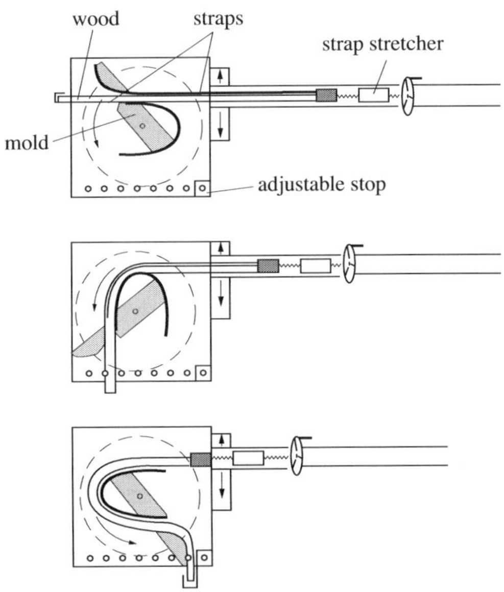

Fig. 10.18 Diagram of a planar bending machine for making S-shaped items.

after which the wood piece is fixed in the mold and the strap stretched with the strap stretcher (locking in the axial direction). The pressure roll is applied, pressing the wood and strap against the mold. The bending starts by rotating the bending table (in the direction of the arrow in Figure 10.17) and, when the material has been bent on the mold, it is fixed with a clamp. The item is then dried while still fixed in the mold. The capacity is 15-25 items per hour and the drying time varies from 1 to 7 days depending on the dimensions of the item. Kivimaa (1948) has investigated such a type of bending machine thoroughly for the bending of birch. A machine for S-shaped items is shown in Figure 10.18. The method is used to manufacture e.g. seat rings, under-rings, arms, S-shaped legs, arm supports etc., as illustrated in Figure 10.19.

Only a few decades ago, planar and lateral bending was carried out manually by two men, but this type of bending is now almost exclusively done in a machine.

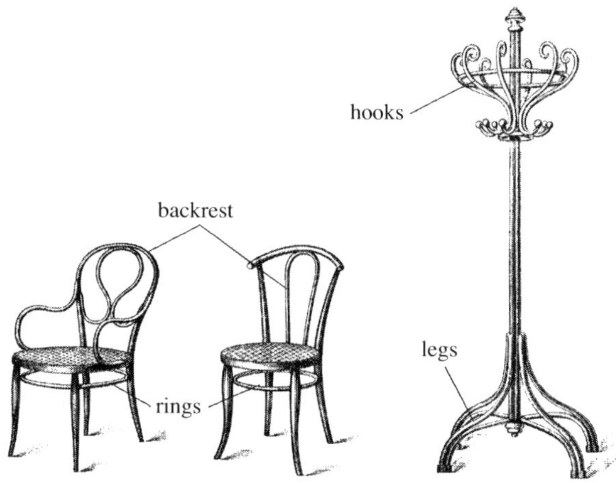

Fig. 10.19 Examples of items that are suitably bent according to the lateral bending technique with a restraining band in one plane.

Bi-planar bending with a restraining band is, however, in the great majority of cases, still carried out manually. There are relatively simple general rules for how much a piece of wood can be bent in uni-planar bending with or without a bending band (see 10.3.1). In bi-planar bending, it is more difficult to theoretically determine the smallest bending radius. Not only the thickness of the material but several interacting factors are also of decisive importance, such as the profile of the turned wood piece, the torque, the length of the torque, the length of the work piece, the distance between the transitions to different planes etc.

Figure 10.20 shows bending in two planes and in this bending operation the restraining bands used must be arranged so that the bent parts of the wood remain covered on the convex face, irrespective of the plane of bending. More than one restraining band with adjustable end stops at a time is, however, never used. The first part of a bend in two planes is made in the normal manner. As soon as this part is completed, it is clamped to the mold and the end-stops removed. It is important to keep these end-stops in position until the fixing clamps are really tight, to prevent the restraining band from slipping along the wood and causing tension failures on the bends. The two loose pieces of restraining band secured by metal fasteners at right angles to the main-restraining band, are next put in position along the wood and the released end-stops fitted to their ends. These restraining bands are pulled tight by manipulation of the adjusting screws, and the two ends of the wood can then be safely bent in a plane in right angles to the center part, as indicated in Figure 10.20.

It is almost impossible to describe the method and technology of bending in two planes. In general, it can be said that the greatest difficulty lies in constantly guiding the restraining band and wood piece to ensure that the axial pressure (locking) is maintained in order for the wood to be forced into compression on the concave side, thus preventing stretching to failure on the convex side. When the wood is taken out of the steaming equipment, it has a temperature close to 100 °C, but this temperature drops quickly. Consequently, the bending must be carried out as quickly as possible

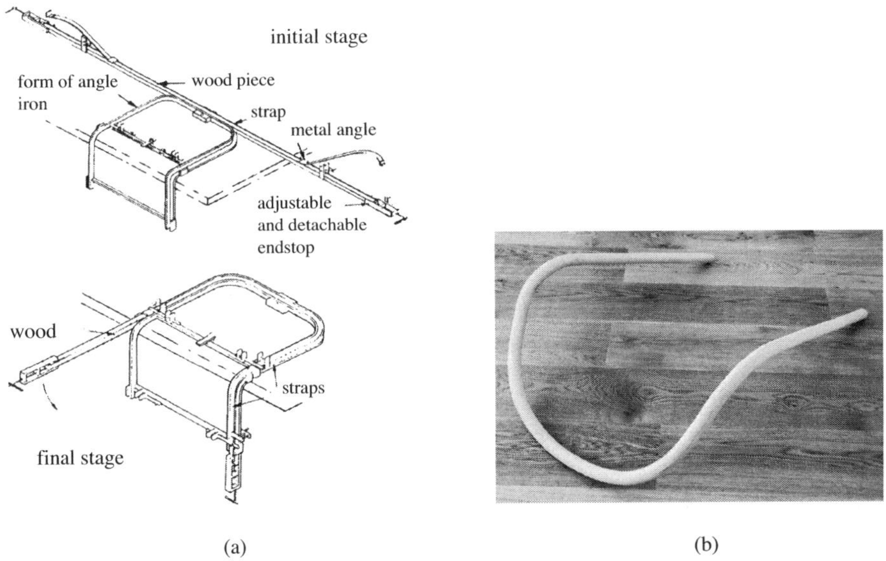

Fig. 10.20 Bending in two planes with two restraining bands (a) and an example of an item bent in two planes (b).

once the process has been started. At the same time, the wood must be given time to become compressed and consideration must be given to the fact that small radii require longer processing times than large radii. A balance must therefore be found between the work rate and the time necessary for the material to be shaped.

In the various bending methods in a single plane, the time for the operator to learn the skills is relatively short and good results can be attained already after a few months of training. Manual bending in several planes is however something of an art, and vocation is a condition for success. Accuracy, feeling and a knowledge of the material as well as of the interplay between "bender" and "restrainer" is necessary. It normally takes 2-3 years to master the most common bending procedures in several planes.

Finishing

After the shaping, the deformations achieved must be "frozen" before the wood piece is released. This is done through the final drying of the material, while it is clamped in its mold or in some other fixture, to a moisture content of 6-10%. The time for drying the curved wood products is long, which is a disadvantage in a productivity context. The drying takes 1 to 3 weeks depending on the species of wood, the drying temperature, the dimensions of the wood and how well the mold encloses the wood pieces and thereby hinders the drying. The drying time can be reduced considerably through the use of dielectric heating in the drying process. Research into the microwave drying of wood has been going on since the early 1960's (Antti, 1999). However, the drying should not be too intense, since there is a risk that the steam formed will rupture the cells.

After the drying, which takes place at 65-70 °C if a restraining band is used, the wood must be left to cool and condition for a few weeks while the bending radius is still under control, often with a clamp or the like. The drying must be carried out carefully to avoid drying-cracks and other damage. After the conditioning, the finishing process is undertaken. This consists of cutting to the desired shape, repairing, polishing, assembly and surface treatment.

It is desirable to finish the wood pieces to an as large an extent as possible before the bending operation. Among the reasons for this can be mentioned:

- Unevennesses in the surfaces giving rise to stress concentrations which can lead to fracture.
- It being easier to bend thin samples.
- It being cheaper to process a straight sample than a bent one.

In those cases where the final product is to have an asymmetric cross section or where the cross section is to vary significantly along the product, additional processing must however be carried out after the bending.

10.4.3 Using dielectric heating in wood bending

Wood at a high moisture content is capable of absorbing a large amount of electromagnetic energy. The amount of energy required to raise the temperature is determined by the specific heat capacity of the material. Low values require less electromagnetic energy to increase the temperature. The specific heat capacity of wood is influenced by its moisture content, its dry density and the temperature.

Electromagnetic radiation is a phenomenon that takes the form of self-propagating waves in a vacuum or in matter. It consists of electric and magnetic field components which oscillate in phase perpendicular to each other and perpendicular to the direction of energy propagation. Electromagnetic radiation is classified into several types according to its frequency. These types include radio waves, microwaves, terahertz radiation, infrared radiation, visible light, ultraviolet radiation, X-rays and gamma rays, Figure 10.21. Electromagnetic radiation carries energy and momentum that may be imparted to matter with which it interacts. This phenomenon is used to heat the wood material during plasticization and drying.

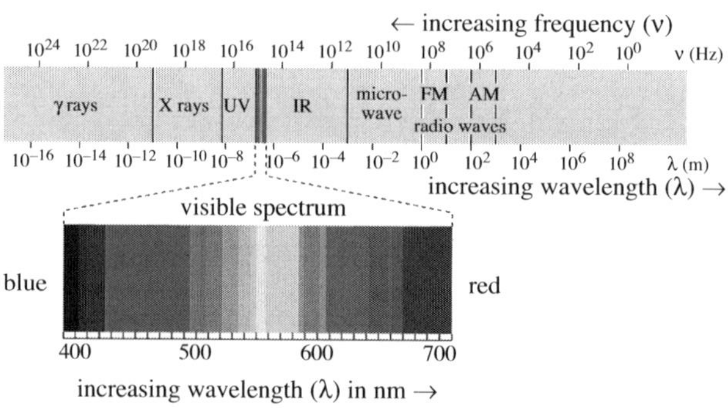

Fig. 10.21 Spectrum of electromagnetic radiation with the visible portion highlighted.

Dielectric heating, i.e., radio-wave heating (or high-frequency heating) and micro-wave heating technologies are the most common techniques used for wood. The literature indicates that there have been problems with uneven heating, but with today's knowledge of dielectric methods, it should be fully possible to achieve a uniform heating.

The plasticization of wood through dielectric heating has been studied for several years (see for instance Norimoto, 1979; Norimoto *et al.*, 1980; Mori *et al.*, 1984; Norimoto & Hasegawa, 1984; Norimoto & Gril, 1989; Eggert, 1995). Moreover, the theoretical approach has been well documented by Torgovnikov (1993). This chapter exemplifies the dielectric heating technique for solid wood bending by a production concept in which the bending and the processing system are integrated together.

Mold for bending

The mold is of the male and female type, made of plywood and sheeted with aluminium and plastic with a thickness of 2 mm, Figure 10.22. It is loaded with a maximum of 21 pieces of wood with dimensions $35 \times 52 \times 452$ mm^3 ($T \times W \times L$). The wood is placed in a metal sheet with metal end-stops (restraining-plate or strap-and-stop) to prevent the tensile strain in the wood during bending from exceeding 1%, Figure 10.22. To make it possible for the vapor in the wood to evaporate in the longitudinal direction of the wood, evacuation holes are drilled in the end stops, Figure 10.23. The concentration of vapor in the mold inevitably leads to flash-over in the electromagnetic field.

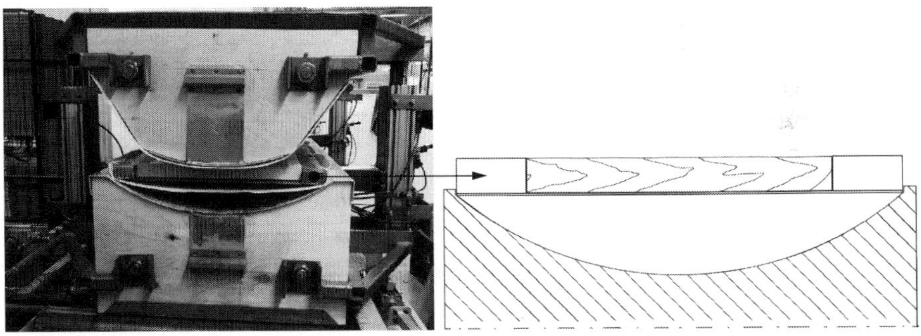

Fig. 10.22 Cross-section of the mold for solid wood bending (left). Wood in the restraining-plate before bending (right).

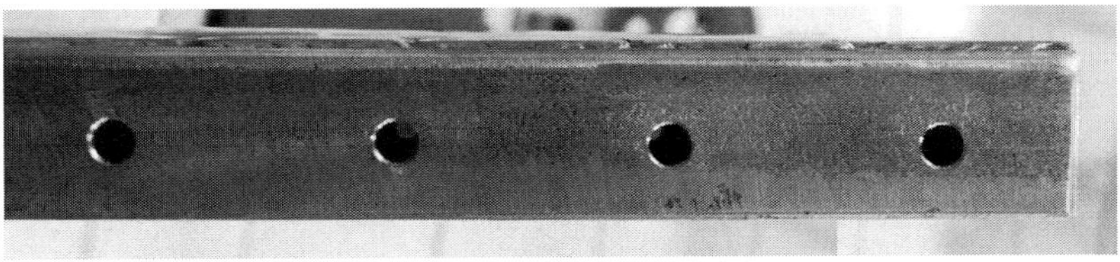

Fig. 10.23 Evacuation holes drilled in the end stops of the restraining-plate.

Press equipment and layout of the process

The pressing equipment is designed to provide an axial compressive force during bending. During bending, the mold is placed in an 80 ton press equipped with a high-frequency (HF) generator, Figure 10.24. The HF-generator (Ib Obel Pedersen T30L) is a single-end system with an electric earth plane. The HF-effect is 33 kW and the frequency 13.56 MHz.

Fig. 10.24 Wood in the mold after bending in the 80-ton press.

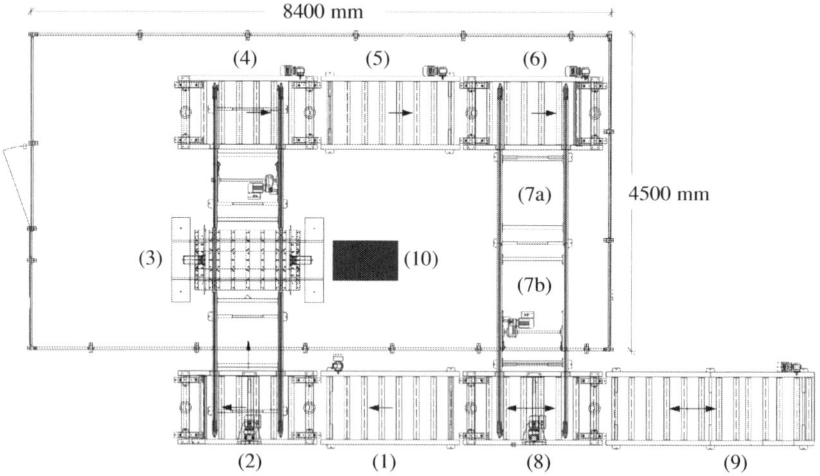

Fig. 10.25 Layout of the line for solid wood bending and the different positions of the mold (1-9). (1) Loading of the mold; (2) Changing the direction of the mold; (3) HF-Press: heating, bending and drying; (4) Changing the direction of the mold; (5-7b) Waiting stations; (8) Unloading of the bent wood; (9) Storage for the molds; (10) HF-generator.

Figure 10.25 shows the layout of the production line for solid wood bending and the different positions of the mold. The mold is loaded with wood pieces at a moisture content of about 25% (1), and is then transported via (2) to the HF-press (3). In the press, the wood is heated, bent and dried to a moisture content of about 6%. After the bending, the mold is transported between the stations (4) and (7b) while the wood cools. At (8), the mold is opened and the bent wood is taken out, verified for damage and stacked for further processing. Table 10.3 shows an example of the processing times in the different positions. The total time from the loading of the unbent wood to the unloading of the bent and dried wood is 26 minutes. The bending takes in all 10 minutes, which means that three molds can be used at the same time.

Table 10.3 An example of the processing time in the different positions, see Figure 10.25.

Position		Cycle time (seconds)	Total time (seconds)
1	Wood pieces in the mold	0	0
2	Transportation to pos. 2	11	11
3	Transportation to pos. 3	40	51
3	The press is waiting for the start of the computer program	20	71
3	Bending process: heating, bending and drying	620	691
3	End of program and start of transportation	21	712
4	Transportation to pos. 4	33	745
5	Transportation to pos. 5	26	771
6	Waiting at pos. 5 and transport to pos. 6	190	961
7a	Waiting at pos. 6 and transport to pos. 7a	200	1 161
7b	Waiting at pos. 7a and transport to pos. 7b	190	1 351
8	Waiting at pos. 7b and transport to pos. 8	190	1 541
8 and 1	Place for loading and unloading of the mold	–	–
	Total process time:		26 minutes

The bending process

An electromagnetic energy field (13.56 MHz) is used to plasticize the wood in a high-speed solid-wood bending process. However, a good control of the moisture content of the wood is needed, and the equipment for bending must be designed to prevent flash-over in the electromagnetic field and to hinder steam in the wood from rupturing the cell walls. Flash-over in the tool or in the wood can have different causes and Figure 10.26 shows some example of flash-over. To reach a low rejection level in the bending process, the process parameters, i.e., the input energy and the time for the different process stages, have to be well controlled since the time-span for optimal bending is very small.

In this process for bending solid beech wood (*Fagus silvatica* L.), wood with a moisture content of 20-25% is heated, bent and dried in a single sequence. The heat is generated by the high frequency equipment in which the wood is placed, and curved in a form tool, as described above. It may be worth noting that when the bending starts the moisture content in the wood is about 15%.

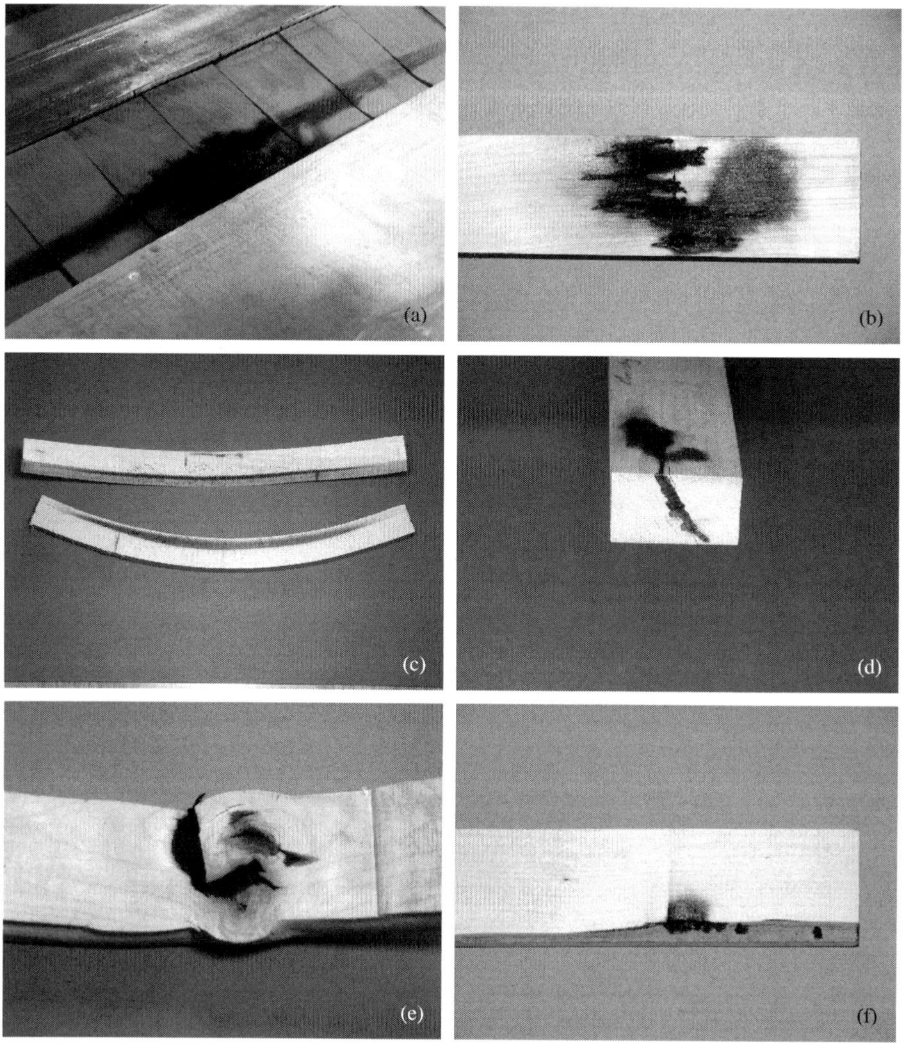

Fig. 10.26 Examples of flash-over in the electromagnetic field: (a) wood pieces with a high average moisture content, (b) high moisture content on the surface of the wood pieces, (c) local high moisture content in the wood pieces, (d) crack in the end-section of the wood, (e) fiber disturbance or a void in the wood, (f) bark left on the wood.

The wood in the process presented here was bent to a radius of 486 mm at a moisture content of about 4%. The curved wood was then processed to an arm-rest, Figure 10.27.

The bending of wood in the HF-press consists of three stages: heating, bending and drying. In each stage, both the power (current intensity) and the time for the current to the mold can be regulated. In the bending stage, the loading speed can be controlled. Each stage is regulated by a control program and it possible to tailor programs for specific purposes.

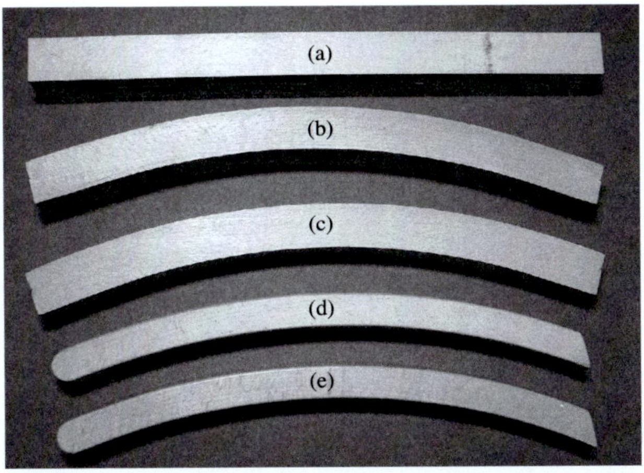

Fig. 10.27 Beech wood in different manufacturing steps to a finished arm-rest. (a) Clear beech wood at 25% moisture content. (b) Curved wood at 6% moisture content. (c) After planing of two sides. (d) After CNC-machining and sanding. (e) The arm-rest after lacquering.

Figure 10.28 show an example of a program for the bending of beech, both the adjusted values in the control program and the actual current from the HF-generator. During the first 30 seconds, the current rises to the adjusted value and is maintained at that level with a tolerance of ± 0.1 A for the rest of the heating stage (0-310 s). The temperature in the wood normally rises to 90-100 °C during this stage. At the end of the heating, it is important that the wood does not have too low a temperature or too low a moisture content, otherwise the plasticizing effect will not be achieved.

During the bending stage, the press is closed and the wood is curved in the mold. The control system then adjusts the current to the control level when the distance between the HF-electrodes (the female and male parts of the mold) changes. At the

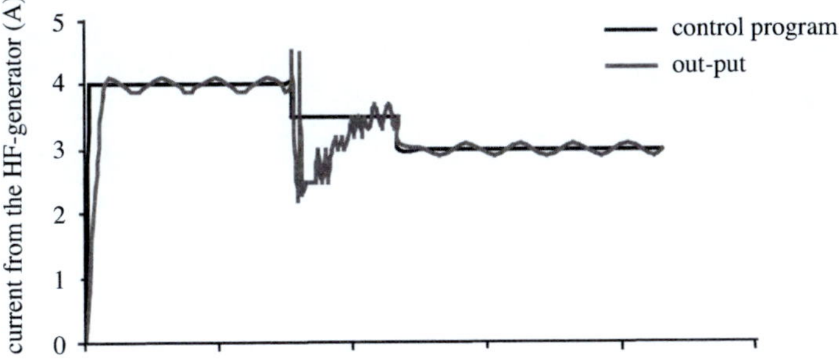

Fig. 10.28 Example of a program for the bending of beech, showing both the adjusted values in the control program and the output current from the HF-generator. The three stages in the sequence are heating 0-310 s, bending 310-470 s, and drying 470-860 s.

end of this stage, the current is at the adjusted value ± 0.2 A. The speed of the press is about 0.04 m/min. In the drying sequence, the output current is stabilized at the adjusted value ± 0.1 A.

Moisture content and temperature in the wood during processing

Figure 10.29 shows the temperature and moisture content in the beech wood during the bending process. The curves are mean values for 21 wood pieces and the temperature was measured at the center of each piece. The current and time for each stage of the bending sequence are listed in Table 10.4.

Table 10.4 Current, time and pressing speed in the control program.

Sequence	Current (A)	Time (s)	Bending speed (m/min)
Heating	4.0	260	0
Bending	3.5	120	0.04
Drying	3.0	360	0

Figure 10.29 also shows the rejection of curved pieces, i.e., curved pieces that do not satisfy the requirements for the bending radius or are not free from tensile or compressive damage, cracks etc. The graph presents the rejected proportion as a function of time.

When the wood is heated, the temperature rises from about 25 °C to 95 °C and the moisture content simultaneously drops from 22% to 14%. During the heating stage, the rejection is by definition 100% as the pieces are not yet curved. During the bending stage, the temperature is almost constant and the moisture content drops rapidly. At the end of this stage the moisture content is approx. 10%. If the process were to be stopped at this time, the rejection would be 100% as a consequence of springback of the curved pieces.

During the drying stage, the moisture content is lowered to a level where the curved shape is fixed at the required radius. In Figure 10.29, it is clear that the mois-

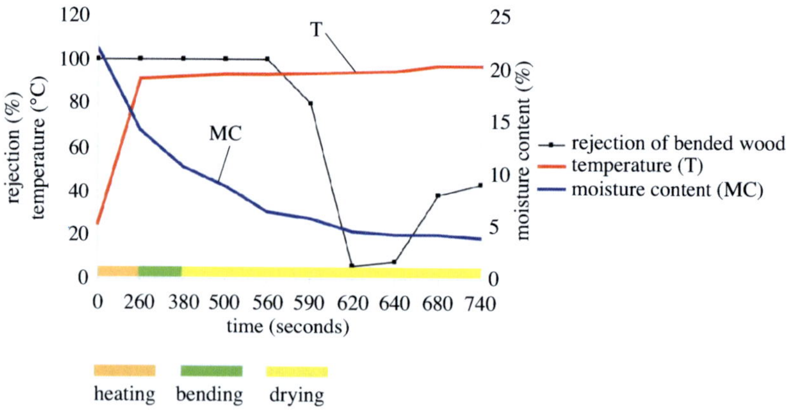

Fig. 10.29 Temperature, moisture content and rejection as a function of time during the bending of solid beech wood.

ture content continues fall during drying until no water remains in the wood. The temperature rises from 95 °C to about 98 °C during 360 seconds. The rejection drops from 100% to about 5% during a time period of about 60 seconds. The process should be stopped within this time period, to obtain a low rejection of wood pieces. If the drying continues, the rejection of bended pieces increases again, and when the bending process is stopped the rejection rate is as high as 40%.

The curves in Figure 10.29 show that there is a relatively short time span during which the process should be stopped to ensure a low rejection of the bent pieces and before the moisture content becomes too low. The optimal processing time is of course dependent on the energy input into the wood.

Figure 10.30 shows an example of different types of damage that may occur in the wood during the bending process. As already mentioned, the wood is not curved during the heating sequence (0-260 s) and is then by definition rejected (100% spring-back). Another reason for rejection during the heating may be a rupture of cell walls if the steam generated within the cells of the wood finds difficulty in escaping, but this was not observed here.

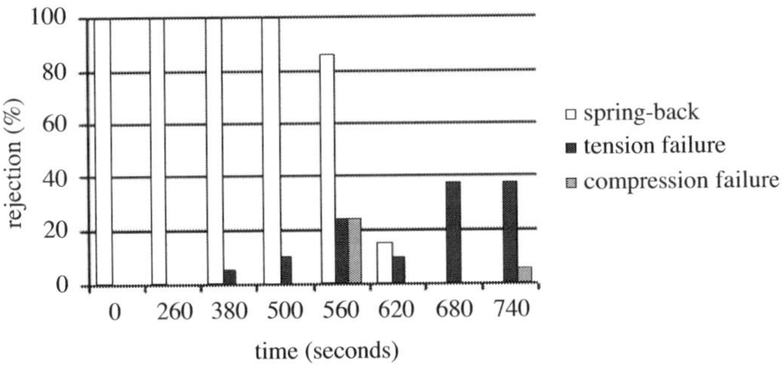

Fig. 10.30 Rejection of wood pieces of beech at different times during the bending process. Current, time and pressing speed in the control program were in accordance with Table 10.4.

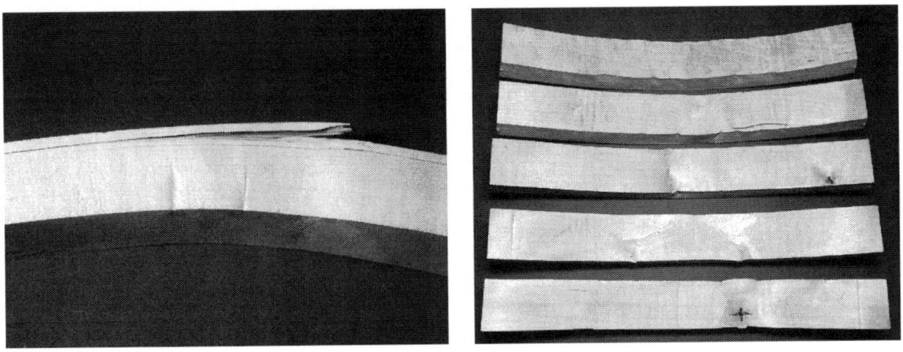

(a) (b)

Fig. 10.31 Examples of a) tensile failure on the stretched side of the wood piece and b) compressive failure on the compressed side.

When the heating sequence (260-380 s) is ended after the bending is finished, the main reason for rejection is spring-back of the pieces which leads to too large radii. Rejection may also occur because of tensile or compressive rupture, especially if the heating stage before bending is too long. Examples of such rupture are shown in Figure 10.31.

During the drying stage (380-740 s), the rejection due to spring-back decreases until it no longer occurs. Continuous drying leads to an increase in tensile damage. Some of the tensile damage, and especially the compressive damage, that may occur in this stage is a result of the earlier process stages.

10.4.4 Pre-compressed wood

Hanemann (1917, 1928) plasticized wood by heating it in steam or water at low pressure, after which the wood was compressed along the fiber direction. In order to avoid bending during compression, the wood was set up in a fixture. The compressed wood was then bent in a cold state and set in the curved shape by heating to 80 °C. The idea of compressing the wood in the longitudinal direction to improve the bending properties was developed further in the 1940s, but a more industrial application did not come until the 1990s, when Compwood AS in Denmark presented a new type of production equipment that had been developed for the pre-compression of wood. Its advantage of which was that whole-wood bending could be achived in a simple manner even for small lot sizes without any great investment costs in steaming equipment and jigs.

The procedure involves the following steps:

1. the steamed wood is longitudinally compressed by ca. 20%;
2. the pressure is then being released, after which the wood has a residual deformation of 4-5%;
3. the wood is allowed to cool to room temperature and is dried to a moisture content of 12%; and
4. the wood piece is polished and processed to the desired cross-sectional shape and surface-finished as a straight piece of material.

The wood piece can thereafter be bent in a cold state. The smallest radius of curvature for 25-mm beech is about 230 mm, which can be compared with the value of 330 mm for plasticized beech without any restraining band, see Table 10.1.

The bendability of the wood can be further increased if, after having been pre-compressed and dried according to items 1-3 in the above list, the material is subjected to a further compression of 20% in the cold state. After the pressure has been released, there is a residual deformation of about 15%. A piece of beech that has been processed in this way can be bent in the cold state to a radius of curvature as small as 100 mm.

10.5 LAMINATED BENDING

Laminated bending is produced by forming and gluing the veneers simultaneously without end pressure against a mold. This method is commonly used in furniture production and can be used for the production of complex forms. Laminated bending can be defined as a process in which a number of concentric veneers are bent and laminated together in such a manner that relative movements are rendered virtually impossible.

Stevens and Turner (1948; 1970) list the following advantages and disadvantages of laminated bending. The advantages are listed below.

- Thick bends of small radius can be built up from thin laminations of any species of wood.
- It is possible to incorporate, within reason, poor quality wood containing knots, splits and other defects which would render the wood quite unfit for solid bending.
- If the lamella are sliced or rotary cut, considerable savings in wood may be effected compared with solid bending.
- Long lengths can be obtained by end-jointing of shorter pieces through scarfing or other methods and, if the joints are properly made and staggered in the completed bend, no appreciable weakening of the cross-section need result. For information about scarfing veneers, see for instance McKean *et al*. (1952).
- Laminated bends can usually be set more readily and made to conform better to the shape of the mold as opposed to similar bends of solid wood.
- No softening treatment is generally required before the pieces are bent.

The disadvantages are that

- more technical skill and better equipment are usually required as opposed to for solid wood bending;
- the presence of glue may be somewhat detrimental to the machines used for the final cleaning of the bent pieces;
- the glue lines, which are usually visible on the sides, may receive objections to for aesthetic reasons; and
- the preparation of the pieces for gluing, the drying of the wood by artificial means, and the cutting of the laminations etc. usually result in a higher cost for the laminated product than for bent products of solid wood.

This comparison by Stevens and Turner between solid wood bending and bending of laminates is not adequate in all situations since solid wood bending is in most cases used for the production of beam-like products such as chair legs etc. whereas laminated wood bending of veneers is widely employed in the manufacture of shells such as chair seats.

The method used in laminated bending is to press glued veneers in a mold made of metal or plywood. The mold is in most cases of the male and female type, but for making a prototype or a partial component, it is possible to use only the male part of the mold and to apply the pressing forces for example through a vacuum bag shaping and veneering press. With the mold, it is comparatively easy to make the bent component. This technique is widely used in the making of chair components for mass production and numerous companies use laminated wood bending technologies worldwide.

The laminated wood bending process consists of

- veneer production or purchase;
- preparation of the veneer, i.e., grading, clipping, sewing, sanding, drying etc;
- gluing and assembling the veneers before pressing;
- pressing, glue-hardening and cooling; and
- final processing to the final shape.

Figure 10.32 shows the principles of the manufacture of laminated products and some distortions that can occur in the product after pressing. An elastic spring-back always occurs when the product is unloaded after pressing and is taken into account when making the mold. Twisting may also occur in the final product, especially when it is subjected to variations of moisture content. Twisting is very troublesome in the production of laminated products.

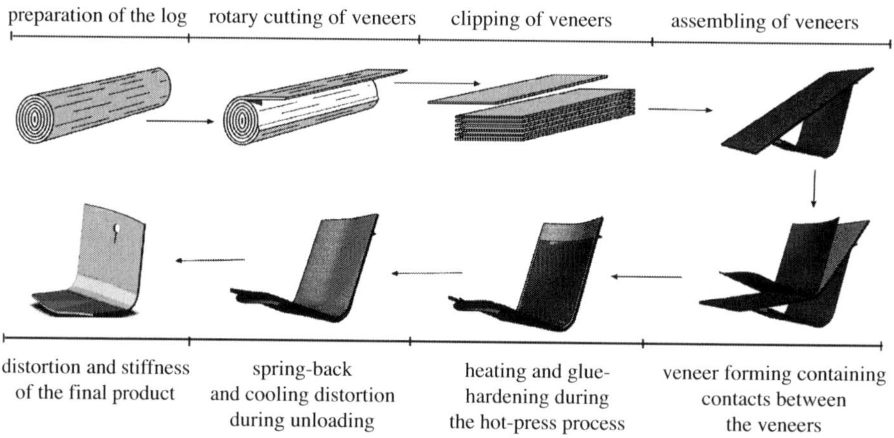

Fig. 10.32 Manufacture of laminated veneer products and distortions of the product after pressing.

10.5.1 Selection and preparation of veneers

In laminated bending, veneers with a thickness of 1 to 5 mm are used. The thickness of the veneers depends on for instance the curvature of the product that is to be produced. In mass production, ready-made veneers are widely employed, i.e. the veneers have been cut and graded for a specific product by the veneer producer or seller.

A veneer is a relatively thin sheet of wood used as surface covering on core stock or other materials, for decorative purposes or for the building-up of laminates for strength. This means that veneers can be divided into two categories: decorative and structural veneers. In laminated wood bending, we usually call these two categories for face veneers and inside veneers.

Making veneer is the most economic way of producing high quality, genuine wood-facing materials. Modern production technologies render possible the exploitation of wood to veneers in numerous ways. Between 800 and 1000 m^2 of veneer can be obtained from one cubic meter of wood. No other kind of wood processing gives such a yield. Veneer can be rotary cut, sliced or sawn from a log. Of the total volume of veneer production, approximately 95% is rotary cut, 3-4% is sliced, and the rest, approximately 1%, is sawn. The first step in veneer production is to decide how to use the veneer log. This is, of course, much more important in decorative veneer production than in structural veneer manufacturing. The properties of the log that are important include the straightness, heart and tension cracking, visible defects such as knots, pin-knots, decay etc., dimensions, color, texture, and shape. To meet present day quality standards for decorative veneers, a veneer surface must as far as is possible be free of defects, uniformly colored and evenly textured.

Before the conversion of the log or parts of the log (sometimes the log is split into two to four pieces prior to the veneer production), it has to be softened. In the most common softening process, the log is heated in water, which leads to two effects.

- The wood is given the suppleness necessary to ensure smooth slicing and thus perfect quality in production.
- The color of the veneer is influenced by the heating time. For example, white beech is changed by the softening process from white to a pink hue.

Depending upon the species and color demands, the cooking period various from one day to a week. Different temperature profiles are recommended for different species of wood during the heating period. This profile has to be precisely maintained to avoid color defects. Since the quality of the water (hardness, chemical composition) plays a considerable role during the course of the heating operation, each veneer manufacturer has developed own time and temperature values which are considered to be the most favorable for the conditions in question.

There are various methods of slicing that give specific textures to the veneer, as indicated in Figure 10.33. For the actual slicing operation, the blocks (flitches) are planed on one or both sides to ensure that the block lies perfectly flat on the slicing bed. Two types of veneer slicing machines are used, i.e.,

- horizontal or vertical slicing machines; and
- peeling machines for rotary cutting, eccentric peeling or stay-log.[1]

In a modern veneer plant, the moist veneers after the slicing or peeling operation are dried in a jet drier which also has a flattening effect. Different species have different drying programs and the program must be run with regard to belt speed and temperature. Bundles containing 24 or 32 sheets are taken out from the drier. Finally the veneers are clipped to the desired dimension.

Before the veneers are bundled and ready to be transformed into a laminated product, they are subjected to various manipulations: cutting, conditioning according to moisture content and temperature, punching, sewing, sorting and organization by grade. The face veneers are also sanded in a sanding machine before pressing to avoid a costly operation after the pressing, or at least to make the final sanding easier and less costly. If the veneers are too buckled to be sanded, they can be plane-pressed with heat to make them flatter. The face veneer is not allowed to have any damage. If

---

1 Stay-log is a special veneer cutting machine in which the line of cut sweeps across the annual rings in a circular direction to give eccentric cutting.

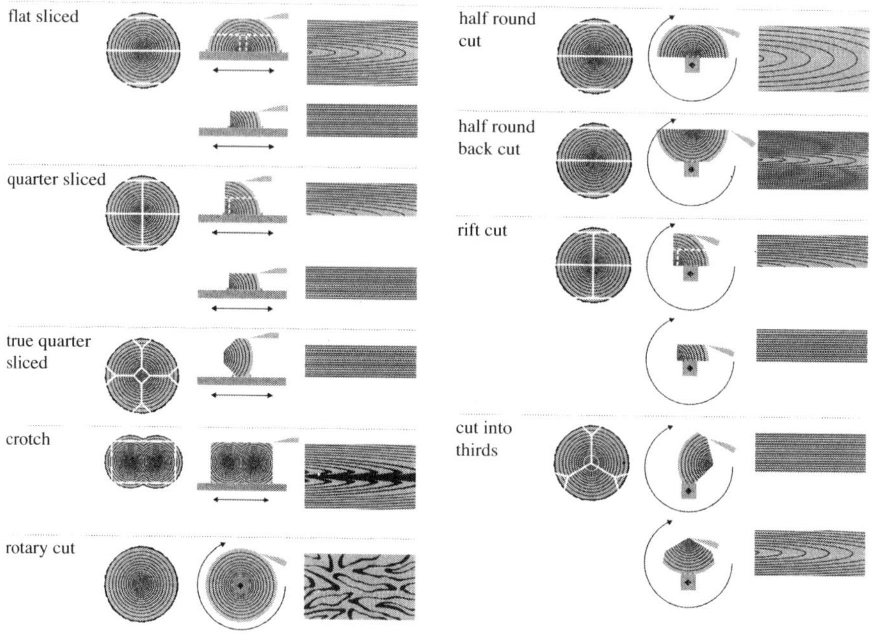

Fig. 10.33 Different ways of slicing the log give different textures on the veneers.

the inside veneer has for instance knots, these must be plugged before use. The time between processing and gluing is important, and different species of wood are more or less sensitive to the time. In order to avoid problems, veneers need to be straight-grained and conditioned to a uniform moisture content before pressing.

It is possible to mix veneers with different thicknesses, qualities and species, as well as with solid wood in a construction. The construction is much easier to handle if the structure is symmetrical. The face veneer, which can be thinner, should always be of high quality, but it is possible to have lower quality materials inside the construction. However, it is important to understand that the inside veneer also affects the entire structure, e.g., the form stability of the final product in use.

The smallest possible radius of curvature during pressing without pre-treating depends on the thickness of the veneer, the manufacturing method and the species. A 1.5-mm birch veneer can be bent to a minimum radius of 40 mm. Experience says that the minimum radius will not be smaller than about 30 times the veneer thickness.

The moisture content of the veneer is determined by the amount of water added with the adhesive and the amount of water that evaporates during heating in pressing. If the required final moisture content is 8-9% for a laminated product, the veneer moisture content should be 4-5% when using the adhesives that is common nowadays.

Bending in more than one direction demands much higher skills, especially when 3D-forms one desired. A major problem in laminated bending is the stretching of the veneer and the risk that the veneers can crack if they are subjected to large three-dimensional deformations. It is well known that the strength of all species is much higher in the longitudinal than in the transverse direction. Therefore, the bending direction of the construction is normally in the longitudinal direction of the external veneers. When the bending also occurs in other directions, the veneers will be stretched in the transverse direction. A seat shell can easily represent a difficult 3D-form, if the seat shell includes convex and concave bends.

A so-called "3D-veneer" is a veneer of wood which can be formed three-dimensionally (like the deep drawing of a metal sheet). This special veneer was developed by Reholz GmbH and has now been successfully introduced onto the market (Müller, 2006). During the production of 3D veneers, narrow grooves spaced 0.1 to 1 mm apart and through the thickness of the veneer are introduced into an ordinary veneer along the fiber direction. To keep the wooden strips together, lines of glue are spread on the rear of the veneer, Figure 10.34.

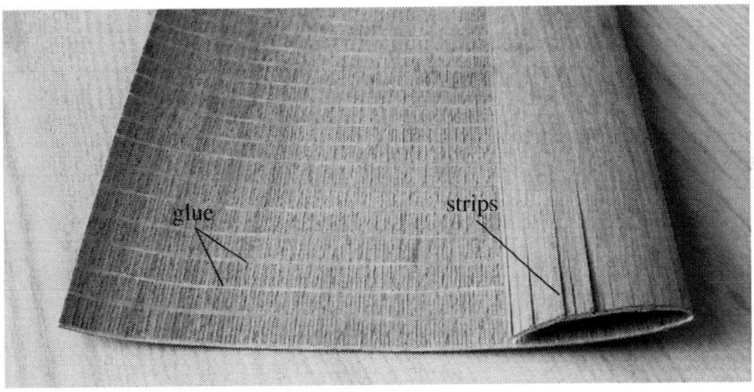

Fig. 10.34 3D veneer showing the strips and the glue keeping the strips together.

3D veneer can be prepared for processing (gluing and packing) by the usual techniques, and 3D forming is performed with a molding tool prepared for the shape required. With 3D veneers, it is possible to achieve form-stable products that are resistant to breakage or distorsion, as shown in Figure 10.35. Neither the structure of the wood nor the texture of the veneer is altered in the process. Another field of application for 3D veneers is coatings for three-dimensional surfaces made of materials such as plastic or metal.

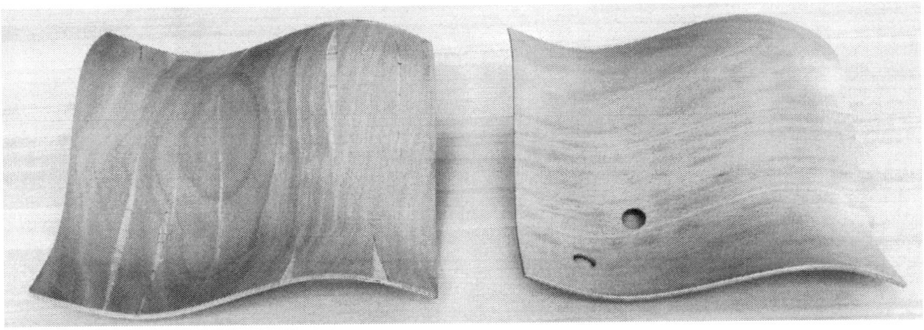

Fig. 10.35 3D-shaped form with five layers manufactured from (left) ordinary beech veneers and (right) 3D veneers.

10.5.2 Adhesives

The laminated product must lock the construction in the desired shape, and a vital factor in the industrial development of laminated wood products has been the development of synthetic adhesives. Examples of adhesives include

- urea formaldehyde (UF) resin;
- melamine urea formaldehyde (MUF) resin, or emulsion;
- polymer isocyanate (EPI) resin;
- polyurethane (PU) resin; and
- epoxy resin.

UF is probably the most common adhesive in the laminated wood-bending industry, but the release of free formaldehyde is a problem. UF adhesive suppliers are therefore developing low-emission UF adhesives. At the same time, suppliers are trying to develop other types of adhesive to replace the UF adhesive. Their goal is to remove formaldehyde without compromising the strength of the construction. Special polyvinylacetat (PVAc) adhesives are used in some cases. EPI, PU and epoxy adhesives are used for laminated products, but these involve problems with the inhalation of isocyanate and with reactive epoxy resins that can cause skin irritation and allergic contact dermatitis.

The adhesive is applied to the veneer using a roller or by spreading with a curtain as shown in Figure 10.36. An adhesive of an appropriate temperature and viscosity is applied evenly on both sides of each veneer. It is not applied directly to the face veneer. The amount of glue needed depends on its type. It is very important to follow the advice of the glue supplier.

Delamination of the glue line after pressing, i.e., separation of surfaces to which the glue does not adhere, is a problem that occurs in laminated bending. A totally delaminated surface feels like an orange peel and is often glossy. Delamination may be due to several factors, e.g., too low pressure or the drying of the glue before the pressing process. A too low relative humidity in the production area or a space between the veneers with adhesive will shorten the "open time" of the adhesive.

Fig. 10.36 Application of adhesive on veneer using rollers. The operator inserts the veneer between the rollers where adhesive is applied (left) and the veneers are subsequently stacked.

10.5.3 Pressing

A very common method for bending laminated assemblies to the desired shape and applying the required pressure to achieve proper bonding of the laminae is to press the assembly in a mold between shaped male and female forms. The mold can be made of plywood, which is most common, or of metal in which case aluminum is normally used, as illustrated in Figure 10.37. In plywood molds, the form is made of several plywood boards that are glued together. The surfaces of the mold in contact with the veneers are covered with an aluminum sheet, which lowers the friction and guides the high-frequency radiation used to heat the glue. Before the 5-axis CNC routers were introduced, more complex forms had to be sculptured, which was a difficult operation.

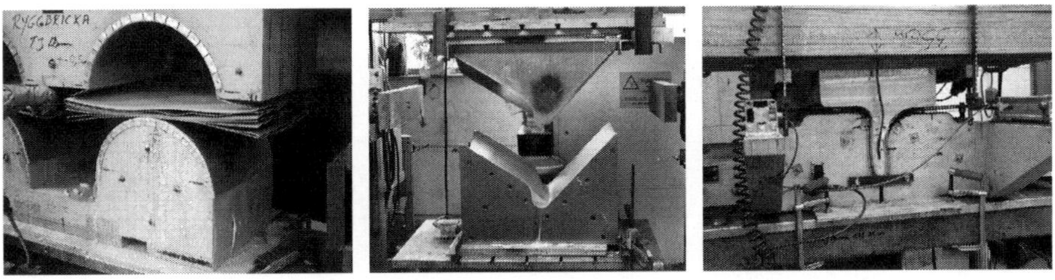

(a) (b) (c)

Fig. 10.37 Molds for pressing in laminated bending: (a) a mold of plywood clothed with aluminum, (b) a mold of aluminum and (c) a mold in which the male part is of aluminum and the female part of plywood.

The pressure must be sufficiently high to conform with the gluing specifications laid down by the producers of the particular adhesive used, making due allowance for the extra pressure required to bend the material and to overcome friction forces into the mold. During pressing, the structure is set under high pressure in the thickness direction. The pressure is usually adjusted so that the volume loss in the thickness direction is 5 to 10%.

The pressing of laminae between male and female forms is limited when the applied force occurs in only one direction in the mold. This is due to the pressure in the mold decreasing as the pressed surface of the lamina deviates to become perpendicular to the applied force, Figure 10.38. To compensate for the decrease in pressure in the mold, pressure is applied from two or more directions, as shown in Figure 10.39. Techniques other than using male and female forms can also be used to apply a pressure: vacuum rubber bags with or without the use of autoclaves, compressed air or water in hoses etc.

It is not necessary to employ heat during pressing, but without it, the curing time for the adhesive is usually long. There are several methods for heating: heating of the mold with hot water or oil, resistive heating of the mold or heating of the laminated assembly with the help of dielectric heating technologies. Ngo & Pfeiffer (2003) describe a method of heating of the veneer assembly by induction, but they do not clearly explain the process.

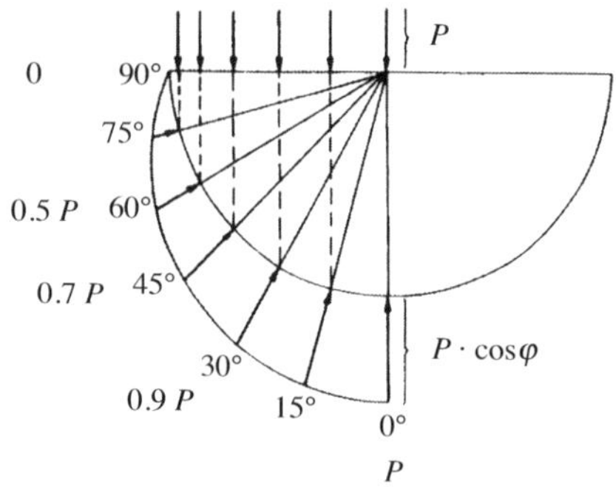

Fig. 10.38 Pressure distribution when a pressure (*P*) is applied on a hemispherical form.

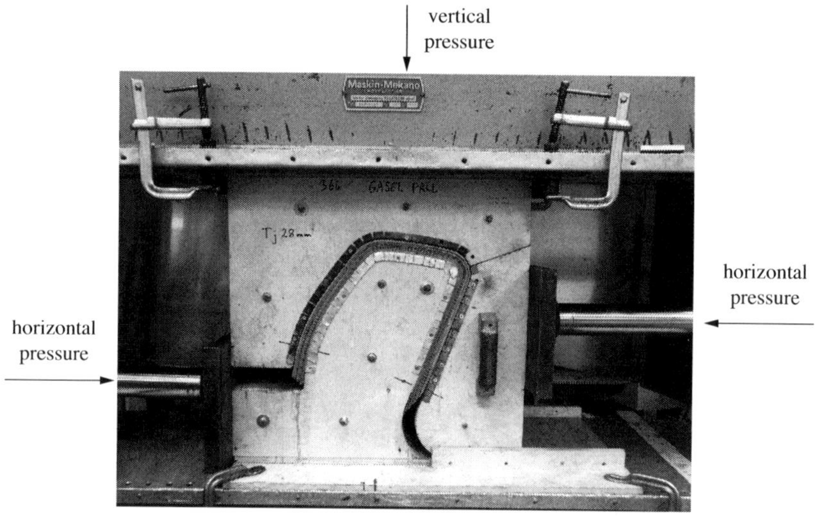

Fig. 10.39 Mold with pressure applied from two directions (horizontal and vertical) to achieve sufficient pressure on the compressed veneers.

For hot water or oil-heated molds, the mold is made of aluminum or stainless steel. This method is used for large production series, with a pressing time of 3-4 minutes.

In resistive heating, the backbone of the mold is of plywood or the like and the mold is covered with metallic and resistance elements that generate heat when a current is passed through them. The press times are about the same as for hot water or oil-heated molds.

In dielectric heating, the workpiece is heated dielectrically in a radio-frequency AC field. The mold is made of plywood and clothed with an aluminum sheet to which the electrodes for the radio-frequency field are attached. This method is more knowledge-intensive than the others, so as not to damage the product during heating. Press times are short and for thin structures they can be about 45 seconds.

After pressing, the product must be stabilized before further processing can take place. Firstly, the temperature is high after the pressing, and the workpiece must cool. Secondly the construction has to be conditioned. This means that the moisture content and the temperature must be in balance. It may be necessary to fasten the construction into a fixture or to stack the material in a balancing climate on each side of the construction. There is no standard time and the minimum time for balancing varies depending on factors such as the dimensions of the product, the temperature, the relative humidity etc. If a non-conditioned construction is further processed, the release of stresses may lead to a distortion of the product.

The material is further manufactured depending on the construction. A seat shell is properly edge processed in a CNC-router.

10.5.4 The movement and distortion of laminated bends

To ensure success in laminated bending, it is necessary to acquire a sound basic knowledge of the behavior of the wood and the adhesives, and to understand the inherent reactions of wood to heat, moisture, strain and stress, which are not constant even within a single species. A good understanding of material behavior during the manufacture of wood veneer products can ensure an efficient wood utilization and promote the development of new processes and products that take advantage of the visco-elastic nature of wood.

Several researchers have indicated that the properties of the final product depend on several material and process parameters, e.g., variations in temperature and moisture content (Hvattum *et al.*, 1978), the design of the pressing mold (Stevens & Turner, 1948, 1970; Lind, 1981; Srinivasan *et al.*, 2008), the applied pressure (Wu *et al.*, 1999) and the quality and properties of individual veneer sheets (Suchsland & McNatt, 1986; Ohya *et al.*, 1989; Zemiar & Choma, 2002; Suchland, 2004; Wagenführ *et al.*, 2006; Srinivasan *et al.*, 2007; Ormarsson & Sandberg, 2007; Buchelt & Wagenführ, 2008). Stresses resulting from the large bending deformations introduced during the molding process can lead to crack formation during the forming and also later when the product is exposed to natural climate variations (Hvattum *et al.*, 1978; Lind, 1981; Cassens *et al.*, 2003).

Ormarsson and Sandberg (2007) have developed a finite element model for studying the distortions of laminated veneer products, Figure 10.40. The simulation model can be used to study how different process parameters, material properties and fiber orientations influence the spring-back and shape stability of the final product. The utilized simulation model consists of three parts: one to study the molding deformations of the veneers, one to study the heating of the veneers, glue-hardening and spring-back deformation during unloading, and one to study the moisture-related distortions of the final product.

A sheet-forming simulation of the curved veneered product was performed with an explicit dynamic analysis for large deformations and contact constraints. The contact algorithm used for the veneer-sheets contacts is based on a so-called penalty-formulation for the mechanical constraints, which means that contact surfaces have some flexibility to penetrate into each other. This reduces the large contact forces that arise in the degrees of freedom normal to the contact surfaces. The interaction properties perpendicular to the contact surfaces allow no separation of the surfaces when they have been in contact. The tangential contact properties assume an isotropic friction behavior. Since the theory of large deformations is used, the orthotropic material directions (l_0, r_0, t_0) are upgraded during the large mold-bending of the veneer sheets (l_t, r_t, t_t). The material directions (l_t, r_t, t_t) in the deformed configuration have been used as initial material directions in the spring-back and distortion simulation (parts 2 and 3 in Figure 10.40).

The stress fields developing during the forming and heating procedures cause the spring-back deformation that occurs when the molding-pressure is removed from the product. The spring-back deformation was simulated by an ABAQUS-standard using the theory for small strain assumptions. The glue-line between the veneers was simulated as a tied constraint condition. This means that no deformation can take place in the glue line.

After the molding procedure, the final chair seat was cooled in a normal indoor climate and with limited external constraints. The distortion simulation was performed in a manner similar to the spring-back analysis; the product was exposed to additional temperature and moisture variations caused by the surrounding climate conditions.

The simulations were performed using the commercial software ABAQUS/Explicit for the veneer molding and ABAQUS/Standard for the spring-back and moisture-related distortions. To describe the wood material, some external subroutines

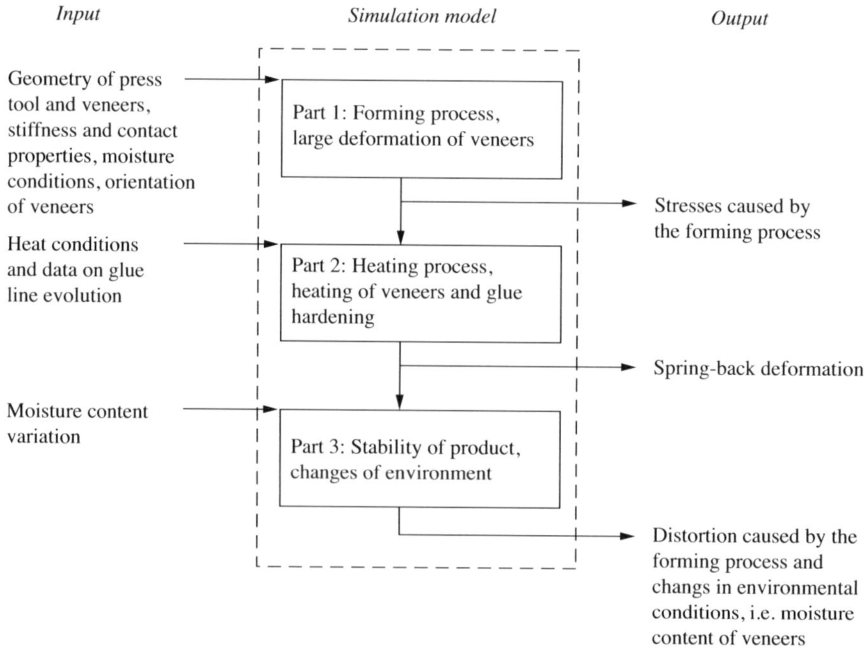

Fig. 10.40 The principal components in the simulation model by Ormarsson and Sandberg (2007) used to study deformations and stresses laminated wood which arise during manufacture and as a result of environmental climate changes.

have been developed and linked to the software. In addition, a few routines have been written for data transfer between the different softwares.

A laminated structure changes its shape when the pressure is released and the product is removed from the mold. This change is often referred to as spring-back. The product will also alter its shape in later stages depending on natural changes in the surrounding environment occuring during the year. For a manufacturer of laminated products, it is important to know how the shape will change during production, and the mold and heating system should be designed to ensure that the product has the desired form for the assembly. The spring-back deformation always occurs during production and this deformation is normally not as great a problem as other modes of distortion. The moisture content is of central importance for the distortion, and should be kept as constant as possible i.e., around 6-9% in the assembly. The parameters that have a significant impact on the spring-back are

- the thickness and number of laminae;
- the elastic modulus and compression strength in the fiber direction;
- the temperature during pressing;
- the moisture content of the veneer; and
- the radius of the bend.

Increasing the number of laminae reduces the contribution of residual stresses, but leads to an increased minimum bending radii and a greater spring-back. However, since in most cases the product has a given final thickness, the use of several thin veneers (rather than a few thick ones) can reduce the spring-back. Increasing the pressing temperature also reduces the spring-back. The stresses in a newly produced laminated product will always tend to reopen the bend, i.e., to increase the radius of curvature.

The temperature change that occurs between the pressing and the fixing of the laminated component also affects the shape. Furniture production often involves high temperatures to cure the adhesive and the subsequent cooling results in a shrinkage in the thickness direction. This leads to a reduction of the radius. The effect of a temperature change is however much smaller than the effect of a change in moisture content. The effect of a temperature change is primarily due to the radius of curvature and the thermal expansion coefficient of the wood. The shape change decreases with an increasing radius of curvature, but it also depends on the number and thickness of the lamellae.

Since the laminated bending using a large amount of adhesive causes a rise in the moisture content, it is important to fully dry the laminate before removing it from the mold. Basically, if the drying process is carried out carefully after applying the adhesive to the material, there should be only small problems of distortion.

Figure 10.41 illustrates deformations resulting from the molding process, unloading (spring-back) and moisture content variations (twist and change in spring-back). The first four parts of the figure display the progressive bending deformation that takes place during the molding process, while the last two show the spring-back deformation after unloading and the twist deformation caused by the moisture changes.

In the bending of elastic materials, compressive and tensile stresses are induced along the concave and convex surfaces respectively. These stresses tend to always

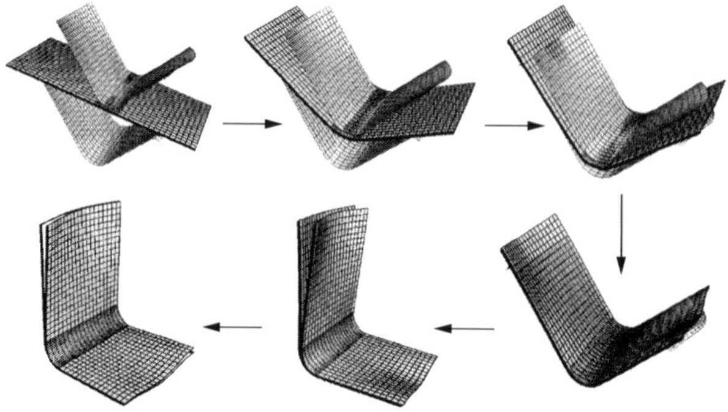

Fig. 10.41 Deformations of a chair seat during the molding process, after unloading (spring-back) and after a change in moisture content (Ormarsson & Sandberg, 2007).

produce strains in their own direction, but at the same time to produce opposite kinds of strain in the perpendicular directions. Thus, the longitudinal shortening of the compressed concave surface is accompanied by a lateral expansion, and similarly the longitudinal stretching of the convex surface is accompanied by a lateral shortening. From thus, it follows that bending the material in one plane will induce a tendency for it to bend in a plane at right angles to the first but in the opposite sense.

A straightness of grain does not preclude the possibility of the laminated assembly twisting as a result of a moisture change unless the grain is absolutely parallel with the sides, or unless a parallel grain exists throughout the assembly. If the fibers in the veneers are in a spiral form or deviate from the main directions of the veneers there will be a tendency for the assembly to twist upon drying. Such issues can obviously be eliminated by selecting only straight-grained veneers and, above all, by ensuring that all the pieces are conditioned to the same moisture content prior to assembly.

Figure 10.42 shows twist deformation in chair seats of beech and birch when the outer veneers are not inclined, when the outer veneers are orientated about 5 degrees in opposite directions (veneer 1: +5° and 7: –5°) and when the fibers in the internal layers are parallel with the veneer edges. In the molding press, all the veneers were heated from 20 to 110 °C and the moisture content increased by about 3 percentage points as a consequence of the water in the glue. The initial moisture content of the veneers was 4.5%. The twist was measured directly after the shell was taken out of the mold, after 14 days' conditioning at 20 °C and 20% RH, and after a further 14 days of conditioning at 20 °C and 90% RH.

In chair seats with no inclination of the outer veneers (test groups 1 and 3 in Figure 10.42), only a small twist was observed after molding and following the subsequent conditioning at low and high relative humidities. The twist that occurred may be a consequence of a variability in the wood properties and the assembly of the veneers, or of an inaccuracy in the deformation measurement.

The chair seats with the outer veneers orientated 5 degrees in opposite directions had a negligible twist after molding. Conditioning at 20 °C and 20% RH led to a drying of the veneers in the glued construction and Figure 10.42 clearly shows that the twist then increased dramatically. A twist greater than 0.8° means that the chair seat must be rejected. When the relative humidity was changed to 90% RH, the twist decreased and in the case of the beech sample, the twist also changed direction. The experimental results showed a good qualitative agreement with the results from finite element simulations by Ormarsson and Sandberg (2007).

To illustrate the effect that the inclination of the fiber may have on the twisting of laminated bends, Figure 10.43 shows an assembly made of two laminae with the fiber direction of the top piece inclined at an angle to the fiber direction of the bottom

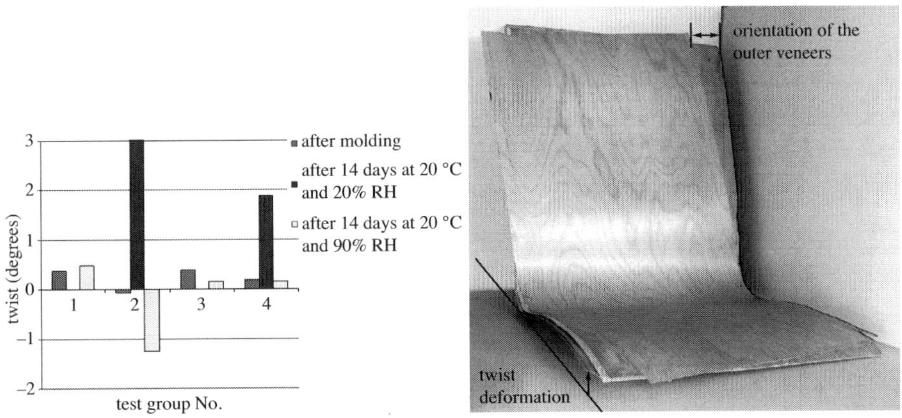

Fig. 10.42 Twist deformation in chair seats of beech and birch directly after molding, and after 14 days conditioning at first 20 °C/20% RH and then 20 °C/90% RH. The chair seat to the right shows twist deformation in the dry state. The test groups are: No. 1: Beech, no or only very small deviation in the fiber orientation of the veneers. No. 2: Beech, outer veneers orientated about 5 degrees in opposite directions. No. 3: Birch, no or only very small deviation in the fiber orientation of the veneers. No. 4: Birch, outer veneers orientated about 5 degrees in opposite directions.

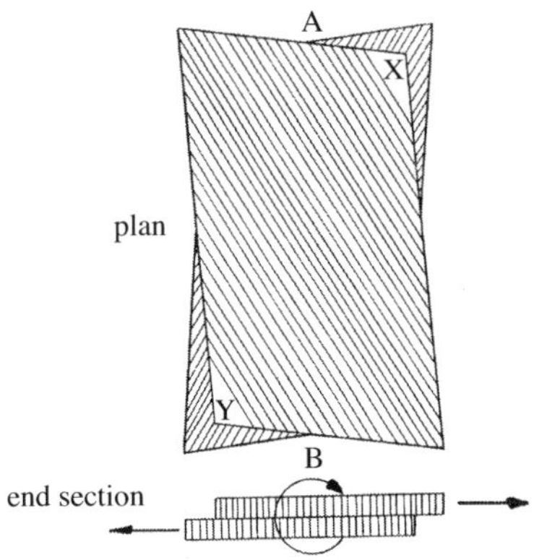

Fig. 10.43 Distortions of veneers when the fiber orientation deviates from an orientation parallel to the veneer edges.

piece. The diagonal lines represent fiber directions as they appeared before assembly of the top and bottom pieces. The inclination of the fibers to the sides of each lamina is assumed to be the same.

If it is assumed that virtually no shrinkage takes place in the fiber direction but that there is shrinkage in the direction normal to the fibers as a result of drying, it is apparent that the top right-hand corner X and the bottom left-hand corner Y of the top lamina will tend to approach one another. The other two corners will also tend to approach one another, but to much less an extent. The face of the lamina will in effect become diamond-shaped, and the longer edges will tend to orientate themselves in the direction of the fiber. The lower lamina will behave in a similar manner, but the orientation will be in the opposite direction.

If they were free to move, the end sections would assume relative positions as illustrated in Figure 10.43. The glue, however, inhibits the relative movement, causing forces to be developed, with the top lamina pulling in one direction with a given force, and the bottom in the other direction with an equal force. These two forces lead a turning moment which tends to twist one end in an anti-clockwise direction and the other in a clockwise direction as viewed from an end. The whole piece is thus subjected to torsion, and twisting is the probable outcome. Stevens and Turner (1948) give advice as to how the laminated bending process can be improved in practice and many of their recommendations have later been verified in a theoretical approach (see for instance Ormarsson & Sandberg, 2007).

It follows that crossing of the fiber direction of two or more lamina in a pack introduces a tendency for the glued assembly to twist as a result of any subsequent moisture change. The intensity of twisting which may develop will be dependent upon a number of factors, such as the magnitude of the moisture change, the angle of crossing of the fibers, the relative positions and thicknesses of the parts having opposed fiber directions, strength characteristics etc.

The best method for avoiding twisting problems would seem to be to select perfectly straight-grained veneers with a fiber direction absolutely parallel to the longitudinal direction of the piece. Failing this, however, it is recommended that the greatest care be taken to ensure that the fiber direction of the various laminae in a pack are parallel throughout; a requirement that can best be fulfilled by assembling the packs from laminae in the order cut.

Twisting may also be minimized or eliminated by an arrangement where any pair of lamina with cross grain, which tend to twist the assembled bend, say, to the right, are counterbalanced by another pair tending to twist it with an equal force to the left. However, such balancing is not easy to accomplish in practice, and it is not recommended as a general method of procedure.

Finally, it is always good to ensure as far as possible that the moisture content of the bent pieces does not become altered after setting, so that shrinkage and swelling can be avoided before incorporation into the finished product. Similarly, bends already twisted may be made serviceable by conditioning them back to their original moisture content, but one should keep in mind that when subsequent moisture changes occur in service, such conditioned bends may cause trouble by twisting out of shape and possibly leading to distortions in the finished article into which they are incorporated.

The following are suggestions for reducing shape-stability problems when gluing laminae:

- Try to maintain the production conditions as constant as possible, e.g., veneer quality, moisture content, glue quality, glue formulation, pressing time, temperature.
- Adapt the moisture content attained by the element after pressing and in the subsequent treatment to the moisture content that the product will have in use. This will avoid any drying or moistening of the element which would lead to distortions. The elastic spring-back can be reduced by using a large number of veneers.
- Balance the elastic spring-back and the change in shape which occur during drying.

When the glued laminate is incorporated into for instance a piece of furniture, it will still tend to change its shape if the ambient relative humidity changes. These alterations in shape and even the functionality of the furniture change. Such effects are primarily due to moisture variations, but the influence of temperature and built-in stresses are often not negligible.

10.6 REFERENCES

AALTO, A. (1936). *Process of bending wood.* U.S. Patent No. 2042976.

ANTTI, L. (1999). *Heating and drying wood using microwave power.* PhD Thesis No. 35, Luleå University of Technology, Sweden.

BACH, L. (1974). Rheological properties of beech wood in the ammonia-plasticized state. *Material Science and Engineering, 15*(2/3):221-220.

BACK, E.L. & SALMÉN, L. (1982). Glass transition of wood components hold implications for moulding and pulp process. *Tappi, 65*(7):107-110.

BARRY, A., PETERSON, F.C. & KING, A.J. (1936). X-ray studies of reactions of cellulose in non-aqueous systems. I. Interaction of cellulose and liquid ammonia. *Journal of the American Chemical Society, 8*(2):333-337.

BUCHELT, B. & WAGENFÜHR, A. (2008). The mechanical behaviour of veneer subjected to bending and tensile loads. *Holz als Roh- und Werkstoff, 66*(4):289-294.

CASSENS, D., LENG, Y. & MCCABE, G. (2003). Face check development in veneered furniture panels. *Forest Products Journal, 53*(10):79-86.

CHOW, S. & PICKLES, K. (1971). Thermal softening and degradation of wood and bark. *Wood and Fiber, 3*(3):166-178.

DAVIDSON, R.W. (1969). Plasticizing – a new process for wood bending. *Furniture Methods and Materials, February*:26-29.

DAVIDSON, R.W. & BAUMGARDT, W.G. (1970). Plasticizing wood with ammonia – a progress report. *Forest Products Journal, 20*(3):19-25.

EGGERT, O. (1995). *Untersuchung der Einflussgrössen beim liegen von Vollholz.* (Investigation of the influencing factors on solid wood.) PhD Thesis, Institut für Werkzeugmaschinen der Universität Stuttgart.

FESSEL, F. (1951). Probleme beim Holzbiegen. (Problems in wood bending.) *Holz als Roh- und Werkstoff, 9*(2):56-62.

HANEMANN, M. (1917). *Holzaufbereitungsverfahren.* (Wood treatment process.) Deutsches Reichs Reichpatentamt, Patentschrift, No. 318197.

HANEMANN, M. (1928). *Verfahren und Vorrichtung zur Herstellung von weichbiegsamem Holz* . (Method and apparatus for producing soft and flexible wood.) Deutsches Reichs Reichpatentamt, Patentschrift, No. 458923.

HON, D.N.-S. & SHIRAISHI, N. (eds.) (1991). *Wood and cellulosic chemistry.* Marcel Dekker, Inc., New York and Basel, ISBN 0-8247-8304-2.

HUTTUNEN, J. (1973). *Method for plasticizing wood.* U.S. Patent No. 3894569.

HVATTUM, O., LIND, P. & OSVIK, H. (1978). *Springback of laminated bends.* Norsk Treteknisk Institut (NTI, Norway), Technical Report 16.

KARMARSCH, K. (1839). *Grundriss der mechanischen Technologie. Zweiter Band.* (Introduction to the mechanical technology. Part 2.) Verlage der Helmingschen, Hannover.

KIVIMAA, E. (1948). *Koivuu taivutuksesta.* (On the bending of birch.) VTT, Technical Research Research Centre of Finland, Report No. 54.

KOLLMANN, F.F.P. & CÔTÉ, W.A. (1968). *Principles of wood science and technology: I. solid wood.* Springer-Verlag, Berlin, Heidelberg.

LIND, P. (1981). *Corner joints and laminated bends – movements causes by moisture changes.* Norsk Treteknisk Institut (NTI, Norway), Technical Report 26.

MCKEAN, H.B., BLUMENSTEIN, R.R. & FINNORN, W.F. (1952). Laminating and steam-bending of treated and untreated oak for ship timbers. *Southern Lumberman, 185*(15):217-222.

MORI, M., NAKAZWAA, S., NORIMOTO, M., YAMADA, T., HIRANO, M., TAKAOKA, M. & MURANAKA, T. (1984). Measurement of internal temperature of wood piece during microwave irradiation by fiber-optical temperature sensor. *Wood Industry, 39*(12):600-603.

MÜLLER, A. (2006). *Method for the production of a three-dimensional, flexibly deformable surface element.* U.S. Patent No. 7131472B2.

NGO, D. & PFEIFFER, E. (2003). *Bent ply – the art of plywood furniture.* Princeton Architectural Press, New York, ISBN 1- 56898-405-7.

NORIMOTO, M. (1979). Wood bending by microwaves. *Wood Research Review, 14*:13-26.

NORIMOTO, M., WADA, H., HASEGAWA, K. & IIDA, I. (1980). Wood bending utilizing microwave heating. *Journal of Society of Rheology, Japan, 8*:166-171.

NORIMOTO, M. & HASEGAWA, K. (1984). *Method and apparatus for shaping wood material into a predetermined configuration.* U.S. Patent No. 4469156.

NORIMOTO, M. & GRIL, J. (1989). Wood bending using microvawe heating. *Journal of Microvawe Power and Electromagnetic Energy, 24*(4):203-212.

OHYA, S., KITAYAMA, S. & KAWAGUCHI, M. (1989). The effect of veneer quality on the bending strength of laminated wood. *Journal of the Japan Wood Research Society, 35*(10):905-911.

OLSSON, A. & SALMÉN, L. (1993). Mechanical spectroscopy: a tool for lignin structure studies. In: *Cellulosics: chemical, biochemical and material aspects,* Kennedy, J.F., Philipps, G.O. & Williams, P.A., (eds.). Ellis Horwood Ltd., Chichester, UK, pp. 257-262.

ORMARSSON, S. & SANDBERG, D. (2007). Numerical simulation of hot-pressed veneer products – moulding – spring-back – distortion. *Wood Material Science and Engineering, 2*(3/4):130-137.

PECK, E.C. (1957). *Bending solid wood to form.* U.S. Department of Agriculture, Forest Service, Agriculture Handbook No. 125.

PENTONEY, R.E. (1966). Liquid ammonia-solvent combinations in wood plasticization. Properties of treated wood. *Industrial and Engineering Chemistry Product Research and Development, 5*(2):105-110.

PRODEHL, A. (1931a). Zur Holzbiegetechnik. (About the wood bending technology.) *Zeitschrift des Vereines deutscher Ingenieure, 75*(39):1217-1222.

PRODEHL, A. (1931b). *Untersuchungen über das Biegen gedämpften Holzes.* (Studies on the bending of steamed wood.) PhD Thesis, Sachsischen Technischen Hochschule zu Dresden.

SALMÉN, L. (1984). Viscoelastic properties in situ lignin under water-saturated conditions. *Journal of Material Science, 19*(9):3090-3096.

SCHUERCH, C. (1963). Plasticizing wood with liquid ammonia. *Industrial Engineering and Chemistry, 55*(10):39.

SCHUERCH, C. (1964). Wood plasticization principles and potential. *Forest Products Journal, 14*(9):377-381.

SCHUERCH, C. (1966). *Method of forming wood and formed wood products.* U.S. Patent No. 3282313.

SCHUERCH, C., BURDICK, M.P. & MAHDALIK, M. (1966). Liquid ammonia-solvent combinations in wood plasticization. Chemical treatments. *Industrial and Engineering Chemistry Product Research and Development*, *5*(2):101-105.

SRINIVASAN, N., BHATTACHARYYA, D. & JAYARAMAN, K. (2007). Thermoforming of wood veneer composite sheets. *Holzforschung, 61*(5):558-562.

SRINIVASAN, N., JAYARAMAN, K. & BHATTACHARYYA, D. (2008). Profile production in multi-veneer sheets by continuous roll forming. *Holzforschung, 62*(4):453-460.

STEVENS, W.C. & TURNER, N. (1948). *Solid and laminated wood bending*. Department of Scientific and Industrial Research, Forest Products Research Laboratory, Great Britain.

STEVENS, W.C. & TURNER, N. (1970). *Wood bending handbook*. Woodcraft supply corp. Woburn, Massachusetts, ISBN 0-918036-06-2.

SUCHSLAND, O. (2004). *The swelling and shrinkage of wood. A practical technology primer.* Forest Products Society, Madison, USA, ISBN 1-892529-38-6.

SUCHSLAND, O. & MCNATT, J.D. (1986). Computer simulation of laminated wood panel warping. *Forest Products Journal, 36*(11/12):16-23.

TEICHGRÄBER, R. (1953). *Über die Spannungszustände bei der Verformung von Holz und die dadurch geänderten Holzeigenschaften*. (On the stress state in the deformation of wood and the resulting modified wood properties.) PhD. Thesis, Hamburg, University.

THUNELL, B. (ed.) (1943). *Verkstadsboken: teknisk handbok för verkstadsindustrien III*. (The workshop handbook: a technical handbook for the engineering industry III.) Lars Hökerbergs Bokförlag, Stockholm.

THUNELL, B. & HONGSLO, A. (1948). *Plasticering av trä*. (Plasticizing wood.) The Swedish Wood Technology Research Institute, Report No. 17.

TORGOVNIKOV, G.I. (1993). *Dielectric properties of wood and wood-based materials*. Springer-Verlag, Berlin.

WAGENFÜHR, A, BUCHELT, B. & PFRIEM, A. (2006). Material behaviour of veneer during multidimensional moulding. *Holz als Roh- und Werkstoff, 64*(1):83-89.

WU, Z.H., FURUNO, T. & ZHANG, B.Y. (1999). Calculation models of pressure and position of curved laminated veneer lumber in molds during pressing. *Journal of Wood Science, 45*(3):213-220.

ZEMIAR, J. & CHOMA, P. (2002). The shape stability of formed laminated wood. *Wood Research (Drevarsky Vyskum), 47*(2):35-43.

CHAPTER 11

RECONSTITUTED WOOD

11.1 INTRODUCTION

The desire to extend and modify the sizes and properties of natural wood, and the need to use manufacturing waste and residues as well as smaller and lower grade trees to produce more versatile and more consistent products, has led to a vast array of materials known as wood composites or reconstituted wood. Wood composites can be broadly grouped into fiber composite products on the one hand and particle wood composites on the other. In fiber composites, wood is broken down to its individual wood cells or fibers and is reformed to the desired shape with or without adhesives.

Particle wood composites, often referred to as particle boards, are manufactured from various wood particles, shavings, chips, flakes, strands and other particle types. The particles are bonded together using polymers or adhesive resins. These materials are classified by particle type, by adhesive type, by density, and by strength. They are more or less homogeneous with uniform properties and they are increasingly being engineered for specific purposes. Another class of wood composites is manufactured with veneers and adhesives.

The term "wood-plastic composite" refers to any number of composites containing wood particles together with a thermosetting or thermoplastic polymer. In contrast to the wood-thermoset composites, wood-thermoplastic composites have seen a phenomenal growth in the United States in recent years and are for this reason often referred to simply as wood-plastic composites (WPCs) with the common understanding that the polymer is always to a thermoplastic (Rowell, 2005).

Heat and mass transfer are involved in various processes in the manufacturing of wood-based composites. These procedures consist of the drying and fabrication of fibers and wood particles, a hot-pressing operation and the conditioning of the mat after it leaves the hot press. Although many researchers have investigated solid wood drying, few have studied the heat and moisture transfer in the drying of wood particles and fibers. Kamke and Wilson (1986a,b) have analyzed the effect of parameters affecting the drying of wood particles within a rotary dryer, in order to develop a computer model to describe the drying phenomena and compute the retention time. Fyhr and Rasmuson (1996, 1997) studied the drying of wood chips under superheated steam.

To consolidate the mat of fibers/particles and adhesives during the manufacture of MDF or particle board, hot pressing is used. The thermal energy applied in the hot press polymerizes the adhesives and softens the wood particles. Mechanical compression increases the contact area between wood particles and adhesive and also promotes bonding, reduces the porosity of the mat, and partially densifies the wood particles. Several external and internal parameters such as temperature, humidity, porosity of the mat, particle dimensions, compressive force, and the characteristics of the wood are involved during hot-pressing of the mat. A careful and exact control of both external and internal variables is necessary to achieve the desired product and minimize the time of pressing. Various physico-mechanical and chemical phenomena also take place during the hot-pressing, which makes the operation very complex.

The fiber/particle mat is a complex heterogeneous structure at multiple levels. It is a porous material with voids in the cellular structure (lumens) and between the fibers/particles. It is thus difficult to describe the transport mechanisms in a mat, especially if the medium is compressible (Aguilar, 2006). Besides heat and mass transfer, other processes such as polymerization of adhesives, chemical degradation of the wood and densification of the wood particles may be taking place. For the last three decades, several researchers have tried to describe these mechanisms and to develop a numerical model for the hot-pressing process (virtual experimentations). These include Bolton and Humphary (1988), Humphary and Bolton (1989), Thoemen (2000), Zombori (2001), Fenton *et al.* (2003), Carvalho *et al.* (2003), Dai and Yu (2004), Pereira *et al.* (2006) and Thoemen and Humphary (2001, 2006). A recently published book on wood-based panels explains different mechanisms of heat and mass transfer in the various processes involved in the manufacture of wood-based composites (Thoemen *et al.*, 2010).

Wood fibers can be used to produce a wide variety of low-density three-dimensional webs, mats, and fiber-molded products. Short wood fibers can be blended with long fibers and formed into flexible fiber mats, utilizing physical entanglement, nonwoven needling, or thermoplastic fiber melt matrix technologies. The most common types of flexible mats are carded, air-laid, needle-punched, and thermo-bonded mats.

Reconstituted wood represents a product range with applications in constructional engineering and in the paper industry. A large number of these products are developed in the form of panels. Their diversity is large and is increasing, taking advantage of scientific knowledge and technological developments in the production field, and taking into consideration the increasing requirements imposed on construction materials. The various types of panels derived from wood are

- fiberboards: insulating fiberboards, hardboards and, low-density and medium-density fiberboards (MDF);
- particle boards: OSB panels (oriented strand board), LSL panels (laminated strand lumber or Intrallam), PSL panels (parallel strand lumber or Parallam); and
- panels made of veneers; LVL Panels (laminated veneer lumber or Microlam), plywood and cylindrical LVL.

The manufacturing processes are divided into two principal classes. In the first class, the cellulose fibers bind together by natural forces (hydrogen bonds). This bonding of the fibers is achieved through drying and compression. In this process, which takes place under wet conditions, no adhesive is used to bind the fibers together. The technique is used to make paper, paperboard, insulating fiberboard, and semi-hard and hardboards.

In the second class, the method involves a dried process where the bonding together of the particles is accomplished by the use of various adhesives. The most widely used adhesives are phenol-formaldehyde and urea-formaldehyde resins. A classification of the various panel materials according to the dimensions of the raw material and the manufacturing process is given in Figure 11.1. A short description as well as a manufacturing diagram of each type of panel material is provided in the following section.

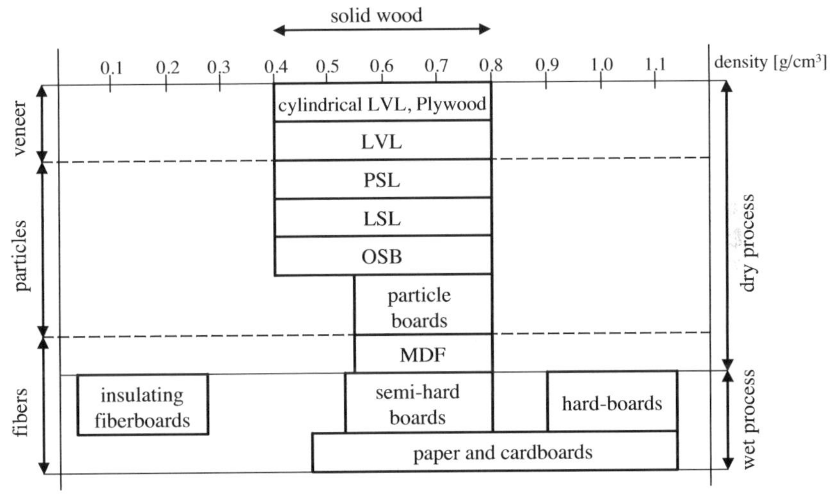

Fig. 11.1 Classification of panel materials according to the process and the raw material (modified from Suchsland & Woodson, 1986).

11.2 MANUFACTURE OF FIBERBOARDS

Fiberboards were first manufactured in the beginning of the 20th century, and the best known products are the hard fiberboards, the insulating fiberboards and the MDF. The raw material used to manufacture these boards is wood fibers and fiber bundles produced in mechanical and thermo-mechanical (pressurized) refiners. Since the manufacture of MDF panels differs from that of the other fiberboards, its fabrication will be discussed separately.

11.2.1 Manufacture of insulating and hard fiberboards

The principal stages in the wet process for manufacturing insulating fiberboards and hard fiberboards are similar to those involved in papermaking and comprise the following steps:

1. Grinding of chips or shavings in mechanical refiners in order to obtain a fibrous pulp.

2. Adding water to the pulp so that a dilute suspension is obtained. There is about 10 kg of fiber in one cubic meter of water, which makes it possible to fix the grammage[1] and the dilution of the pulp.
3. Transforming the pulp into a felted and homogeneous web by draining, aspiration and pressing between rollers. At the end of this stage, the fibers are closely bonded and tangled and a fiber board is obtained.
4. Pressing the fiber panel under heating (approximately 50 kg/cm^2 compressive stress at a temperature 200 °C) to obtain hardboard. This reduces the pore volume of the fiber panel (fibers are usually also densified) and dries the material. It should be noted that no resin is used to bind the fibers together.

It has been suggested by Suchland *et al.* (1983, 1987) that, under wet conditions and at a high temperature, certain phenolic substances may be produced from lignin, and that these act as adhesives establishing strong connections between fibers in the hardboard. On the other hand, to obtain insulating boards, these are completely dried by hot air without compression after the third stage.

A diagram of the manufacture of the insulating and hard fiberboards is given in Figure 11.2.

After machining, these boards are transformed into acoustic insulating panels, thermo-insulating boards or boards for other applications. Figure 11.3 presents a photographs of different board materials.

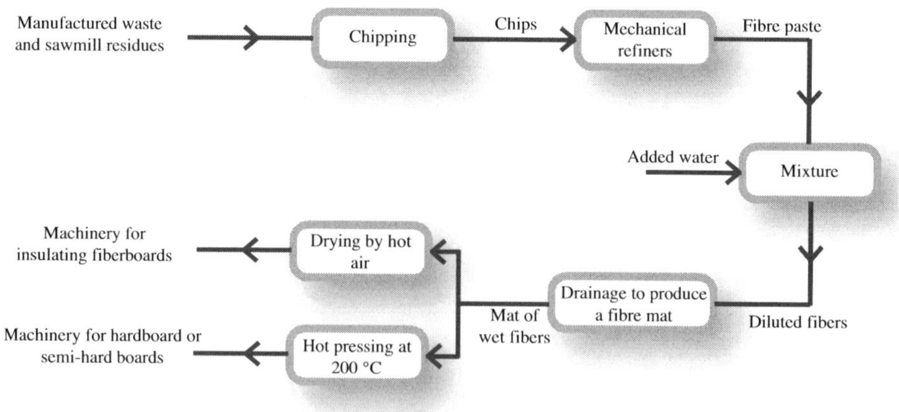

Fig. 11.2 Diagram of the manufacture of insulating and hard fiberboards.

1 In the metric system, the density of all types of paper and paperboard is expressed in terms of grams per square meter (g/m^2).

Fig. 11.3 Photographs of different board materials (a) insulating fiberboard, (b) hardboard, (c) medium density fiberboard (MDF), (d) particle board.

11.2.2 Manufacture of medium density fiberboard (MDF)

The manufacture of MDF panels started in the USA around 1950. The essential difference between hard fiberboards and MDF is that adhesives are used as binders in MDF, whereas in hard fiberboards the lignin, under the effect of pressure and temperature, is transformed into a kind of adhesive. The manufacturing process was first developed as a semi-dry process and this led to a fully dry-process method. A diagram of the manufacture of MDF is given in Figure 11.4. In this method, less water is used than in the wet process and thus less polluted water is produced. In addition, this technique allows the fabrication of thicker panels, from 2 mm up to 100 mm, as opposed to hard fiberboard where the thickness is limited to 12.5 mm due to the wet process. A photograph of an MDF board is shown in Figure 11.3.

A uniform distribution of fibers during manufacture ensures that the MDF has a homogeneous structure. It is possible to manufacture MDF boards with specific characteristics to suit particular applications. The MDF is a homogeneous and dense product, with a density from 600 to 800 kg/m^3. An MDF board is easy to machine and its regular surface lends exceptionally well to painting or the application of decorative coatings. It is this quality that has given MDF the place that it occupies in the furniture industry. The thick MDF is used in joinery and for the doorframes, window frames, etc. MDF fibers are usually prepared by the thermo-mechanical pulping process.

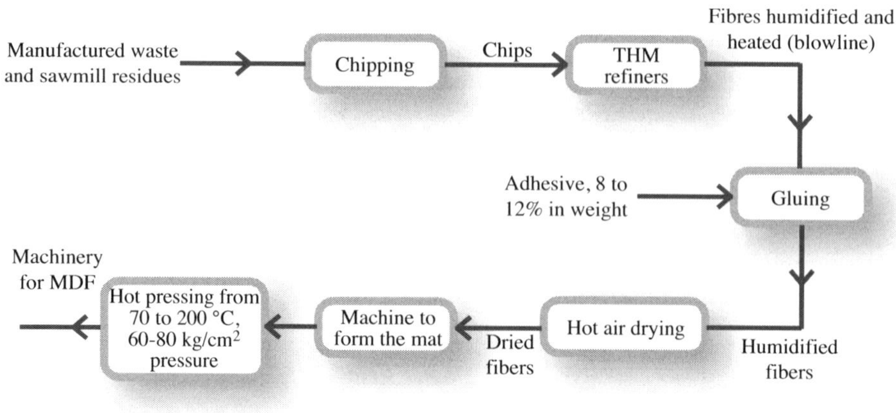

Fig. 11.4 Diagram of the manufacture of MDF by the dry-process.

11.2.3 Manufacture of particle boards

The manufacture of particle boards was started in Germany around 1940. For various reasons, this industry was commercialized successfully throughout the world:

- The process makes it possible to use wood trunks of small diameters and wood residues in the manufacture. Historically, the availability of raw materials was good and the cost was relatively low. However, this relationship changed drastically in resent years as a result of competition from biomass fuel use.
- New high quality synthetic resins are cheap and affordable.
- The particle boards are very useful in furnishing and construction industries. The production of particle boards also promotes an effective use of wood, with a very small percentage of waste, 10% to 25%, instead of 50% in the sawing of logs.
- The mechanical properties of the boards depends on the dimensions as well as on the orientation and arrangement of the wood particles used in the manufacture of the board. The wood particles employed in the construction of panels have various dimensions, cut mechanically by chippers, flakers etc. The particles are divided into chips, splinters and wood meal, as shown in Figure 11.5 (Tsoumis, 1991).

(a) (b) (c)

Fig. 11.5 Particle types: (a) chips, (b) splinters, (c) wood meal (Tsoumis, 1991).

The manufacture of particle board is a dry process and there are two different methods of production; flat hot pressing and extrusion, which give two types of boards with two particle orientations. In the first method, the particles are parallel to the panel surface, whereas in the second technique the particles are oriented perpendicularly to the surface. Figure 11.6 presents a diagram of two types of particle board.

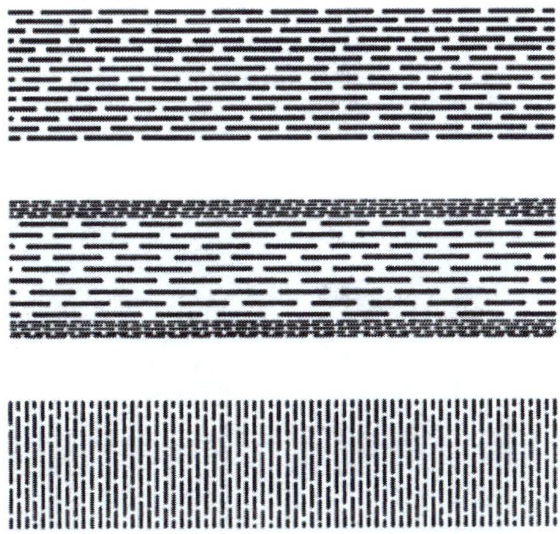

Fig. 11.6 Diagrammatic presentation of the cross-section of particle boards, (a) manufactured by flat hot pressing; the orientation of the particles is parallel to the surfaces and the board is homogeneous (b) manufactured by flat pressing, a board with three layers (c) manufactured by extrusion; the orientation of the particles is perpendicular to the board surfaces.

The various stages of the production of particle boards are illustrated in Figure 11.7.

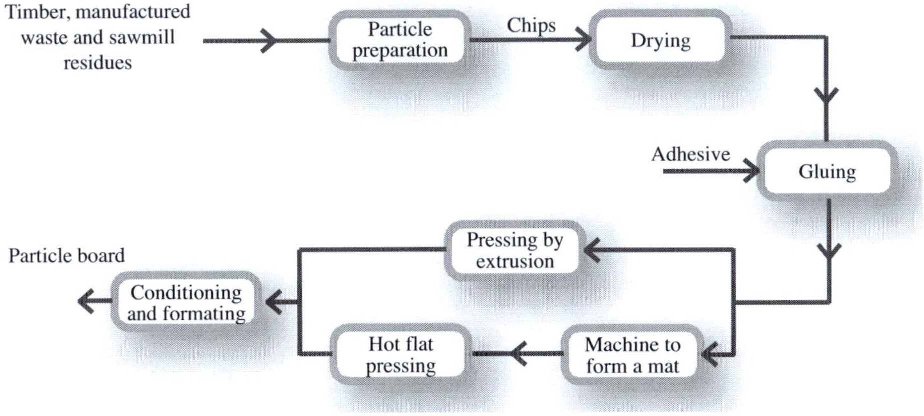

Fig. 11.7 Diagram of the various stages of two methods for manufacturing particle boards.

11.2.4 Manufacture of oriented strand board (OSB)

OSB is developed from strands of wood, which are typically 15-25 mm wide, 75-150 mm long and 0.3-0.7 mm thick, cut from logs of small diameters, Figure 11.8. A resin is used to bond the strands together. Such boards are fabricated under pressure at high temperature. The strands in the outside layers are aligned parallel to the length whereas the internal strands are deposited randomly or perpendicularly to the face layers. OSB is therefore a multi-layer board, used in a large variety of structures of

Fig. 11.8 Illustration of an OSB and its fine strands.

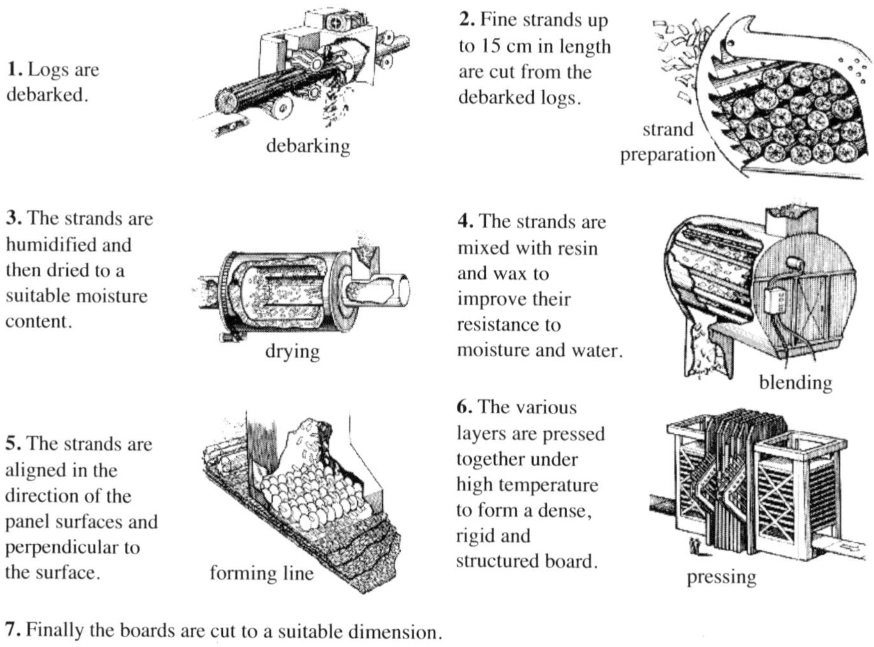

Fig. 11.9 Stages in the manufacture of OSB.

Fig. 11.10 Segments of (a) OSB and (b) plywood.

an industrial or decorative nature. The boards are used for covering floors, ceilings and sometimes walls. The OSB are also being utilized more and more in packaging, for cases, pallets, etc. Stages of the manufacture and a photograph of a segment of an OSB are shown in Figures 11.9 and 11.10, respectively.

11.2.5 Manufacture of laminated strand lumber (LSL) or Intrallam

LSL or Intrallam is a product where fast growing poplar is converted into large-sized panels. The manufacture of LSL are explained briefly in Figure 11.11 and a photograph of an LSL section is presented in Figure 11.12.

The manufacture of Intrallam or LSL involves the following stages:

1. Cleaning and debarking of logs with small cross-sections (rapid growing poplar, length: 2.5 m).

2. Cutting the logs into strands (30 cm long and 3 cm wide) and drying; then treating with polyurethane resin to obtain hydrophobic properties.

3. Stacking a mattress of the strands placed parallel to each other in the longitudinal direction and pressing under vapor injection.

4. Cutting to a large-sized panel (11 × 2.5 m, thickness: 3 to 14 cm).

Fig. 11.11 Stages in the manufacture of LSL.

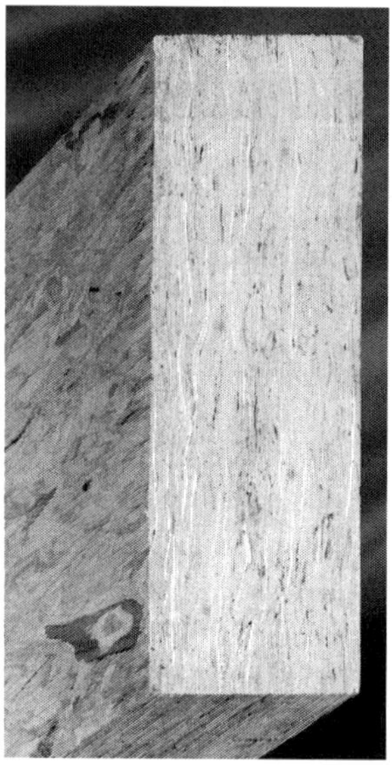

Fig. 11.12 Section of a LSL or Intrallam.

1. Peeling of the logs of small sections to obtain veneers. All the veneers, even those with irregular dimensions obtained at the beginning of the peeling are used.

2. Drying (moisture content < 6%) and cutting veneers into strands 2.4 m long and 3 mm thick. Elimination of defects and treating of the strands with an adhesive with hydrophobic properties.

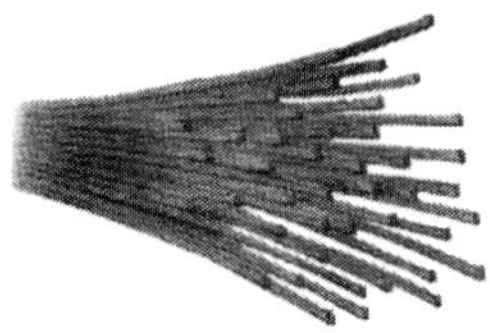

3. Introduction of the strands in the longitudinal direction into a press (continuous process) and polymerization of the adhesive by microwaves.

4. Obtaining a rectangular beam of section (28 × 49 cm^2) and sanding. After cutting into sections, it is ready for use.

Fig. 11.13 Stages in the manufacture of PSL.

11.2.6 Manufacture of parallel strand lumber (PSL) or Parallam

The manufacture of PSL is based on a technology wich makes it possible to convert small trees into elements of large cross-sections (28×49 cm^2) and considerable lengths (up to 20 m). The manufacture of PSL is explained briefly in Figure 11.13. A photograph of a section of PSL is presented in Figure 11.14.

These products are used in building construction as elements in compression, large trusses, beams or posts, Figure 11.15.

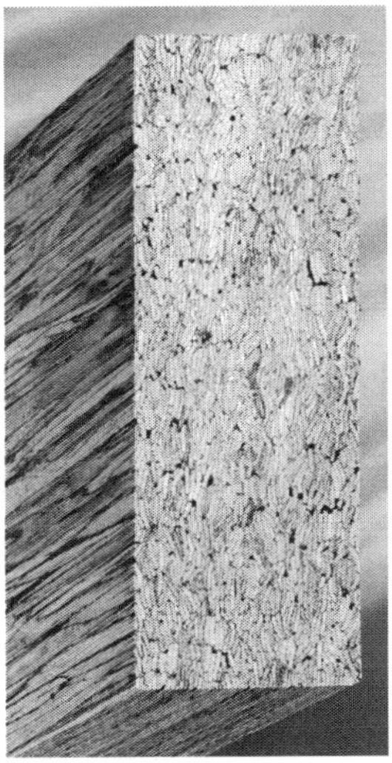

Fig. 11.14 Photograph of a section of PSL.

Fig. 11.15 Various PSL products are used in building construction: elements in compression, large trusses, beams and posts.

11.2.7 Manufacture of laminated veneer lumber (LVL) or Microllam

LVL is a wooden product intended for structural applications. It is manufactured from veneers of coniferous trees obtained by peeling the logs. The different stages in the

1. The logs are subjected to thermo-hydrous treatment to facilitate the barking and peeling of the logs. The logs are placed for 1 to 3 days in a hot bath or under saturated vapor between 90 °C and 120 °C.

2. Peeling of the logs into veneers, and cutting into small veneers 1.5 × 2.5 m^2 in size and a thickness of 2.5 mm up to 4.5 mm. The sheets are dried to a moisture content below 8%, and then classified according to their density and the slope of the fibers. Any veneers containing major defects are rejected or obtain a lower grade.

3. Gluing and stacking of the veneers. Treatment of the veneers with a hydrophobic adhesive (approximately 30 kg/m^3) and stacking them with the same orientation.

4. Continuous thermo-pressing of the mattress of veneers to obtain panels 1.22 m wide and a maximum in thickness of 9 cm. The temperature and the time of pressing are adapted to the species and the thickness of the LVL panels. Sawing to obtain beams with a rectangular section.

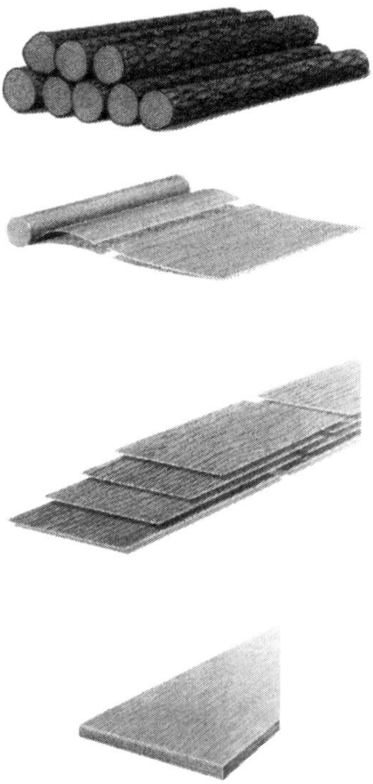

Fig. 11.16 Stages in the manufacture of LVL.

Fig. 11.17 Section of an LVL panel.

manufacture of LVL are intended to eliminate the defects inherent in the ordinary sawn wood and to develops a higher strength, a good dimensional stability and a homogeneity of physical and mechanical properties along the beam or elements. The various stages in the manufacture of LVL are shown in Figure 11.16. A photograph of a rectangular LVL section is presented in Figure 11.17. LVL products are used in structural applications like purlins, lintels and beams, Figure 11.18.

(a)

(b)

Fig. 11.18 LVL products are used in structural applications such as (a) purlins (b), lintels.

11.2.8 Manufacture of plywood

Plywood is manufactured by gluing together relatively thin veneers (0.5 to 6 mm). The principal stages in the manufacture of plywood are similar to those of LVL, i.e., production of veneers, stacking and gluing of veneers, pressing at high temperature, cutting and storage. The various types of plywood are characterized by the wood species, the number of layers and the orientation of the veneers. Figure 11.19 shows three examples of plywood composed of veneers superimposed and glued to others of various orientations.

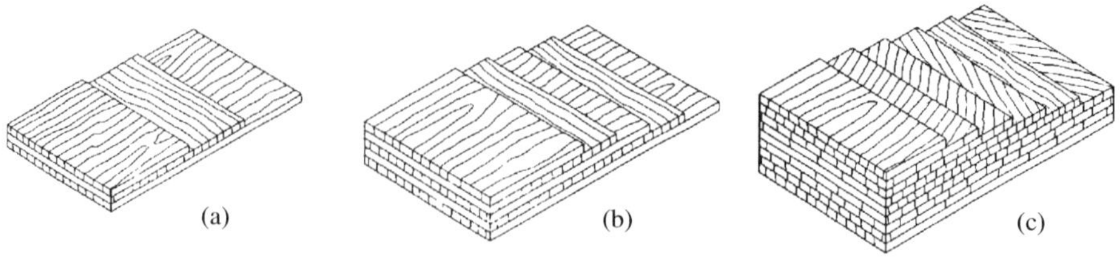

Fig. 11.19 Examples of plywood, (a) three veneers crosswise glued at 90°, (b) five veneers crosswise glued at 90°, (c) nine veneers crosswise glued at 22.5°.

11.2.9 Manufacture of cylindrical LVL

The manufacture of the cylindrical LVL was developed in Japan, at the Wood Research Institute, Kyoto University in 1994 (Sasaki *et al.*, 2000). The concept is based on the technology of manufacturing paper tubes by winding sheets around a cylinder reproducing the natural winding of the microfibrils on a macroscopic level. The raw material used to manufacture the cylindrical LVL consists of veneer tapes, as shown in Figure 11.20.

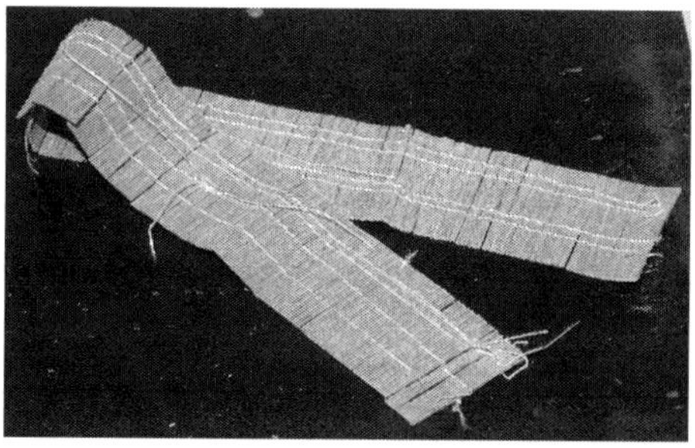

Fig. 11.20 Veneer tapes used in the manufacture of cylindrical LVL.

Fine strands are cut from a veneer with a thickness of 1 to 4 mm. These fine strands are sewn with plastic thread to hold them together as flexible veneer tapes. The dimension of these tapes is selected according to the diameter of the cylinder and the mechanical performance required. The various stages of winding the veneer tapes on a mandrel to manufacture the cylindrical LVL are given in Figure 11.21.

The veneer tapes are wound up in spiral of several layers around a mandrel to form the cylindrical LVL. The first layer is wound without adhesive in a spiral on a casing, whereas the subsequent layers contain adhesive. To prevent the unwinding of the cylindrical LVL when in use under the action of a longitudinal force, the direction of winding is alternated from one layer to the next. This gives the wall of the cylindrical LVL an orthotropic behavior. Figure 11.21 shows the winding of the veneer tapes in spiral.

A resorcinol or isocyanate adhesive is spread on the back of the veneer tapes in an amount of approximately 25 g/m^2. At the end of this winding process, a rubber belt is wound to apply a pressure of approximately 0.4 MPa on the LVL cylinder. To

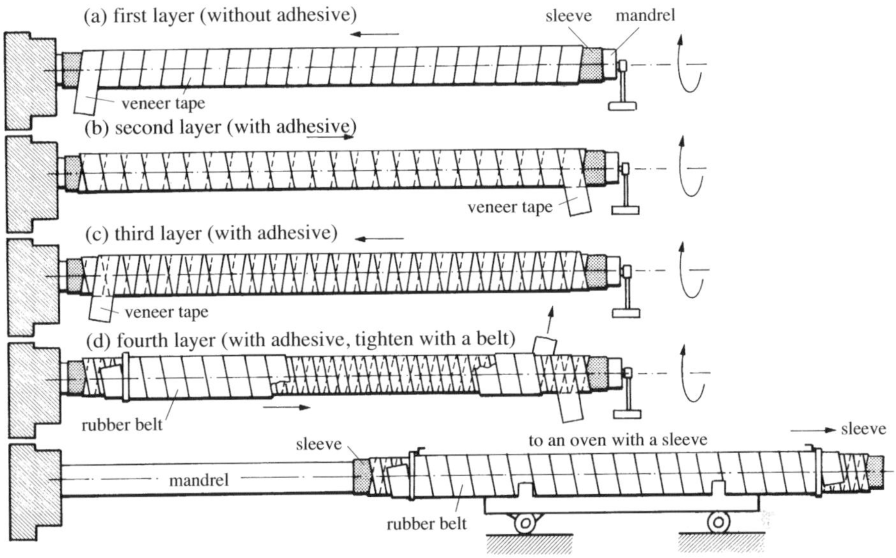

Fig. 11.21 Winding of the "veneer tapes" to produce cylindrical LVL, (Sasaki *et al.*, 2000; Yamauchi *et al.*, 2001; Kawai *et al.*, 2001).

polymerize the adhesive, an oil circuit inside the sleeve heats the cylindrical LVL. After curing of the adhesive, the sleeve is withdrawn from the LVL. Two photographs of cylindrical LVL sections are given in Figure 11.22.

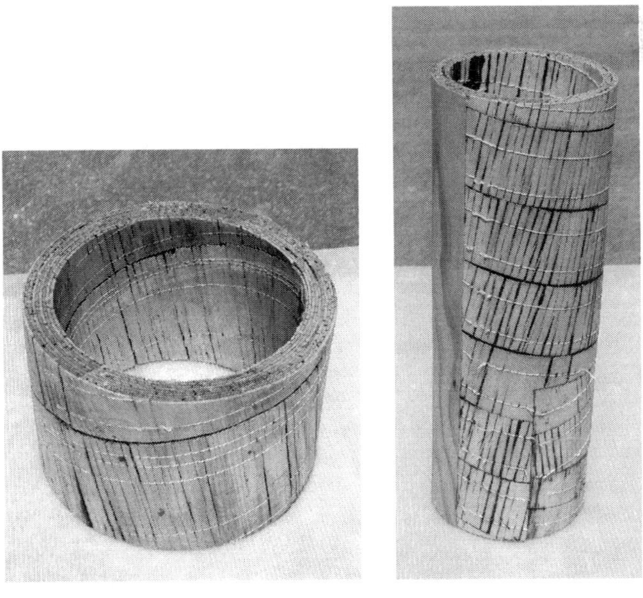

Fig. 11.22 Segments of two cylindrical LVLs.

The cylindrical LVL is used in various ways, as a structural element in building construction, as a pipe for the transport of fluid or as a guide for electric wiring. The technology for its implementation is recent, and these composite cylinders are being used more and more in the construction industry. It is possible to manufacture hollow columns with lengths of 12 m and a diameter of 1.2 m using veneer tapes fabricated from relatively small diameter logs. The possibility of producing cylindrical LVL with various layer orientations gives it excellent mechanical properties making it very suitable as a construction material.

11.3 REFERENCES

AGUILAR, H. (2006). *Study of the rheology of deformable porous media: application to paper production.* PhD Thesis in Chemical Engineering, Faculty of Engineering, University of Porto, Portugal.

BOLTON, A.J. & HUMPHREY, P.E. (1988). The hot pressing of dry-formed wood-based composites. Part I. A review of the literature, identifying the primary physical processes and the nature of their interaction. *Holzforschung, 42(6)*:403-406.

CARVALHO, L.M.H., COSTA, M.R.N. & COSTA, C.A.V. (2003). A global model for the hot-pressing of medium density fiberboard (MDF). *Wood Science and Technology, 37*(3/4):241-258.

DAI, C. & YU, C. (2004). Heat and mass transfer in wood composite panels during hot-pressing: Part I. A physical-mathematical model. *Wood and Fibre Science, 36*(4):585-597.

FENTON, T., BUDMAN, H., PRITZKER, M., BERNARD, E. & BRODERICK, G. (2003). Modeling of oriented strandboard pressing. *Industrial and Engineering Chemistry Research, 42*(21):5229-5238.

FHYR, C. & RASMUSON, A. (1996). Mathematical model of steam drying of wood chips and other hygroscopic porous media. *American Institute of Chemical Engineers (AIChE) Journal, 42*(9):2491-2502.

FHYR, C. & RASMUSON, A. (1997). Some aspects of the modelling of wood chips drying in superheated steam. *International Journal of Heat Mass Transfer, 40*(12):2825-2842.

HUMPHREY, P.E. & BOLTON, A.J. (1989). The hot pressing of dry-formed wood-based composites. Part II. A simulation model for heat and moisture transfer and typical results. *Holzforschung, 43*(3):401-405.

KAMKE, F.A. & WILSON, J.B. (1986a). Computer simulation of a rotary dryer. Part I: Retention time. *American Institute of Chemical Engineers (AIChE) Journal, 32*(2):263-268.

KAMKE, F.A. & WILSON, J.B. (1986b). Computer simulation of a rotary dryer. Part II: Heat and mass transfer. *American Institute of Chemical Engineers (AIChE) Journal, 32*(2):269-275.

KAWAI, S., SASAKI, H. & YAMAUCHI, H. (2001). Bio-mimetic approaches for the development of new wood composite products, In: *Proceedings of the First International Conference of the European Society for Wood Mechanics*, Navi, P., (ed.). Lausanne, Switzerland, pp. 503-514.

PEREIRA C., CARVALHO, L.M.H. & COSTA, C.A.V. (2006). Modeling the continuous hot-pressing of MDF. *Wood Science and Technology, 40*(4):308-326.

ROWELL, R.M. (ed.) (2005). *Handbook of wood chemistry and wood composites.* Taylor & Francis, CRC Press, Boca Raton, USA, ISBN0-8493-1588-3.

SASAKI, H., YAMAUCHI, H., MA, L-F., PULIDO, O.R., KATAYA, M. & KAWAI, S. (2000). Technology for the production of cylindrical LVL. In: Fifth Pacific Rim Bio-based Composite Symposium, Cambera, Australia, pp. 215-219.

SUCHSLAND, O., WOODSON, G.E. & MCMILLIN, C.W. (1983). Effects of hardboard process variables on fiberbonding. *Forest Products Journal, 33*(4):58-64.

SUCHSLAND, O. & WOODSON, G.E. (1986). *Fiberboard manufacturing practices in the United States.* USDA Forest Service Agriculture Handbook, 640, US Government Printing Office, Washington, DC.

SUCHSLAND, O., WOODSON, G.E. & MCMILLIN, C.W. (1987). Effect of cooking conditions on fiber bonding in dry-formed binderless hardboard. *Forest Products Journal*, 37(11/12):65-69.

THOEMEN, H. (2000). *Modelling the physical processes in natural fiber composites during batch and continuous pressing.* PhD. Thesis, Oregon State University, USA.

THOEMEN, H. & HUMPREY, P.E. (2001). Hot pressing of wood-based composites: selected aspects of the physics investigated by means of simulation. In: *Proceedings of the Fifth European Panel Products Symposium*, Llandudno, Wales, UK, pp. 38-49.

THOEMEN, H. & HUMPREY, P.E. (2006). Modeling the physical processes relevant during hot pressing of wood-based composites: Part I. Heat and mass transfer. *Holz als Roh- und Werkstoff, 64*(1):1-10.

THOEMEN, H., IRLE, M. & SERNEK, M. (eds.) (2010). *Wood-based panels –An introduction for specialists.* Brunel University Press, London, England. ISBN 978-1-902316-82-6.

TSOUMIS, G. (1991). *Science and technology of wood: structure, properties, utilization.* Van Nostrand Reinhold, New York, ISBN 0-442-23985-8.

YAMAUCHI, H., SASAKI, H., PULIDO, O.R. & KAWAI, S. (2001). Manufacture an application of cylindrical LVL for building materials. In: *Proceedings of the First International Conference of the European Society for Wood Mechanics*, Navi, P., (ed.). Lausanne, Switzerland, pp. 519-525.

ZOMBORI, B. (2001). *Modelling the transient effects during the hot-pressing of wood-based composites.* PhD. Thesis, Faculty of the Virginia Polytechnic Institute and State University, Blacksburg, Virginia, USA.

INDEX